Unifying Microbial
Mechanisms

Unifying Microbial Mechanisms

Shared Strategies of Pathogenesis

Michael F. Cole

CRC Press
Taylor & Francis Group
Boca Raton London New York

CRC Press is an imprint of the
Taylor & Francis Group, an **informa** business

CRC Press
Taylor & Francis Group
6000 Broken Sound Parkway NW, Suite 300
Boca Raton, FL 33487-2742

© 2020 by Taylor & Francis Group, LLC
CRC Press is an imprint of Taylor & Francis Group, an Informa business

No claim to original U.S. Government works

Printed in Canada acid-free paper

International Standard Book Number-13: 978-0-8153-4540-4 (Paperback)
978-0-367-20664-2 (Hardback)

Visit the Taylor & Francis Web site at
http://www.taylorandfrancis.com

and the CRC Press Web site at
http://www.crcpress.com

I dedicate this book to Professor George H. W. Bowden and the late Professor William H. Bowen. George is my oldest and dearest friend who started me on my adventure in microbiology when I was a first year student at The London Hospital Dental Institute in Whitechapel, East London. A microbiologist of international stature I cannot begin to describe how much I have learned, and continue to learn from him both as a scientist and a human being. The late Professor Bill Bowen, another dear friend and my Ph.D. mentor, instilled in me a can do attitude. Bill brought me to the United States and, in so doing, expanded my horizons.

Contents

Preface   **xv**

Acknowledgement   **xvii**

Chapter 1: Introduction to Pathogenesis   1

Introduction

Henle-Koch postulates and evolving
 views of infectious disease causation

Experimental models of pathogenicity
 The ethics of using humans, animals and cell lines in
  pathogenics research   6
 Advantages and disadvantages of human experimentation
  in the study of the pathogenesis of infectious diseases   8
 Animal models in the study of the pathogenesis of
  infectious diseases   8
 Other considerations in experimental models of the
  pathogenesis of infectious diseases   10
 Human cell lines as a surrogate for microbe–host interactions   12

Bibliography

Chapter 2: Normal Microbiotas of the Human Body   17

Introduction
 Microbiota of the skin   20
 Microbiota of the vagina   22
 Microbiota of the urinary tract   25
 Microbiota of the conjunctiva   31
 Respiratory tract microbiota   31

Microbiota of the alimentary canal
 Mouth   34
 Oesophagus   40
 Stomach   40
 Small intestine   40
 Large intestine   44

Key concepts

Bibliography

Chapter 3: Biofilms   59

Introduction

Biofilms structure and properties
Mucosae versus skin   60
Initial steps in biofilm formation   61
Biofilm development and the climax community   63
Quorum sensing in biofilms   65
Biofilm dispersal   67

Biofilms in human infections
Peripheral and central i.v. catheters   67
Urinary catheters   68
Bladder biofilms   69
Endotracheal tubes   69
Peritoneal cavity dialysis catheters   70
Prosthetic joints   71
Heart valves   71
Chronic wounds   71
Otitis media   73
Cystic fibrosis   74

Biofilm formation by filamentous fungi

Biofilm formation by viruses

How biofilms are studied

Key concepts

Bibliography

Chapter 4: Adhesion to Host Surfaces   87

Introduction

Barrier epithelia
Skin   88
Mucous membranes (mucosae)   89
Blood and lymphatic vessels   91
Blood-brain barrier   92
Foeto-placental interface   93

The extracellular matrix and intercellular adhesion molecules
Abiotic surfaces   94
Initial adhesion events   95

Adhesin–receptor interactions
Protein–carbohydrate (lectin) interactions   97
Bacteria   98
Fungi   102

Viruses 102
Parasites 106

Protein–protein interactions
Microbial surface components recognising adhesive matrix
  molecules 107
Fibrinogen-binding MSCRAMMs 115
Vitronectin-binding MSCRAMMs 116
Proteoglycan-binding adhesives 117
Bacteria 117
Viruses 117
Parasites 118
Anchorless adhesins (Moonlighting proteins) 119
Bacteria 119
Fungi 120
Protozoa and multicellular parasites 120
Cell wall glycopolymers 121
Capsules 121

Galectins as bridging molecules in microbial adhesions

Adhesion to other barriers
Endothelium of blood vessels and lymphatics 124

Key concepts

References

Chapter 5: Facilitated Cell Entry 129

Introduction

Crossing intact skin
Enzymatic degradation 131

Crossing intact mucosal epithelium
Entry *via* microfold (M) cells 132
Enzymatic degradation 133
Polar tube formation 134
Moving junction 134
Paracytosis 135
Endocytosis 138
Reorganisation of the actin cytoskeleton and endosomal
  trafficking 141

Exploitation of endocytosis pathways by pathogens
Bacteria 145

Zipper mechanism
Trigger mechanism 148
Viruses 156
Fungi 156

Microtubule reorganisation

Transcytosis

Key concepts

Bibliography

Chapter 6: Exotoxins and Endotoxins 167

Bacterial exotoxins
Introduction 167
Membrane-acting toxins 168
   Superantigens (SAs) 168
   Heat-stable exotoxins (STs) 170

Membrane-damaging exotoxins
α-helical pore-forming exotoxins 172
β-barrel pore-forming exotoxins 173
RTX exotoxins 174
MARTX exotoxins 175

Intracellular exotoxins
AB exotoxins 176
AB_5 exotoxins 178
AB exotoxins 181

Fungal toxins

Parasite exotoxins

Endotoxins

Key concepts

Bibliography

Chapter 7: Extracellular Degradative Enzymes 187

Introduction

Proteases

Potential roles of microbial proteases in pathogenesis
Tissue destruction and cell internalisation 188
Inactivation of plasma protease inhibitors 189
Activation of bradykinin-generating and blood-clotting
   cascades 190
Protease-activated receptor 190
Chemoattractant molecules 191
Immunoglobulins 191

Microbe and parasite glycosidases
Deglycosylation of immunoglobulins 194
Adhesion 194

Microbe and parasite phospholipases
Bacterial phospholipases · 196
Fungal phospholipases · 198
Parasite phospholipases · 198

Key concepts

Bibliography

Chapter 8: Evasion of the Human Innate Immune System · 201

Introduction

Antimicrobial peptides
Overview · 201
Bacterial evasion of AMPSs · 204
Fungal evasion of AMPs · 205
Virus evasion of AMPs · 205
Parasite evasion of AMPs · 205

The complement system
Recruiting and mimicking RCAs · 210
Destroying complement components · 213
Microbial envelope/wall components that inhibit complement · 214
Evasion resulting from cell wall structure · 214
Consuming complement in the fluid phase · 214

Circumvention of phagocytosis
Chemoattraction · 216
Regulation of chemokines · 218

Circumventing pattern recognition receptors
Subversion of PRR crosstalk · 223
Targeting cytosolic PRRs, IPS-1, RIG-I and MDA5 · 224
Masking microbe-associated molecular patterns · 225

Manipulating host inhibitory signaling

Pathogen survival inside host cells
Bacteria · 228
Arresting the phagosome/endosome · 229
Diverting the endosomal/phagosomal pathways · 232
Survival in the endolysosome/phagolysosome · 233
Fungi · 235
Parasites · 237

Escape to the cytosol

Virus interactions with intracellular vacuoles

Cytosolic motility of intracellular pathogens

Escape of intracellular pathogens from host cells
Cytolysis   245
Actin-mediated cell-to-cell spread   245
Protrusion into the extracellular environment (extrusion)   247
Induction of programmed cell death   247
Preventing programmed cell death   248
Interference with the host cell cycle   248
Reprogramming the host cell   249

Evading autophagy
Preventing the induction of autophagy   250
Preventing the maturation of the autophagosome into an
    autolysosome   251
Avoiding pathogen capture by the autophagosome   251
Utilising the autophagosome as a habitat for survival,
    replication, or escape from the host cell   253
The role of autophagy in eukaryotic pathogens   253

Evading natural killer cells
Evasion of the natural killer group 2D receptor   257
Evasion of natural cytotoxicity receptors   258

Key concepts

Bibliography

Chapter 9: Evasion of the Human Adaptive Immune System   263

Introduction

Antigen presentation
Linking sensing of MAMPs by pattern-recognition
    receptors with antigen processing   264
Activating naïve T cells by licenced dendritic cells   265
Follicular helper CD4$^+$ T cells help B cells make
    high-affinity, class-switched antibodies   267

Inhibition of antigen presentation by
MHC class I and class II pathways
Viral subversion of the MHC class I antigen-processing
    pathway   270
Bacterial subversion of the MHC class I antigen-processing
    pathway   270
Viral subversion of the MHC class II pathway
    antigen-presenting pathway   271
Bacterial subversion of the MHC class II antigen-processing
    pathway   273
Parasite evasion of the MHC class II antigen-processing
    pathway   273

Manipulation of co-stimulatory molecules 275
Manipulation of regulatory receptors and ligands 276
Up-regulation of IL-10 276

Evasion of antibody

Antigen modulation

Antigenic and phase variation

Subverting B lymphocytes (B Cells)

Subverting T lymphocytes (T Cells)

Key concepts

Bibliography

Chapter 10: Persistent and Latent Infections 295

Introduction

Persistent bacterial infections
Introduction 296
Helicobacter pylori 296
Treponema pallidum subspecies pallidum 301
Mycobacterium tuberculosis 302
Salmonella typhi serovar Typhi 307

Persistent virus infections
Introduction 308
Herpesviruses 309
Hepatitis B, C and D viruses 313
Measles virus 316
Adenoviruses 317
Human papilloma viruses 317
Human polyomaviruses 319
Human immunodeficiency virus 320
Human T-cell lymphotropic virus type 1 321

Persistent parasite infections
Introduction 322
Helminths 322
Plasmodium 323
Leishmania 325
Trypanosoma cruzi 327
Toxoplasma gondii 329
Myeloid-derived suppressor cells in chronic infections 330

Key concepts

Bibliography

Index 337

Preface

The idea for this book came from team teaching a course in Microbial Pathogenesis in which bacterial pathogenesis, viral pathogenesis, fungal pathogenesis and parasite pathogenesis were taught sequentially. When sourcing textbooks for the course it was necessary to have four, one for each class of pathogen. Moreover, almost half of most pathogenic textbooks are devoted to basic immunology and microbiology before considering pathogenic mechanisms. This is understandable since pathogenics is at the confluence of microbiology and immunology and, in my opinion, is best understood by students who have already taken courses in both of these disciplines. For this reason and because of the availability of many excellent microbiology and immunology textbooks I chose not to include basic microbiology and immunology in this book beyond a minimum necessary to understand certain concepts of pathogenics. Since human pathogens of all types adhere to, and invade, the same cells and evade the same immune system it seemed to me more logical to approach pathogenesis from the perspective of common themes employed by pathogens during their infectious cycle in humans. Most is known about the pathogenic mechanisms of bacteria and viruses as compared to fungi and parasites so it will come as no surprise that the former classes of pathogen have more coverage. I hope that by presenting common themes used by pathogens, whatever their type, students will gain an integrated perspective of the constant battle between pathogens and the immune system of their human hosts.

Michael F. Cole, B.D.S., M.Sc., Ph.D.
Professor of Microbiology & Immunology
Georgetown University School of Medicine
Washington, DC, U.S.A.

Acknowledgement

It gives me much pleasure to acknowledge the help of Patricia Sikorski, B.S., a graduate student in the joint Georgetown University–NIH program who helped me so much to find the numerous research papers that were the source of the information found in this book. Elizabeth Owen at Taylor and Francis was invaluable in the creation of this book and providing constructive criticism as it progressed.

Chapter 1: Introduction to Pathogenesis

INTRODUCTION

The aim of this text, is to examine and discuss common themes in microbial pathogenesis. Despite the class of pathogen – whether virus, bacterium, fungus or parasite – they all cause tissue injury and evade the host immune system in much the same ways and attack the same targets. Therefore, it seems only sensible to discuss pathogenic mechanisms in the context of all of these classes of pathogens. With rare exception, texts on pathogenesis have focused on a single class of pathogen, such as bacteria, without conveying any sense that the pathogenic mechanisms described can just as easily be applied to any other class of pathogen. Pathogenics is at the interface of microbiology and immunology because it considers the properties of a microorganism that harm the host and the innate and acquired host defence mechanisms that can neutralise them. The relationship between the pathogen and host is dynamic and reflects thrust and counter thrust. The fulcrum of this interaction can be moved to benefit either combatant by, for example, enhancing or compromising the host immune system or by the acquisition of new genes or loss of existing genes by the microorganism. We will return to this concept later. Because one cannot consider the microbe in the absence of the host, most texts of microbial pathogenesis feel obliged to devote a considerable amount of the text to basic immunology and basic microbiology as a prelude to the consideration of the mechanisms of pathogenesis of microbes. However, there is a plethora of excellent immunology and microbiology texts, both concise and comprehensive, that are superior to the coverage of these disciplines in pathogenesis texts. It is reasonable to say that students should be competent in immunology and microbiology before embarking on a course in microbial pathogenesis. For these reasons, basic microbiology and immunology are not covered in this book beyond that essential to understand particular pathogenic mechanisms.

There are three goals in the study of the pathogenesis of microbes. The first is to identify the aetiological agent of a particular infectious disease, the second is to determine what property or properties of the microorganism and the host allow it to cause disease, and the third is to find methods of neutralising and eradicating the agent by the use of antimicrobial agents and by harnessing the host immune system.

HENLE-KOCH POSTULATES AND EVOLVING VIEWS OF INFECTIOUS DISEASE CAUSATION

Criteria for identifying the aetiology of infectious diseases first were presented in an address given by Robert Koch to the Tenth International Congress of Medicine in Berlin in 1890. They are as follows:

1. The parasite occurs in every case of the disease in question and under circumstances which can account for the pathological changes and clinical course of the disease.

2. The parasite occurs in no other disease as a fortuitous and non-pathogenic parasite.

3. After being fully isolated from the body and repeatedly grown in pure culture, the parasite can induce the disease anew.

These postulates were based on Koch's study of tuberculosis and anthrax caused by the bacteria *Mycobacterium tuberculosis* and *Bacillus anthracis*, respectively. Later others added a fourth postulate, *viz*:

4. The parasite should be re-isolated from the susceptible host (postulate 3) and shown to be identical to that isolated originally (postulate 1).

Although these postulates are adequate to describe some infectious agents, Koch and others realised that failure to fulfil the postulates does not exclude a microbe from being the aetiologic agent of an infectious disease. For example, viruses and some bacteria that cannot be grown on artificial media fail postulate 3, and resident microbes that cause disease in immunocompromised individuals, so-called *opportunistic pathogens*, fail postulate 2. The advent of the application of molecular biological techniques to the study of microbial pathogenesis resulted in a revised set of molecular Koch's postulates formulated by Falkow (1988) that took advantage of these new techniques. Falkow's molecular postulates are listed below.

1. The phenotype or property under investigation should be associated with the pathogenic members of a genus or pathogenic strains of a species.

2. Specific inactivation of the gene(s) associated with the suspected virulence trait should lead to a measurable loss in pathogenicity or virulence, or the gene(s) associated with the supposed virulence trait should be isolated by molecular methods. Specific inactivation or deletion of the gene(s) should lead to loss of function in the clone.

3. Reversion or allelic replacement of the mutated gene(s) should lead to restoration of pathogenicity, or the replacement of the modified gene(s) for its allelic counterpart in the strain of origin should lead to loss of function and loss of pathogenicity or virulence. Restoration of pathogenicity should accompany the reintroduction of the wild-type gene(s).

To allow sequence-based methods to establish causal relationships between microbes and disease in cases where the suspected aetiologic agent cannot be cultivated, Fredricks and Relman (1996) formulated a set of guidelines that are listed below:

1. A nucleic acid sequence belonging to a putative pathogen should be present in most cases of an infectious disease. Microbial nucleic acids should be found preferentially in those organs or gross anatomic sites known to be diseased, and not in those organs that lack pathology.

2. Fewer, or no, copy numbers of pathogen-associated nucleic acid sequences should occur in hosts or tissues without disease.

3. With resolution of disease, the copy number of pathogen-associated nucleic acid sequences should decrease or become undetectable. With clinical relapse, the opposite should occur.

4. When sequence detection predates disease, or sequence copy number correlates with severity of disease or pathology, the sequence-disease association is more likely to be a causal relationship.

5. The nature of the microorganism inferred from the available sequence should be consistent with the known biological characteristics of that group of organisms.

6. Tissue–sequence correlates should be sought at the cellular level: Efforts should be made to demonstrate specific *in situ* hybridization of microbial sequence to areas of tissue pathology and to visible microorganisms or to areas where microorganisms are presumed to be located.

7. These sequence-based forms of evidence for microbial causation should be reproducible.

What is notable about the three sets of criteria listed above is their focus on the microbe without regard to the role of the host immune system in pathogenesis. However, the immune status of the host clearly modulates pathogenicity. *Pathogenicity* is defined in this text as the capacity of a microbe to cause damage to the host. *Virulence* is defined as the *relative* capacity of a microbe to cause damage to the host. In a series of papers, Casadevall redefined the basic concepts of virulence and pathogenicity by integrating microbe-centric and host-centric approaches to pathogenesis into a damage–response framework based on the following three tenets:

1. Microbial pathogenesis is the outcome of an interaction between the host and a microbe and is attributable to neither the microbe nor the host alone.

2. The pathological outcome of the microbe–host interaction is determined by the amount of damage to the host.

3. The damage to the host can result from microbial factors and/or the host response.

Casadevall's damage–response framework can be described by the parabolic curve shown in **Figure 1.1**. The position of the base of the parabola is variable and, whereas host damage can occur throughout the continuum of the host response (x axis), it is maximal at both extremes. Both *weak* and *strong* host responses are defined by both qualitative and quantitative aspects of the host immune response. **Figure 1.2** shows six classes of pathogenic microbes fitted to the damage–response framework. **Figure 1.2a** shows situation in which a microbial factor, for example, an exotoxin, completely responsible for host damage and in which it binds to its receptor before the host can produce antibodies to neutralise it. The exotoxin in this example is, by definition, a *true* virulence determinant/factor in that the exotoxin-producing microbe is rendered avirulent if the gene encoding the exotoxin is inactivated or if the exotoxin

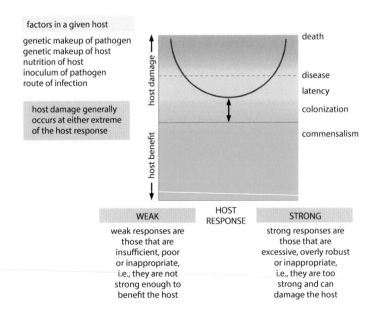

Figure 1.1 Basic parabolic curve of the damage–response framework. (From Casadevall, A. and Pirofski, L.A., *Nat. Rev. Microbiol.*, 2003, 1, 17–24.)

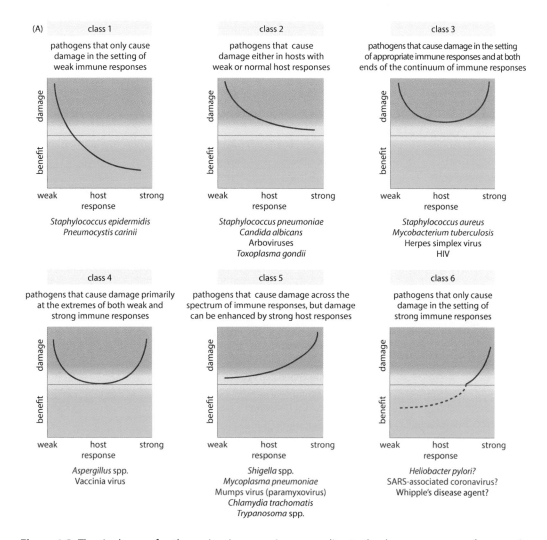

Figure 1.2 The six classes of pathogenic microorganisms according to the damage–response framework. (From Casadevall, A. and Pirofski, L.A., *Nat. Rev. Microbiol.*, 2003, 1, 17–24.)

is neutralised by host antibodies. It may surprise the reader to realise how few microbial components fulfil these criteria. There is but one more microbial component that fulfils these criteria and it is the capsule. All other components/molecules that are thought to contribute to pathogenicity but have not fulfilled the criteria stated above are more accurately termed *determinants of pathogenicity*. **Table 1.1** lists vaccines currently approved for use in the United States. It will be noted that vaccines exist for certain pathogenic bacteria and viruses only. Currently there are no antifungal or anti-parasite vaccines. All of the vaccines against virus infections employ either killed or attenuated virus. All but two of the bacterial vaccines are toxoids (toxins that have been rendered nontoxic by chemical means but retain their antigenicity) or capsular polysaccharides. Only the pertussis vaccine and the meningococcus serogroup B vaccine contain defined vaccine antigens other than exotoxins or capsules.

microbe	vaccine antigen(s)
adenovirus	live adenovirus type 4 and type 7
anthrax*	protective antigen exotoxin
diphtheria (*Corynebacterium diphtheriae*)*	exotoxin**
hepatitis A	inactivated/recombinant virus
hepatitis B	inactivated/recombinant virus
Haemophilus influenzae type B*	capsule**
human papilloma virus	inactivated/recombinant virus
seasonal influenza	inactivated/recombinant virus
Japanese encephalitis	inactivated/recombinant virus
measles	live, attenuated virus
Neisseria meningitidis ACWY*	capsules**
Neisseria meningitidis B*	factor H binding protein, *Neisseria* heparin binding antigen, *Neisseria* adhesin A, outer membrane vesicles
mumps	live attenuated virus
pertussis (*Bordetella pertussis*)*	chemically or genetically detoxified pertussis toxin, filamentous haemagglutinin (FHA), 69-kDa outer-membrane protein (also known as pertactin), fimbrial-2 and fimbrial-3 antigens
*Streptococcus pneumoniae**	capsule**
polio	inactivated or attenuated virus
rabies	killed virus
rotavirus	live attenuated virus
rubella	live attenuated virus
shingles	live attenuated virus
smallpox	live vaccinia virus
tetanus (*Clostridium tetani*)*	exotoxin**
tuberculosis (*Mycobacterium tuberculosis*)*	bacille Calmette-Guerin (live attenuated strain of *Mycobacterium bovis*)
typhoid fever (*Salmonella enterica*) serotypes typhi and paratyphi A, paratyphi B (tartrate negative), and paratyphi C*	Capsular polysaccharide** and live-attenuated Ty21a strain of *Salmonella* serotype typhi
varicella	live attenuated virus
yellow fever	live attenuated virus

Table 1.1 Vaccines currently approved for use in the United States and those for which the vaccine antigens are capsule or exotoxins. Bacterial vaccines shown in *; vaccine antigens that are either capsule or an exotoxin shown in **.

EXPERIMENTAL MODELS OF PATHOGENICITY

The ethics of using humans, animals and cell lines in pathogenics research

Research using human subjects: Clearly, human beings are the model with the highest fidelity to study microbes that infect them. However, because of the virulence and pathogenicity of certain microbes, it is unethical to examine these agents in human beings and experiments must be conducted in appropriate animal models or *ex vivo*. That does not mean that humans cannot serve as experimental models for certain infectious agents under tightly controlled and ethically approved conditions.

Informed consent: The Office for Protection from Research Risks, U.S. Department of Health and Human Services, defines informed consent as follows: It is a process, not just a form. Information must be presented to enable persons to voluntarily decide whether or not to participate as a research subject. It is a fundamental mechanism to ensure respect for persons through provision of thoughtful consent for a voluntary act. The procedures used in obtaining informed consent should be designed to educate the subject population in terms that they can understand. Therefore, informed consent language and its documentation (especially explanation of the study's purpose, duration, experimental procedures, alternatives, risks, and benefits) must be written in 'lay language,' (i.e., understandable to the people being asked to participate). The written presentation of information is used to document the basis for consent and for the subjects' future reference. The consent document should be revised when deficiencies are noted or when additional information will improve the consent process. Unfortunately, there are appalling examples where informed consent was not sought for the study of infectious agents in human beings. Informed consent is a central tenet in the ethics of human experimentation, but it by no means is the only consideration. There are other requirements that apply not only to human research but to animal research, too.

Significance: The research must test a hypothesis that allows acquisition of data that has scientific, clinical or social value.

Sound experimental design and methods: Generally, in large multicentre studies biostatisticians are an integral part of the research team that design the experiments and ensure that they are adequately powered and that data are obtained in a form conducive to statistical analysis. Determination of adequate sample size should take into account subject attrition and withdrawal from the study. Unfortunately, all too often, investigators fail to consider how they will analyse their data when they are designing experiments, and statistical analysis is relegated to an afterthought once the experiments have been completed. As a result, often it becomes necessary to change the initial hypothesis to fit the results of the experiment. This is termed *post hoc* analysis.

Subject selection: Human subjects should be selected based on the goals of the research and not because of convenience. Women, minorities and children should be included if appropriate in order to maximise the benefits and value of the research. In general, the individuals selected for study should come from populations likely to benefit from the results of the study.

Risk–benefit ratio: Potential risks to individual subjects should be minimised and benefits maximised, and benefits should outweigh risks.

Independent review: Protocols involving either human subjects or animals must be approved by institutional review boards.

Advantages and disadvantages of human experimentation in the study of the pathogenesis of infectious diseases

As stated above, humans are the experimental subject with the greatest fidelity when the infectious agent is solely a human pathogen and the experiments meet ethical requirements. However, because of high inter-subject variability, large numbers of subjects must be enrolled with due consideration to the inclusion of women and minorities and children if necessary. The experiments must be adequately powered to take account of the potential requirement for subset analysis and the inevitable attrition and dropouts.

Animal models in the study of the pathogenesis of infectious diseases

The use of animals in research does not relieve ethical requirements. The same requirements exist for independent review, significance, sound experimental design and methods to ensure that the number of animals used is adequate to support or refute the hypothesis. As always, biometry should be an integral part of the experimental design. Many species have been and are utilised as models of human infectious disease, and these include yeast, fruit flies, the nematode *Caenorhabditis elegans*, zebra fish, mice, rats, guinea pigs, rabbits, cats, dogs and pigs, among others. Consideration of the advantages and disadvantages of each species of experimental animal is beyond the scope of this book, so discussion will be limited to subhuman primates and rodents, particularly mice. The ongoing question remains whether any of these species are valid models of human infectious diseases.

Because nonhuman primates are closely related to humans and susceptible to human infectious agents, it is logical that they should be the experimental model of choice if human beings cannot be employed. However, it must be realised that even primates differ considerably from human beings. The high cost and limited availability of primates from the wild restrict their use largely to regional primate centres that maintain breeding programs which allow the provision of animals breed in captivity. Captive breeding also allows animals of known pedigree. In addition, specific pathogen-free (SPF) primates are available in these centres. Other than the concern that primates are not a perfect substitute for human beings, the major limitation of research with nonhuman primates is adequately sizing the groups to provide sufficient power to allow for robust statistical analysis of data taking into account the extensive genetic variation of outbred primates.

In contrast to primates, rodents – particularly mice – are far easier and far cheaper to house and maintain and are readily available. An additional advantage to the use of mice is the availability of many inbred strains that serve to minimise inter-animal variation. This permits the use of smaller group sizes than is possible with outbred animals. In addition to inbred strains, transgenic knock-in, knock-out, congenic, chimeric and 'humanised' mice are available. Furthermore, stains are available that are reared germfree (axenic) or SPF. Clearly, the mouse and rat are the preferred laboratory animals and account for more than 95% of all animals used in medical research. The most commonly used mouse strains are C57BL/6 and BALB/c; the most commonly used rat strains are Sprague-Dawley and Wistar.

There is, among many researchers, the tacit assumption that data obtained from experiments conducted in mice and rats are extrapolatable to human beings. There are many instances where this has been shown not to be the case and which have led to questioning whether rodent models – or indeed other animal models – are relevant to human disease. Some of the limitations/disadvantages of rodent models are discussed below.

- **Age:** Most rodents used in research are young (weanlings). Mice are weaned at 21 days of age, that is, their mothers are removed from the cage to prevent them from nursing. Mice are considered juveniles between the ages of 3 and 8 weeks and adults after that. The life span of laboratory mice is about 18 months. Therefore, in many cases the age of the mice used in experiments does not reflect the age of human beings when they are most susceptible to disease.
- **Environment:** Rodents used in research are generally housed in 'shoe box' cages and, therefore, do not mimic human beings as far as the effects of stress and exercise levels, for example.
- **Lack of genetic variability:** As stated above, almost all rodent strains used in research are inbred. Although this allows a level of homogeneity that minimises variance in the data, it does not replicate the situation in human beings. Rarely are data obtained using one strain of mouse confirmed in other inbred strains.
- **Use of animals of one sex:** It is customary to use rodents of a single sex in experiments. Again, this does not reflect the distribution of infectious diseases in human beings.
- **Anatomical and physiological differences:** Mice have a greater ratio in the length of the small intestine to the colon than do humans. In addition, they have a prominent cecum and lack an appendix. Rats lack a gall bladder. Mice have a considerable amount of bronchus-associated lymphoid tissue, whereas humans do not.
- **Different microbiotas and pathogens:** Mice and rats harbour different microbiotas compared with humans and are susceptible to different

pathogens. Thus, a human pathogen implanted into a rodent exists in a different ecological relationship than it does in its human host. Often implantation of human pathogens in rodents requires antibiotic suppression of the endogenous microbiota. Nowhere is the consideration of the ecological relationship between the microbe and its host more significant than when human commensal bacteria and fungi are 'implanted' into conventional and germfree rodents with the expectation that data relevant to the autochthonous microbe will result.

Coprophagy: Coprophagy is the practice of eating faeces, and it is performed by rodents and many other animals, including subhuman primates. Mice and rats will eat faecal pellets from the anus of another animal and from the floor of the cage. The significance of coprophagy is that it recycles the endogenous gut microbiota and microorganisms that have been experimentally implanted. In fact, because of coprophagy, a microbe cannot be said to have been established in the alimentary tract of rodents, rather its presence may reflect continuous recycling. Coprophagy can be eliminated by housing rodents individually in screen-bottom cages or by tail cupping but these are rarely done because of added cost and inconvenience.

Other considerations in experimental models of the pathogenesis of infectious diseases

Choice of microbial strain: All too often strains of pathogenic bacteria used in animal and other models of infectious diseases are obtained from type culture collections such as the American Type Culture Collection (ATCC) or the British National Collection of Type Cultures (NCTC) or they come from laboratory strain collections. In these cases, the strains are many generations removed from their human hosts and their properties may be different. Moreover, to compound this problem, only a single such strain may be used repeatedly. The possibility exists that data obtained with such a strain will not predict the performance of wild-type strains of the pathogen in its host. Therefore, it is preferable to use several fresh clinical isolates whenever possible.

Mucosal surfaces: The vast majority of infectious agents invade the human body *via* mucosal surfaces, particularly those of the respiratory and gastrointestinal tracts. Thus, it is frequently necessary to study, in humans or experimental animals, attachment to and entry into mucosal epithelial cells by pathogenic microbes (discussed below) and to detect and quantitate exocrine antibodies notably secretory immunoglobulin A (SIgA) directed against these pathogenic organisms. Collection of mucosal secretions such as tears, nasal secretions, saliva, respiratory secretions, milk, gastrointestinal, genitourinary secretions and sweat, present difficulties that are not encountered when collecting blood, cerebrospinal fluid, and peritoneal fluid. Regardless of the type of external or internal secretion, if repeated sampling is part of the experimental design, then samples should be collected at the same time of the day on each occasion to control for diurnal variation. This is particularly important when collecting saliva from both humans and other mammals because there is little salivary flow during sleep so the concentration of analytes is highest upon waking. With the exception of the collection of whole-mouth saliva, urine and milk from human subjects, it is necessary either to stimulate the flow of secretions or to employ a lavage. Both methods come with their own set of difficulties. The cholinergic agent, pilocarpine, is frequently used to stimulate salivation, lacrimation and sweating in experimental mammals. Oxytocin is used to induce lactation. Unfortunately, there is an inverse relationship between flow rate and antibody concentration in external secretions. Because of this relationship, it is essential to control for differences in flow rate both within and between animals and humans. In the respiratory tree (including the nasal cavity), gastrointestinal tract and vagina there are no secretions to speak of, simply a layer of mucus overlying the epithelium. In these cases, it is necessary to lavage or use an absorbent wick/sponge to recover the mucus. More often than not, the full volume of the lavage solution is not recovered. For example, in the case of nasal lavage, some of the lavage solution invariably runs down the back of the throat into the mouth. Whatever the reason, the inability to recover the full lavage volume affects the quantitation of analytes. Because the function of the lavage is to recover the mucus overlying the epithelium, it is difficult to determine how much mucus has been recovered, that is, how much the lavage has been diluted by the volume of mucus. This difficultly can be overcome by including an internal standard in the lavage solution so that the volume of mucus recovered can be determined by the reduction in concentration of the internal standard. An alternative approach is to use an absorbent sponge/wick. It is claimed that macromolecules such as antibodies can be completely eluted from the sponge/wick. This is accomplished by suspending the sponge/wick in buffer followed by centrifugation and collection of the supernatant. With the exception of milk, the concentration of antibodies in external secretions is low, so it becomes necessary to concentrate lavage samples, and perhaps the supernatants, obtained by sponge/wick elution. Concentration is generally accomplished by lyophilisation (freeze-drying). However, this process results in protein denaturation, so a variable fraction will not be recovered.

In secretions, SIgA exists predominantly as a dimer covalently linked to a joining chain (J-chain) and a secretory component (SC). SIgA forms Ca^{2+} ion-dependent complexes with high-molecular-weight mucins and other factors found in secretions. Therefore, if secretions are frozen without the prior addition of EDTA, a variable fraction of SIgA is trapped in the precipitate that forms when the secretion is thawed. All too often, this precipitate is removed by centrifugation and discarded, taking with it a variable amount of analyte. There is a high level of proteolytic activity in external secretions derived from the endogenous microbiota that colonise the barrier epithelia. The addition of EDTA will also inactive metalloproteases, but the addition of other classes of protease inhibitors are indicated.

The need for normalisation: It is obvious from the foregoing that the data obtained from external secretions must be normalised to control for differences in flow to enable meaningful data to be obtained both within and between experiments. One way in which this can be accomplished is to express data based on the concentration of total protein in each sample. Thus, the analyte can be expressed per milligram of total protein. The enzyme-linked immunosorbent assay (ELISA) is the most widely used assay to measure antimicrobial antibodies in both internal and external secretions. One limitation of the measurement of antimicrobial antibodies by ELISA is that the data are output as optical density units. Although this may be satisfactory for making comparisons within individual laboratories, it makes comparisons with data from other laboratories difficult because optical density is affected by incubation time, source of antibody reagents, and nature of the plastic plate, to name but a few possible confounding factors. However, this limitation can be overcome by combining the measurement of antimicrobial SIgA antibodies and total SIgA immunoglobulin on a single 96-well microtitration plate. In this manner, the read-out is in primary units, *viz.*, micrograms of specific SIgA antibody per milligram of total SIgA immunoglobulin.

Human cell lines as a surrogate for microbe–host interactions

Because of the complexity of animal models and continued concern about their relevance to human disease, human cell lines have been used as more simple, more controlled and more reproducible models of microbe–host interactions. The aim of experimental models using cell lines ought to be to attempt to reproduce, as far as possible, the anatomy and physiology of the cells in the organs from which they were derived and to reflect the nature of the cells that are encountered by the pathogen in the natural infection. Fresh-frozen sections obtained from biopsy material can be used to study adherence of bacteria and fungi. Alternatively, biopsy tissue can be digested with enzymes and individual cells isolated. The advantage of primary cells is that they are physiologically normal and, thus, a better correlate of *in situ* microbe–host interactions than immortalised cells. However, primary cells have a limited life span in culture and die after a limited number of population doublings. The transfer of an exogenous human telomerase reverse transcriptase complementary DNA (hTERT cDNA), that encodes the catalytic subunit of human telomerase, can prevent telomere shortening and immortalise primary human cells. Significantly, hTERT-immortalised cells have the same phenotypic properties as the primary cells from which they were derived. Other methods of immortalisation are to select malignant cell that have lost growth control or to transform cells with oncogenic viruses. In fact, the HeLa cervical adenocarcinoma cell line, the first immortal cell line selected from biopsy tissue, contains human papillomavirus (HPV) 18 DNA. The availability of immortal cells of various types has dramatically expanded their use in research. However, as with all models, immortalised cells have their limitations. For example, if the cell line is derived from malignant cells, these cells will continue to accrue additional mutations as they replicate. This means that not only will a cell line used over time in a single laboratory change, but the same cell line used in different laboratories will also change independently in each laboratory. *In situ* cells, particularly epithelial cells, are polarised, that is, they have apical, basal and lateral surfaces. Neighbouring cells are attached to each other by junctions that consist of multi-protein complexes. In addition, the cells are connected to underlying connective tissue *via* a basement membrane comprising a scaffold consisting of glycoproteins, including type IV collagen. The basement membrane plays a role in cellular functions such as signalling and epithelial cell heterogeneity. Cell lines are usually used in suspension or are allowed to attach to the surface of plastic petri dishes or multi-well plastic dishes. Tissue culture cells growing in suspension or in nonconfluent surface growth lack polarity and molecules that are localised to a particular cell surface on cells *in situ* are distributed over the entire cell surface. Growing a cell line to confluence does not overcome the lack of polarisation. In addition, certain receptors and other molecules that are expressed *in situ* may not be expressed in cell lines. In an attempt to more closely approximate a basement membrane, the plastic substratum can be coated with a basement membrane surrogate such as Matrigel. Matrigel is a solubilised basement membrane preparation derived from the Engelbreth-Holm-Swarm mouse sarcoma and contains copious amounts of extracellular matrix proteins such as laminin, type IV collagen, heparin sulphate, proteoglycans, nidogen-1, and several growth factors. However, this tumour-derived basement membrane is a poor substitute for natural basement membranes. Cells *in situ* are bathed by tissue fluid or secretions, however, cell lines are suspended in or submerged below tissue culture medium or buffer. Furthermore, the apical surfaces of mucosal epithelial cells are covered with mucus and are flushed by mucus-containing secretions (**Figure 1.3**).

Neither of these features are replicated in experiments employing epithelial cell lines. Finally, it is not uncommon that the cell lines used to study adhesion and/or 'invasion' are not the type of cells for which the pathogen has tropism in the human host. This is because either the appropriate cell line does not exist or it is simply a matter of convenience. Regardless of reason, it calls into question the extent to which the resulting data are relevant to the relationship between the microbe and its intact host.

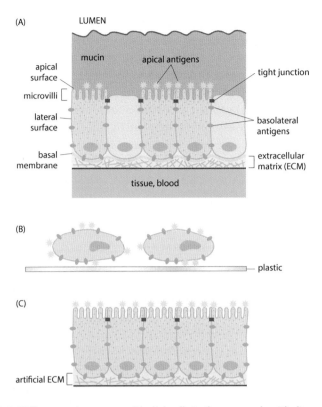

Figure 1.3 Differences between epithelial cells in the mucosal epithelium and cell lines in culture. (A) Epithelium *in vivo*; (B) nonconfluent, nonpolarised tissue culture cells; and (C) polarised monolayer of tissue culture cells attached to a semipermeable membrane. (From Wilson, B.A. et al., *Bacterial Pathogenesis: A Molecular Approach*, 3rd ed., ASM Press, Washington, DC, 2011.)

BIBLIOGRAPHY

Bliven KA, Maurelli AT. 2016. Evolution of bacterial pathogens within the human host. *Microbiology Spectrum*, 4(1). doi:10.1128/microbiolspec.VMBF-0017-2015.

Calderone RA, Cihlar RL. 2002. *Fungal Pathogenesis: Principles and Clinical Applications* (Mycology Book 14). CRC Press, Boca Raton, FL.

Casadevall A, Pirofski L-A. 2003. The damage-response framework of microbial pathogenesis. *Nature Reviews Microbiology*, 1: 17–24.

Cihlar RL, Calderone RA. 2009. *Methods in Molecular Biology 499:* Candida albicans *Methods and Protocols.* Humana Press, New York.

Didelot X et al. 2016. Within-host evolution of bacterial pathogens. *Nature Reviews Microbiology*, 14(3): 150–162.

Falkow S. 1988. Molecular Koch's postulates applied to microbial pathogenesis. *Reviews of Infectious Diseases*, 10: S274–S276.

Fredricks DN, Relman DA. 1996. Sequence-based identification of microbial pathogens: A reconsideration of Koch's postulates. *Clinical Microbiology Reviews*, 9: 18–33.

Fux CA et al. 2005. Can laboratory reference strains mirror 'real-world' pathogenesis? *Trends in Microbiology*, 13(2): 58–63.

Guidance for applicants: Informed consent. http://ec.europa.eu/research/partici-pants/data/ref/fp7/89807/informed-consent_en.pdf.

Informed consent tips. 1993. https://www.hhs.gov/ohrp/regulations-and-policy/guidance/informed-consent-tips/index.html.

Katze MG, Korth MJ, Law GL, Nathanson N. 2016. *Viral Pathogenesis: From Basics to Systems Biology.* 3rd ed. Academic Press, Elsevier, London, UK.

Koch R. 1884. Die Atiologie der Tuberculose. In: Schwalbe J (Ed.), *Gesammelte Werke von Rober Koch,* Vol. 1, pp. 467–565.

Lemichez E, Lecuit M, Nassif X, Bourdoulous S. 2010. Breaking the wall: Targeting of the endothelium by pathogenic bacteria. *Nature Reviews Microbiology,* 8: 93–104.

Loker ES, Hofkin BV. 2015. *Parasitology: A Conceptual Approach.* Garland Science, New York.

Nash AA, Dalziel RG, Fitzgerald JR. 2015. *Mim's Pathogenesis of Infectious Disease.* 6th ed. Elsevier, London, UK.

Nathanson N. 2007. *Viral Pathogenesis and Immunity.* 2nd ed. Academic Press, Elsevier, London, UK.

Robbins JR, Bakardjiev AI. 2012. Pathogens and the placental fortress. *Current Opinion in Microbiology,* 15(1): 36–43.

Wilson BA, Salyers AA, Whitt DD, Winkler ME. 2011. *Bacterial Pathogenesis: A Molecular Approach.* 3rd ed. ASM Press, Washington, DC.

Wilson M, McNab R, Henderson B. 2002. *Bacterial Disease Mechanisms: An Introduction to Cellular Microbiology.* Cambridge University Press, Cambridge, UK.

Chapter 2: Normal Microbiotas of the Human Body

INTRODUCTION

The purpose of this chapter is to provide an overview of the composition of the resident microbiotas of the barrier epithelia of the human body based on the use of modern molecular techniques that employ deep sequencing of 16S rRNA. This review is not intended to be exhaustive and is limited almost entirely to microbiotas in healthy humans. For more detailed information, the reader is referred to several excellent texts and many excellent journal review articles that consider individual microbiotas or the microbiome in both health and disease. We begin with terms used in microbial ecology.

Microbiome: The ecological community of commensal, symbiotic and pathogenic bacteria that share our body space (as defined by Joshua Lederberg, Nobel Prize-winning molecular biologist). Others have defined the microbiome as the collective genomes of the microorganisms in an environment or community.

Ecosystem: A natural unit that includes living and non-living parts interacting to produce a stable system.

Microbiota: The community of microorganisms in a particular environment, for example, the small intestine.

Symbiosis: A close, prolonged association between two or more different organisms of different species that may or may not benefit each member.

Commensalism: A symbiotic relationship between two species in which one species benefits and the other neither benefits nor harmed.

Mutualism: A type of symbiosis between two species of organisms in which both benefit from the association.

Habitat: The natural home of an organism.

Niche: The *function* of an organism in a community, for example, the consumption of lactic acid produced by streptococci and other homolactate fermenters in the presence of fermentable carbohydrate by *Veillonella* species.

The barrier epithelial surfaces of the human body are colonised with complex communities of microorganisms that have co-evolved with us. These communities are termed endogenous, resident or autochthonous. In addition, the microbiotas contain transients, allochthonous, members that are not permanent residents but come from other sites or other individuals. Immediately *postpartum*, the barrier epithelia of the neonate become colonised with microorganisms from the vagina of the mother, the immediate family and perhaps from the environment. Colonisation begins with a nucleus of organisms termed pioneers that are specifically adapted for residence at the many body habitats that vary in their physiological properties. These initial colonisers are unique to their habitat.

Resident microbiotas are generally commensal or mutualistic and exist in a state of dynamic equilibrium with the human host. However, the balance between the body surface microbiotas and the host may shift in favour of a particular microorganism or microorganisms as the result, for example, of a breach in the barrier epithelium, impairment of the immune system or the chronic use of broad-spectrum antibiotics. Under these circumstances, commensal microorganisms may become opportunistic pathogens. Antibiotics can shift the balance of the endogenous microbiota, allowing either the outgrowth of particular endogenous bacteria or the establishment of extrinsic pathogens. This is termed dysbiosis. An example of dysbiosis is the outgrowth or acquisition of ToxA$^+$ *Clostridium difficile* that causes pseudomembranous colitis that may follow the chronic use of broad-spectrum antibiotics. In an attempt to restore balance to the gut microbiota, faeces from a healthy individual may be transplanted (faecal transplant) into the gut of the affected person (recipient).

With the exception of the alimentary canal where diet is a factor in the composition of the microbiota, the endogenous microbiotas at other epithelial surfaces obtain nutrients from the host and there is no requirement for exogenous nutrient sources. Even in the mouth, the entrance of the alimentary canal, the only bacteria affected by hyperalimentation (nutrients supplied intravenously) are highly aciduric (acid tolerant) and acidogenic (acid-producing) bacteria such as *Streptococcus mutans*.

Our understanding about the composition of the endogenous microbiotas of the human body resulted from a wealth of studies conducted over many decades by isolating and culturing bacteria and fungi from the barrier epithelia. However, recent advances in high-throughput molecular techniques such as sequencing, proteomics, and metabolomics with metabolic network modelling have extended knowledge obtained through culture by not only enabling identification of microorganisms – almost exclusively bacteria – that could not be cultured, but determining the functions of these bacteria in their ecosystems. Exploration of the human microbiome has been driven by the Human Microbiome Project funded by the U.S. National Institutes of Health, the European MetaHit Project and the creation of the International Microbiome Consortium. These molecular techniques now are being extended to fungi (the mycobiome) and viruses (the virome). However, although these new molecular techniques have resulted in a more comprehensive cataloguing of microorganisms on body surfaces at the phylum level and below, they have not altered our fundamental understanding of the predominant members of human microbiotas established by culture. The efficiency of molecular techniques has the capacity to permit examination of the microbiotas at a larger number of sites and in a larger number of individuals at more time points than was generally possible using culture techniques. Significantly, the sensitivity of molecular techniques has led to the detection of bacterial populations on surfaces of the human body that were formerly believed to be sterile. The availability of these new data have allowed attempts to correlate population shifts in commensal microbiotas with health and disease. However, to date, with few exceptions, studies have examined the microbiotas of relatively few subjects. Far too few, in fact, to allow such generalisations to be made. Moreover, the sweeping surveys of resident microbiotas allowed by recent molecular techniques rarely consider community structure at the level of the clone (strain). This is important because each member of a microbial community occupies a unique niche. For example, although the human pioneer commensal oral bacterial species *Streptococcus mitis* is the core component of the oral microbiota among individuals, each individual, even monozygotic twins, harbour unique clones. Moreover, these clones turn over and are replaced over time, perhaps as a result of the selective pressure of the mucosal immune system. Furthermore, within this species different physiological variants occupy distinct habitats in the oral cavity, and this is likely the case for other resident bacteria.

Microbial communities on the external and internal epithelial surfaces of the human body exist as biofilms (see Chapter 3, *Biofilms*) and many are not amenable to non-invasive sampling. Such is the case for the urinary, gastrointestinal and respiratory tracts, so data obtained from sampling these areas utilise surrogates such as secretions/lavage samples, urine and faeces. The extent to which these samples reflect the composition of surface residents rather than planktonic microorganisms remains unclear.

Using advanced molecular methods, the microbiota of faeces – usually in small numbers of subjects – has been surveyed with the assumption that it represents the adherent colonic microbiota. These studies have revealed a previously unknown complexity of the faecal microbiota that is highly variable across time and human populations. Such findings have led some to propose that every individual harbours their own unique *taxonomic* microbiotas on their barrier epithelia and reject the notion of core microbiotas, that is, a set of microorganisms common to a particular habitat among subjects. Rather, they argue that commonality lies with the functions of the bacteria in these communities so that different individuals share a core microbiome but not a core microbiota. However, others argue that while it may be true that different proportions of phyla and different clones are found in different individuals, a core microbiota that is common to each barrier epithelium exists. Data showing that multiple lineages of the predominant bacterial taxa in the gut of modern humans arose *via* co-speciation with humans, chimpanzees, bonobos, and gorillas over the past 15 million years supports the notion of the existence of core components of all barrier epithelial microbiotas common to all individuals.

Microbiota of the skin

The barrier epithelium of the skin is second in area only to the mucosal epithelia. The skin is covered with hair except for the glabrous areas comprising the ventral portion of the fingers, palms, soles of feet, lips, labia minora, and glans penis. Some areas of the skin are more moist (axillae, groin, and perineum) than others (dorsal and ventral surfaces of the thorax, arms and legs). The skin contains eccrine glands concentrated on the forehead, axillae, palms and soles and apocrine sweat glands located in the eyelids, axillae and perineum that secrete a variety of antimicrobial substances (reviewed later in the text) and sebaceous glands connected to hair follicles. There is a higher concentration of sebaceous glands in the skin of the head, neck and shoulder girdle than other areas of the skin. Certain skin bacteria break down sebum, the product of sebaceous glands, to fatty acids that contribute to the acid pH of the skin surface, which can be as low as pH 4.8. Common to all barrier epithelia, the surface cells (squames) of the skin desquamate continuously at a rate estimated to be approximately 1,000 cells/cm^2/h. Furthermore, the skin surface is subject to frictional forces as well as the shear forces of airflow, bathing and dressing and undressing. In addition, the temperature of the skin varies by about 10°C depending on location and is lower than core body temperature. It is clear that the skin presents surprisingly varied and complex environments, underlying the need to sample multiple sites to gain a comprehensive understanding of microbial diversity. Factors contributing to variation in the skin microbiota are shown in **Figure 2.1**.

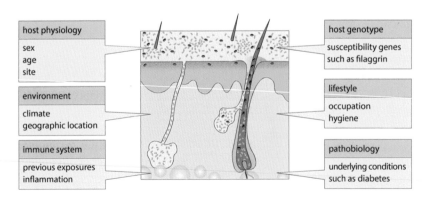

Figure 2.1 Factors contributing to variation in the skin microbiome. (From Grice, E.A. and Segre, J.A., *Nat. Rev. Microbiol.,* 9, 244–253, 2011.)

Human skin harbours bacteria, fungi, viruses and archaea. Bacterial cell densities vary from approximately $10^2/\text{cm}^{-2}$ on the fingertips and back to $10^6/\text{cm}^{-2}$ on the forehead and axilla.

As defined by 16S ribosomal RNA metagenomic sequencing, the microbiotas of the skin, oral cavity and GI tract fall into four phyla, Actinobacteria, Firmicutes, Bacteroidetes and Proteobacteria, although they are present in different proportions. Recent inventories of the skin bacterial microbiota put its diversity at some 19 phyla and 205 genera. However, the principal genera are but three, *Corynebacterium*, *Propionibacterium*, and *Staphylococcus*. Other genera present are *Streptococcus* and *Lactobacillus*. A larger number of species within these genera have been found by molecular methods than had previously been detected by classical culture techniques (**Figure 2.2**). *Corynebacterium* species predominate at moist sites, and *Propionibacterium* and *Staphylococcus* species predominate at sebaceous sites and *Staphylococcus* – generally *S. epidermidis* – at dry sites. Studies have shown the most diverse site to

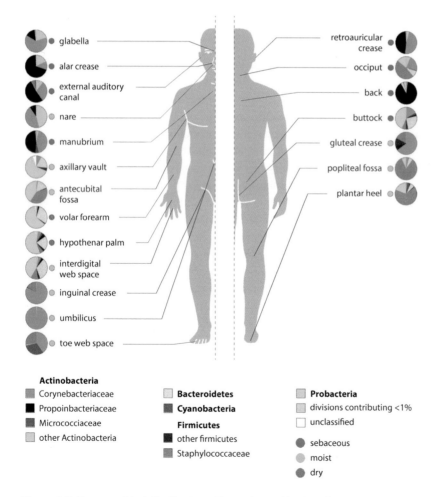

Figure 2.2 Topographical distribution of bacteria on skin sites. (From Grice, E.A. and Segre, J.A., *Nat. Rev. Microbiol.*, 9, 244–253, 2011.)

be the volar forearm and the least diverse site the retroauricular crease (Figure 2.2). Overall, sebaceous sites tend to be less diverse than moist and dry sites. Palmer surfaces of the hands have been shown to contain as many as 158 bacterial species, perhaps because hands are in frequent contact with non-sterile objects.

In contrast to bacterial composition and diversity, relatively little is known about the fungal, viral or archaean contribution to the skin microbiome. Species of yeasts belonging to the genus *Malassezia* appear to dominate the skin fungi with greater diversity only observed in some samples from the foot where *Aspergillus, Cryptococcus, Rhodotorula* and *Epicoccum* have been identified. The relative abundance of fungal genera and *Malassezia* species at different human skin sites is shown in **Figure 2.3**. However, it is difficult to determine which fungi are human commensals and which are transients from the environment. Both *Archaea* and DNA viruses have also been detected on the skin but, as with fungi, whether viruses are transients or commensals remain to be determined.

Microbiota of the vagina

The environment of the vagina is profoundly affected by the level of the hormone oestradiol, which is low in neonatal and postmenopausal females and high in pubertal and premenopausal females. The level of oestradiol correlates with the thickness and glycogen content of the epithelium, acidity and oxidation/reduction potential of the vagina. These variables influence the composition of the vaginal microbiota. The principal source of nutrients for the vaginal microbiota is cervical and vaginal mucus. Recent studies of the vaginal bacterial microbiota of healthy premenopausal women have lead to the definition of six distinct communities termed 'vagitypes' or community state types (CSTs), some of which are dynamic, whereas others are relatively stable. One of four species of *Lactobacillus*, *L. crispatus, L. inners, L. jensenii*, and *L. gasseri*, dominate four of the CSTs whereas the remaining two lack significant numbers of lactobacilli and contain anaerobic gram-negative rods such as *Prevotella, Megasphaera, Gardnerella, Sneathia*, and *Atopobium* (**Figure 2.4**). The forces driving variability are likely the menstrual cycle, sexual activity and the dynamics of the communities themselves. In some studies samples are self-collected, thus there may be inherent variability in the collection site. Other studies have sampled various sites in the vagina. As expected, based on cultural studies, *Lactobacillus* species are a core component of the microbiota but they do not always dominate. For example, some communities are dominated by a single species of *Lactobacillus* such as *L. crispatus, L. gasseri, L. inners* or *L. jensenii*, whereas others contain low numbers of several *Lactobacillus* species complemented by consortia comprising the genera *Anaerocoocus, Corynebacterium, Fingoldia* and *Streptococcus* or *Atopobium, Prevotella, Parvimonas, Sneathia, Gardnerella, Mobiluncus* or *Peptoniphilus* (**Figure 2.5**). Over 250 species have been identified in the vagina.

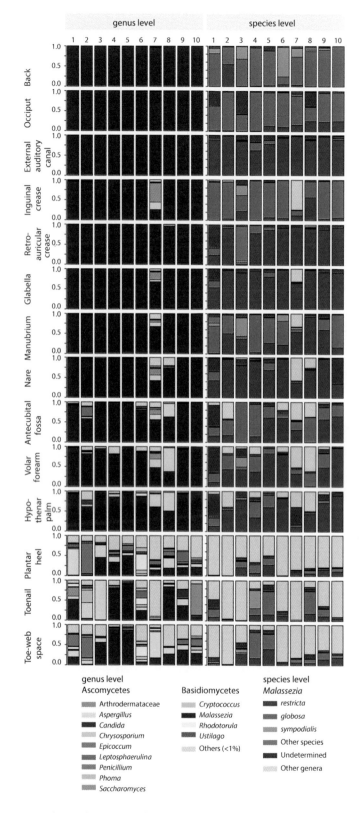

genus level

species level

Back
Occiput
External auditory canal
Inguinal crease
Retro-auricular crease
Glabella
Manubrium
Nare
Antecubital fossa
Volar forearm
Hypo-thenar palm
Plantar heel
Toenail
Toe-web space

genus level
Ascomycetes
- Arthrodermataceae
- Aspergillus
- Candida
- Chrysosporium
- Epicoccum
- Leptosphaerulina
- Penicillium
- Phoma
- Saccharomyces

Basidiomycetes
- Cryptococcus
- Malassezia
- Rhodotorula
- Ustilago
- Others (<1%)

species level
Malassezia
- restricta
- globosa
- sympodialis
- Other species
- Undetermined
- Other genera

Figure 2.3 Relative abundance of fungal genera and *Malassezia* species at different human skin sites. (From Findley, K. et al., *Nature*, 498, 367–370, 2013.)

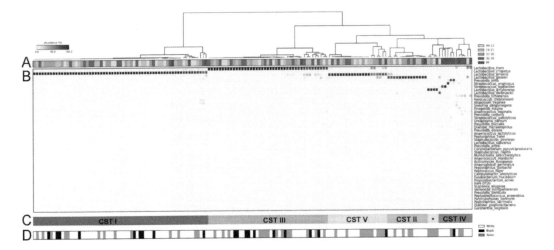

Figure 2.4 Bacterial species composition of vaginal community state types (CST) throughout pregnancy and postpartum. (A) Hierarchical clustering analysis of microbial species data shows that vaginal microbiomes from a UK cohort can be clustered into five major groups consistent with vaginal CSTs identified in non-pregnant and pregnant North American populations. Around 75% of all postpartum samples were found to cluster into CST-IV. (B) Heat map of relative abundance of bacterial species characterising the CSTs. (C and D) CSTs, I, III and IV were represented by similar proportions of white, Asian and black ethnicities; however, CST II and V were almost void of representation from black women. CST*, *Lactobacillus amylovorous*–dominated microbiome. (From MacIntyre, D.A. et al., *Sci. Rep.*, 5, 8988, 2015.)

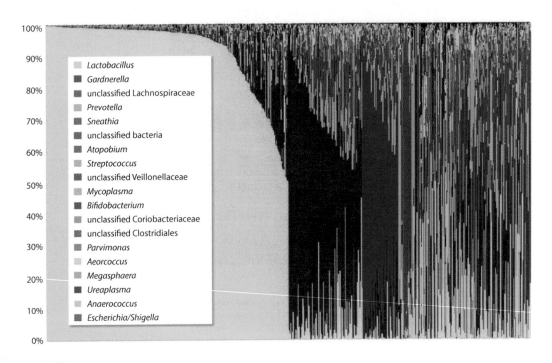

Figure 2.5 Mid-vaginal microbiome profiles using genus-level taxonomic classification. A total of 650 microbiome profiles are represented. Each bar represents the genus-level microbiome profile for one mid-vaginal swab sample, and each colour represents a distinct genus. The V1–V3 hypervariable region of the 16S rRNA gene was targeted and, on average, *ca.* 30,000 reads per sample were generated. (From Fettweis, J.M. et al., *Chem. Biodivers.*, 9, 965–975, 2012.)

It has been generally accepted that a high level of lactobacilli in the vagina is consistent with health and that a low abundance of lactobacilli and high microbial diversity is indicative of disease, but this is not necessarily so. This is because other lactic acid–producing species maintain a low physiological pH. However, the protective role of lactobacilli likely extends beyond fermentation of glycogen to lactic acid to include competition for adhesion to epithelium, and production of hydrogen peroxide and bacteriocins.

Premenstrual girls: The microbiota of premenstrual girls is also dominated by species of *Lactobacillus*. Although associated with bacterial vaginosis in women, *Gardnerella vaginalis* is a prominent component of the microbiota in about one-third of premenstrual girls.

Postmenopausal women: The decline in oestrogen associated with menopause results in a thinning of the vaginal epithelium, a reduction in its glycogen content, blood flow and secretions. These changes are associated with a reduction in the numbers of lactobacilli as the dominant genus and the appearance of *Bacteroides, Prevotella, Escherichia coli, Gardnerella, Mobiluncus*, Lancefield group B, C and G streptococci and *Staphylococcus aureus*.

Vaginal microbiota during pregnancy: The vaginal microbiota during pregnancy varies in composition by site and gestational age. The overall diversity and richness of the microbiota is less, but more stable, than that observed in the vagina of non-pregnant women. The microbiota remains dominated by *Lactobacillus* species and the predominant species appears to be related to ethnicity. Clostridiales, Bacteroidales, and Actinomycetales are also prominent. Indeed there is a reduction in many of the usual members of the vaginal microbiota as a result of the dominance of lactobacilli. Differences between the relative abundance of bacterial phylotypes between pregnant and non-pregnant women are shown in **Table 2.1** and **Figure 2.6**.

Microbiota of the urinary tract

The urinary tract was formerly considered to be sterile, but recently a resident microbiota has been detected in the bladder. The source of samples has been urine obtained from healthy individuals and those with urinary tract disorders. Clearly, urine obtained by transurethral catheterisation of the bladder does not permit examination of other areas of the urinary tract. Moreover, the microbiota of urine reflects planktonic (free-swimming) bacteria and not those adherent to the bladder and urethral epithelia. Thus, it is not surprising that the microbiota of voided urine is different from that collected by catheterisation or suprapubic aspiration (**Table 2.2**). The number of bacteria in urine is low ranging from 100 to perhaps 10,000 colony-forming units (CFUs) per millilitre. In the female, the urinary microbiota resembles the vaginal microbiota. The richness of the microbiota in urine obtained by transurethral catheterisation is less than that of many other barrier epithelia but still some

phylotypes
phylotypes less abundant in pregnancy
Clostridiales Family XI Incertae Sedis
Anaerococcus vaginalis
Anaerococcus
Prevotella genogroup 2
Peptoniphilus
Streptococcus anginosus
Actinomycetales
Leptotrichia amnionii
Finegoldia magna
Prevotella
Clostridiales
Atopobium
Bacteria
Prevotellabivia
Eggerthella
Gardnerella vaginalis
Dialister
Ureaplasma
Lactobacillus
Atopobium vaginoe
Parvimonas micra
Bifidobacteriaceae
phylotypes more abundant in pregnancy
Lactobacillus vaginalis
Lactobacillus jensenii
Lactobacillus crispatus
Lactobacillus gasseri
non-significantly different phylotypes
Lactobacillus iners
Aerococcus christensenii

Table 2.1 Differential relative abundance of microbial phylotypes between pregnant and non-pregnant women. (From Romero, R. et al., *Microbiome*, 3, 4, 2014.)

97 operational taxonomic units (OTCs) have been detected in urine. The female urinary microbiota obtained by transurethral catheter has been shown to comprise several clusters of bacteria termed 'urotypes', which are dominated for the most part by the genera *Lactobacillus* and *Gardnerella* (**Figure 2.7**). *L. crisptatus* is the dominant lactobacillus species in health. Studies of mid-stream clean catch urine from healthy women show a dominance of *Lactobacillus* and *Gardnerella* but also of *Prevotella* and a greater level of richness in a range similar to that of the vaginal microbiota (~1,500 OTCs). Species identified in one study of mid-stream clear catch urine from healthy women are shown in **Figure 2.8**.

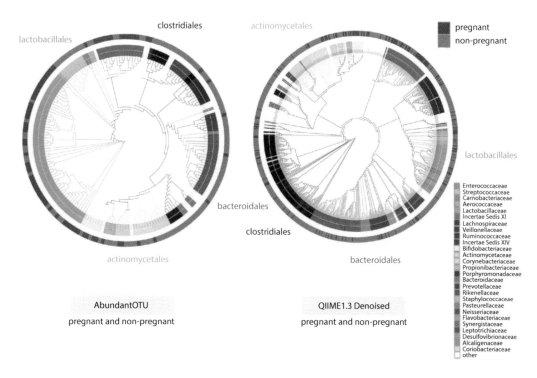

Figure 2.6 Global phylogenetic trees show microbial distribution among all the pregnant and non-pregnant subject samples. The internal cluster dendrograms are coloured by taxa family level projections (see figure key), while the mid-circle is coloured by the majority origins of OTUs from pregnant or non-pregnant subject samples (magenta, pregnant; brown, non-pregnant; red, Bacteroidales; yellow, Actinomycetales; green, Lactobacillales; blue, Clostridiales). (From Aagaard, K. et al., *PloS One*, 7, e36466.)

As might be expected from the anatomic difference between males and females, one of which is the significantly greater length of the male urethra, voided urine obtained from healthy men has a different microbiota than that of women. One significant difference is the replacement by *Corynebacterium* of *Lactobacillus* in urine from healthy males as the dominant genus (**Figure 2.9**). Furthermore, the microbiota of urine from males is far less diverse than that from females. Genera of bacteria identified in the urine of healthy males and females of various ages are shown in Table 2.2. First-catch urine, urethral swab and coronal sulcus (CS) specimens from sexually active men have been observed to contain what are generally regarded as vaginal taxa. The microbiota of the CS is more stable than the urine microbiota of the same person and is strongly affected by circumcision (**Figure 2.10**). Circumcision appears to result in a decrease in anaerobic bacteria such as members of the genera *Anaerococcus*, *Finegoldia*, *Peptoniphilus*, and *Prevotella*. Interestingly *Atopobium*, *Megasphaera*, *Mobiluncus*, *Prevotella* and *Gemella* that are generally associated with bacterial vaginosis are found in the CS of both sexually active and inactive males.

patients (n)	notable taxa[a]	sample collection method
men with STI (10) men without STI (9)	*Lactobacillus, Sneathia, Gemella. Aerococcus, Corynebacterium, Streptococcus, Veillonella, Prevotella, Anaerococcus, Propionibacterium, Atopobium, Staphylococcus*	first-void urine
men with STI (10) men without STI (22)	*Lactobacillus, Sneathia, Veillonella, Corynebacterium, Prevotella, Streptococcus, Ureaplasma, Mycoplasma, Anaerococcus, Atopobium, Aerococcus, Staphylococcus, Gemella, Enterococcus, Finegoldia, Neisseria, Propionibacterium, Ralstonia*	first-void urine
healthy women (8)	*Lactobacillus, Prevotella, Gardnerella, Peptoniphilus, Dialister, Finegoldia, Anaerococcus, Allisonella, Streptococcus, Staphylococcus*	clean-catch midstream urine
healthy controls (26; 58% women) patients with NBD (27; 48% women)	Orders: Lactobacillales, Enterobacteriales, Actinomycetales, Bacillales, Clostridiales, Bacteroidales, Burkholderiales, Pseudomonadales, Bifidobacteriales, Coriobacteriales	midstream urine, intermittent catheterization, Foley catheter
healthy adolescent men (18)	*Lactobacillus, Streptococcus, Sneathia, Mycoplasma, Ureaplasma*	first-void urine
women with IC (8)	*Lactobacillus, Gardnerella, Corynebacterium, Prevotella, Ureaplasma, Enterococcus, Atopobium, Proteus, Cronobacter*	clean-catch midstream urine
healthy women (12) women with POP or UI (11)	*Lactobacillus, Actinobaculum, Aerococcus, Anaerococcus, Atopobium, Burkholderia, Corynebacterium, Gardnerella, Prevotella, Ralstonia, Sneathia, Staphylococcus, Streptococcus, Veillonella*	clean-catch midstream urine, suprapubic aspirate, transurethral catheter
healthy men (6) healthy women (10)	*Jonquetella, Parvimonas, Proteiniphilum, Saccharofermentans* Phyla: Actinobacteria, Bacteroidetes	clean-catch midstream urine
patients receiving first renal transplant (60; 37% women)	*Lactobacillus, Enterococcus, Pseudomonas, Streptococcus* Families: Bifidobacteriaceae, Corynebacterineae	not described
healthy women (24) women with OAB (41)	*Lactobacillus, Corynebacterium, Streptococcus, Actinomyces, Staphylococcus, Aerococcus, Gardnerella, Bifidobacterium, Actinobaculum*	transurethral catheterization
healthy women (58) women with urgency UI (60)	*Gardnerella, Lactobacillus, Actinobaculum, Actinomyces, Aerococcus, Arthrobacter, Corynebacterium, Oligella, Staphylococcus, Streptococcus*	transurethral catheterization
patients with acute uncomplicated UTI (50; 76% women)	*Anaerococcus, Peptoniphilus, Streptococcus, Lactobacillus, Staphylococcus, Escherichia, Pseudomonas*	midstream urine

Table 2.2 The microbiome of the urinary urinary tract. Pooled data from several studies listed as genera, unless otherwise noted. Abbreviations: IC, interstitial cystitis; NBD, neurogenic bladder dysfunction; OAB, overactive bladder; POP, pelvic organ prolapse; STI, sexually transmitted infection; UI, urinary incontinence. (From Whiteside, S.A. et al., *Nat. Rev. Urol.*, 12, 81–90, 2015.)

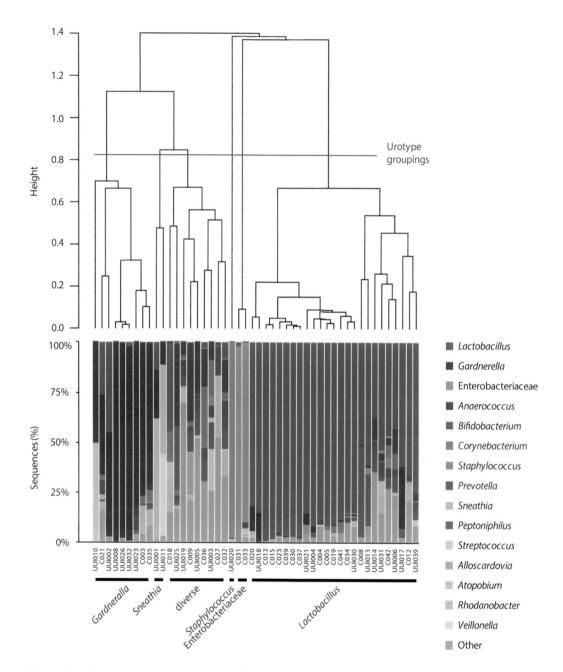

Figure 2.7 Clustering of the urinary microbiome into urotypes. The dendrogram is based on hierarchical clustering of the Euclidean distance between samples in combined urgency urinary incontinence (UUI) and non-UUI cohorts. The dashed line depicts where the clades were divided into six urotypes: *Gardnerella, Sneathia,* Diverse, *Staphylococcus, Euterobacteriaceae,* and *Lactobacillus.* The stacked bar plot below the dendrogram depicts the sequence abundances of the overall most abundant taxa. (From Pearce, M.M. et al., *MBio,* 5, e01283–14, 2014.)

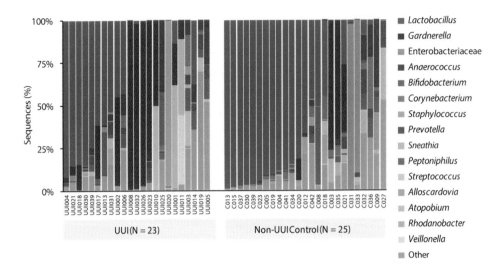

Figure 2.8 Urinary microbiome profile by cohort based on 16S rRNA gene V4 sequencing. Stacked bar plots depict the sequence abundances of the 15 most abundant genus- or family-level taxa in urgency urinary incontinence (UUI) and non-UUI cohorts. Taxa are ranked according to mean abundance across all samples. The y-axis represents the percentage of sequences for a particular bacterial taxa; the x-axis represents the study participants separated by cohort. The family *Enterobacteriaccae* could not be classified to the genus level. The remainder of sequences were combined in the category labelled 'Other.' (From Pearce, M.M. et al., *MBio*, 5, e01283–14, 2014.)

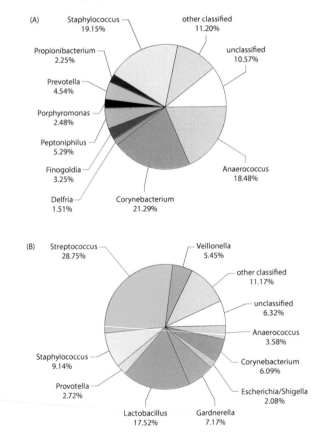

Figure 2.9 Distribution of major taxa in coronal sulcus (CS) and urine specimens. Sanger data set. (A) CS specimens; (B) urine specimens. (From Nelson, D.E. et al., *PloS One*, 5, e36298, 2012.)

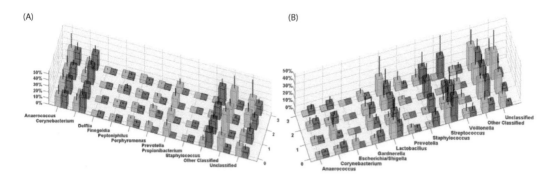

(A)   (B)

Figure 2.10 Impact of circumcision on the coronal sulcus (CS) and urine microbiotas. Relative normalised abundance of major (A) CS and (B) urine taxa in circumcised (red) and uncircumcised (blue) participants at four sampling points (z-axis). Bars indicate 95% confidence intervals. (From Nelson, D.E. et al., *PloS One*, 5, e36298, 2012.)

Microbiota of the conjunctiva

The conjunctiva is reported to harbour about 59 genera although some one-third of all DNA reads represented unclassified or novel bacteria (**Figure 2.11**). There appears to be a core conjunctival microbiota comprising the genera *Pseudomonas, Propionibacterium, Bradyrbizobium, Corynebacterium, Acinetobacter, Brevundimonas, Staphylococcus, Aquabacterium, Sphingomonas, Streptococcus, Strepophyta*, and *Methylbacterium* that account for approximately 96% of the classified bacteria (**Figure 2.12**).

Respiratory tract microbiota

The respiratory tract can be divided into upper and lower regions. The upper respiratory tract comprises the nose, nasal passages, paranasal sinuses, pharynx, and the larynx above the vocal cords. The lower respiratory tract comprises the larynx below the vocal cords, trachea, bronchi and bronchioles. The upper tract can be sampled using nasopharyngeal swabs, whereas a bronchoscope is required to obtain samples from the glottis and below. Two methods are used, bronchiolar lavage and the protected brush method. The protected brush method allows sampling of a specific area of the lower respiratory tract without contamination from the upper respiratory microbiota and so is the more discriminating method of the two. As might be expected, the number of microorganisms are 100 to 10,000 times greater in the upper respiratory tract than in the lower respiratory tract. Although the nostrils harbour bacteria common to the skin such as staphylococci, propionibacteria, and corynebacteria, it appears that the composition of the microbiota of the upper and lower

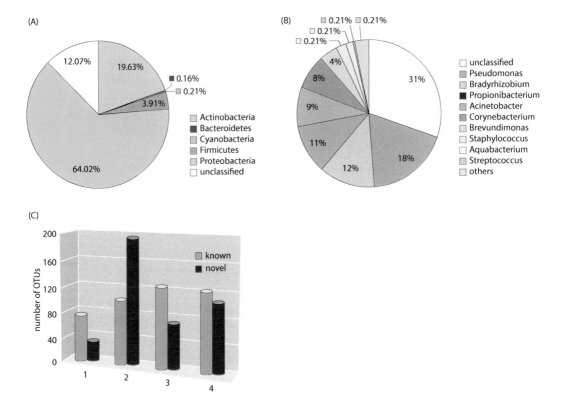

Figure 2.11 Relative abundance of bacterial taxa in the conjunctiva. (A) Phylum-level representation of the bacteria at the ocular surface (OS) of four subjects calculated according to relative abundance of classified 16S rRNA gene reads. The percentage of reads that failed to classify to known bacterial phyla is indicated as 'Unclassified,' shown in orange. The circular diagram presents average values calculated for all analysed subjects. Color-coding key on the right shows taxonomic identities of the classified bacteria. (B) Genus-level representation of the bacteria at the OS. Unclassified reads (31% of the total 115,003 sequences) are shown in dark blue. (C) Relative abundance of known (I6S-classified) and novel (unclassified) bacterial phylotypes at the conjunctiva of the individual subjects. All percentages were calculated relative to the total number of qualified DNA reads for each individual. (From Dong, Q. et al., *Invest. Ophthalmol. Vis. Sci.*, 52, 5408–5413, 2011.)

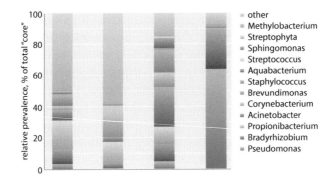

Figure 2.12 Putative core microbiome of the ocular surface (OS). Genus-level representation of the 12 most prevalent bacteria at the OS of four subjects calculated according to relative abundance of classified DNA reads. Percentage of DNA reads that fail to classify beyond bacteria is indicated as 'Other'. Color-coded key on the right side of the panel designates each genus. (From Dong, Q. et al., *Invest. Ophthalmol. Vis. Sci.*, 52, 5408–5413, 2011.)

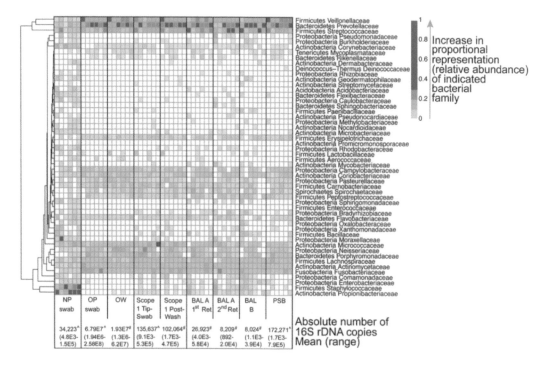

	NP swab	OP swab	OW	Scope 1 Tip-Swab	Scope 1 Post-Wash	BAL A 1st Ret	BAL A 2nd Ret	BAL B	PSB

Firmicutes Veillonellaceae
Bacteroidetes Prevotellaceae
Firmicutes Streptococcaceae
Proteobacteria Pseudomonadaceae
Proteobacteria Burkholderiaceae
Actinobacteria Corynebacteriaceae
Tenericutes Mycoplasmataceae
Bacteroidetes Rikenellaceae
Actinobacteria Dermabacteraceae
Deinococcus–Thermus Deinococcaceae
Proteobacteria Rhizobiaceae
Actinobacteria Geodermatophilaceae
Actinobacteria Streptomycetaceae
Acidobacteria Acidobacteriaceae
Bacteroidetes Flexibacteraceae
Proteobacteria Caulobacteraceae
Bacteroidetes Sphingobacteriaceae
Firmicutes Paenibacillaceae
Actinobacteria Pseudonocardiaceae
Proteobacteria Methylobacteriaceae
Actinobacteria Nocardioidaceae
Actinobacteria Microbacteriaceae
Firmicutes Erysipelotrichaceae
Actinobacteria Promicromonosporaceae
Proteobacteria Rhodobacteraceae
Firmicutes Lactobacillaceae
Firmicutes Aerococcaceae
Actinobacteria Mycobacteriaceae
Proteobacteria Campylobacteraceae
Actinobacteria Coriobacteriaceae
Proteobacteria Pasteurellaceae
Firmicutes Carnobacteriaceae
Spirochaetes Spirochaetaceae
Firmicutes Peptostreptococcaceae
Proteobacteria Sphingomonadaceae
Firmicutes Enterococcaceae
Proteobacteria Bradyrhizobiaceae
Bacteroidetes Flavobacteriaceae
Proteobacteria Oxalobacteraceae
Proteobacteria Xanthomonadaceae
Firmicutes Bacillaceae
Proteobacteria Moraxellaceae
Actinobacteria Micrococcaceae
Proteobacteria Neisseriaceae
Bacteroidetes Porphyromonadaceae
Firmicutes Lachnospiraceae
Actinobacteria Actinomycetaceae
Fusobacteria Fusobacteriaceae
Proteobacteria Comamonadaceae
Proteobacteria Enterobacteriaceae
Firmicutes Staphylococcaceae
Actinobacteria Propionibacteriaceae

1
0.8 Increase in proportional
0.6 representation (relative abundance)
0.4 of indicated bacterial family
0.2
0

Absolute number of 16S rDNA copies Mean (range)

34,223^	6.79E7^	1.93E7#	135,637^	102,064#	26,923#	8,209#	8,024#	172,271^
(4.8E3-	(1.94E6-	(1.3E6-	(9.1E3-	(1.7E3-	(4.0E3-	(892-	(1.1E3-	(1.7E3-
1.5E5)	2.58E8)	6.2E7)	5.3E5)	4.7E5)	5.8E4)	2.0E4)	3.9E4)	7.9E5)

Figure 2.13 Proportions of bacterial taxa in each sample type inferred from 16S rDNA pyrosequence data. Each column corresponds to an individual respiratory tract sample. The type of sample in each group is indicated in the boxes at the bottom of each group of columns. Each row corresponds to a specific bacterial family. OTUs were collected into families, so that some rows harbour multiple operational taxonomic unit - clusters of bacteria grouped by DNA sequence similarity. OTUs that were not assigned at the family level are omitted. Rows were subjected to hierarchical clustering to emphasise families that show similar abundance patterns. The proportional representation (relative abundance) of each family is represented by the colour code (key to the right). The absolute number of 16S copies determined from quantitative PCR of genomic DNA extracts are shown along the bottom. #, copies per millilitre; ^, copies in the total volume eluted from the swab or brush. NP = Nasopharyngeal swab; OP = Oropharyngeal swab; OW = Oral wash; Scope 1 tip wash and Scope 1 post-wash = bronchoscope inserted to above the glottis; BAL A1 and BAL 2 = Bronchoscope introduced into bronchus and lavaged twice; BAL B = Bronchoscope introduced into bronchus and sample collected with a brush. (From *Am. J. Respir. Crit. Care. Med.*, 184, 957–963, 2011.)

respiratory tract is largely similar to that found in oral cavity (**Figure 2.13**). It has been proposed that microaspiration is the source of bacteria in the lower respiratory tract. Microaspiration is the asymptomatic aspiration of small volumes of oropharyngeal secretions or gastric fluid into the lungs. Because of the similarity between the oral microbiota and the upper and lower respiratory tract microbiotas, it has been proposed that the entire respiratory tract from mouth and nose to the alveoli be considered a single ecosystem. However, although the mouth is the major contributor to the lower respiratory microbiota, bacteria such as *Ralstonia*, *Bosea*, *Haemophilus*, *Enterobacteriaceae* and *Methylobacterium* have been found to be disproportionally represented in the lower tract, arguing that there may be a lung-specific bacterial microbiome. Environmental fungi have

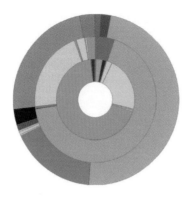

Ascomycota (1.101%; 5.14%; 0.33%)
Taphrinomycetes (1.97%; 0%; 0.09%)
Saccharomycetes (48.13%; 67.28%; 2.60%)
Pezizomycetes (0.11%; 0%; 1.53%)
Leotiomycetes (0.016%; 0.15%; 0.012%)
Sordariomycetes (0.16%; 0.08%; 0.012%)
Dothidemycetes (16.59%; 0.45%; 0.26%)
Eurotiomycetes (0.23%; 20.99%; 2.23%)
Basidiomycota (0%; 0.44%; 0%)
Cystobasidiomycetes (0.55%; 0.011%; 0.058%)
Tritirachiomycetes (0%; 0%; 0.023%)
Microbotryomycetes (0%; 0.13%; 0%)
Pucciniomycetes (0%; 0%; 0.006%)
Malasseziomycetes (0%; 2.21%; 21.66%)
Exobasidiomycetes (0%; 0.15%; 0.62%)
Tremellomycetes (0%; 0.35%; 0.46%)
Agaricomycetes (0.27%; 2.36%; 68.21%)
Fungi incertae sedis (1.14%; 0%; 0.006%)
Neocallimastigomycetes (0%; 0.012%; 0%)
Chytridiomycetes (2.98%; 0%; 0.064%)
Microsporidia (23.58%; 0%; 0%)
Cryptomycota (3.01%; 0%; 0.19%)
Basidiobolomycetes (0.125%; 0%; 1.60%)
Glomeromycetes (0.011%; 0%; 0.006%)

Figure 2.14 Distribution of fungal classes (percentage of relative abundance) in the sputum of healthy individuals (outer ring) and patients with cystic fibrosis (CF; middle ring) and asthma (inner ring). The percentages on the key correspond to each class identified in healthy, CF, and asthma populations (from the outer to inner rings, respectively). Reads that were not identified as class level are group at phylum levels (Ascomycota, Basidiomycota). Classes less than 0.1% are not represented in the rings; the class named 'Fungi incertae sedis' refers to unclassified fungi. (From Nguyen, L.D.N. et al., *Front. Microbiol.*, 6, 2015.)

been detected in the lungs of healthy persons, persons with cystic fibrosis and with asthma (**Figure 2.14**) probably as the result of inhalation of spores. The principle taxa found in healthy persons are *Davidiellaceae*, *Cladosporium*, *Eurotium*, and *Penicillium*. To date the only attempts to examine the lung virome have been in patients with cystic fibrosis and, likely, do not reflect the situation in healthy lungs.

MICROBIOTA OF THE ALIMENTARY CANAL

Mouth

It is clear that the alimentary canal *in toto* supports the richest and most diverse microbiota found on the barrier epithelia (see the Human Oral Microbiome Database [HOMD]). The microbiota of the mouth, the entrance of the alimentary tract, is complex, harbouring over 700 species from at least 12 phyla. Notably, the mouth provides more discrete and diverse habitats than any other barrier epithelium. Some of the oral epithelium is keratinised, including the hard palate, gingivae and dorsum of the tongue, whereas other epithelial surfaces are non-keratinised. These various epithelia provide different ligands for adhesion and contribute to the tropism of particular bacteria for these different surfaces. The different surfaces in the mouth are subject to different shear forces resulting from the flow of saliva, intake of fluids, mastication and different oxygen levels. The mouth of the neonate and pre-dentate infant consist of shedding epithelial surfaces. These epithelial surfaces might be thought to be well oxygenated; however, the oxidation–reduction potential (Eh) of areas of the mouth in the pre-dentate infant is low enough to support anaerobic bacteria such as *Prevotella* species. The eruption of teeth at about 6 months of age provides non-shedding surfaces that allow formation of a complex biofilm community termed dental plaque. Furthermore, the erupted teeth are surrounded by a 2-mm deep cuff of keratinised epithelium termed the gingival sulcus, which provides an environment with reduced oxygen tension. In persons with periodontal disease, plaque extends into the sulcus, resulting in loss of epithelial attachment and apical migration of the bottom of the sulcus to form periodontal pockets. The redox potential of periodontal pockets can reach −400 mV, which is lower than that found in the large bowel.

It is clear that the mouth cannot be considered as a single ecosystem and, in order to adequately determine the richness and diversity of the oral microbiota, it is necessary to sample the various habitats that exist there. Site-specific diversity of oral bacteria revealed by the use of DNA probes is shown in **Figure 2.15** and that revealed by sequencing 16S rRNA is shown in **Figure 2.16**. It is clear from these data that saliva samples or oral rinses – reflecting as they do planktonic microorganisms that have been differentially released from the surface-adherent communities – do not adequately represent the complexity of the site-specific oral microbial communities. A comprehensive understanding of the composition of the human oral microbiota from birth to old age has come about from a wealth of cultural studies conducted over the past several decades, so it is fair to say that the use of advanced molecular techniques has not significantly altered our understanding beyond identifying species that cannot be cultivated. Moreover, these un-culturable bacteria have not been found to constitute major fractions of the oral microbiota. Comparisons between older culture data and recent molecular studies are complicated by taxonomic revisions that have created new genera out of old. For example, the nutritionally variant streptococci now have been placed in the genera *Abiotrophia* and *Granulicatella*.

It is assumed that the oral microbiota of the neonate is acquired from the mother during passage through the birth canal, from the mothers' skin and saliva and those of the immediate family. However, maternal–infant and intra-family transmission have only been convincingly demonstrated for *Streptococcus mutans*, a bacterium that establishes only after the eruption of teeth provides non-shedding surfaces for its colonisation.

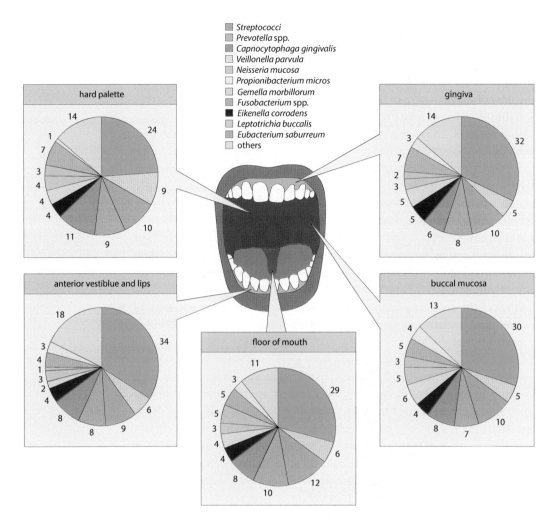

Figure 2.15 Results of a culture-independent analysis of the microbiotas of various mucosal sites (the floor of the mouth, buccal mucosa, hard palate, lip, and the attached gingiva) in 225 adults. DNA was extracted from samples from each site and probed with whole-genome DNA probes specific for 40 oral bacterial taxa. Results are shown for only the 11 most plentiful species present in the communities. Each figure represents the percentage of the total DNA probe count, and so does not necessarily represent its proportion of the microbiota of the site because of the presence of organisms other than those recognised by the probes used. (From Wilson, M., *Bacteriology of Humans: An Ecological Perspective*, Blackwell Publishing, Malden, MA, 2008.)

However, studies conducted in Denmark and in the United States on the acquisition of the pioneer streptococcus *S. mitis* by neonates could not confirm this assertion. The establishment of the oral microbiota is first influenced by whether the neonate/infant is breast- or formula-fed. For example, a higher prevalence of *Actinobacteria* and *Proteobacteria* have been found in breast-fed infants and a higher prevalence of *Bacteroidetes* in formula-fed infants (**Figure 2.17**). The complexity of the oral microbiota increases with age and the eruption of teeth. Tooth eruption has a profound effect on microbial succession (**Figure 2.18**). Accordingly, the loss of non-shedding surfaces in edentulous individuals modifies the oral microbiota.

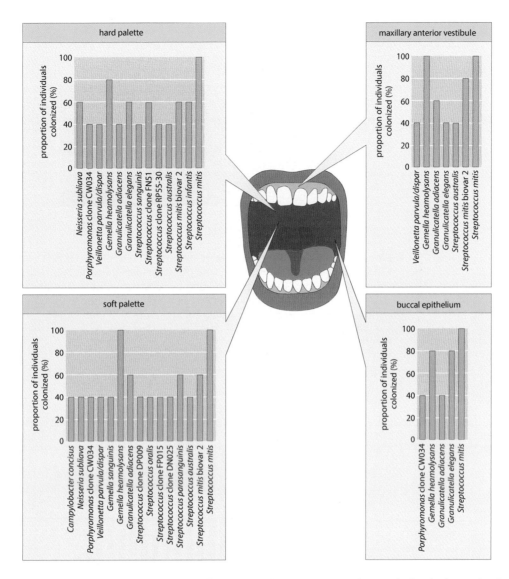

Figure 2.16 Culture-independent analysis of the microbial communities residing on the hard palate, soft palate, buccal epithelium and maxillary anterior vestibule of five healthy individuals. Identification of the resident microbes involved the amplification, cloning, and sequencing of the 16S rRNA genes present in DNA extracted from the communities. Results are shown for only those taxa that were detected in at least two individuals. (From Wilson, M., *Bacteriology of Humans: An Ecological Perspective*, Blackwell Publishing, Malden, MA, 2008.)

In contrast to bacteria, there is relatively little information about the mycobiome and virome of the mouth. The mouth – reflected by oral rinses obtained from adults of various ethnicities – supports a diverse mycobiome with as many as 85 genera and approximately 100 species (**Figure 2.19**). However, it is difficult to determine which are residents and which are transients that originate from inhaled air and/or food. Based on frequency of isolation *Candida* species are predominant members of the oral mycobiome, followed by *Cladosporium*, *Aureobasidium*, Saccharomycetales, *Aspergillus*, *Fusarium*, and *Cryptococcus*, although not all are harboured by all individuals (**Figure 2.20**).

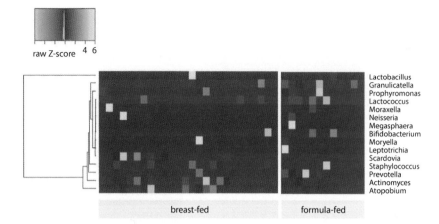

Figure 2.17 Heat map of oral bacterial community compositions within breast-fed and formula-fed infants and their hierarchical clustering. The bacterial genera are represented as heat maps corresponding to a phylogenetic tree. The heat map demonstrates the higher abundance of *Prevotella* spp. within the oral samples from breast-fed compared to formula-fed infants. (From Al-Shehri, S.S. et al., *Sci. Rep.*, 6, 2016.)

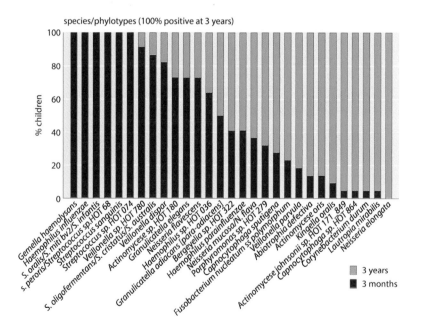

Figure 2.18 Prevalence at 3 months and 3 years of age of species/phylotypes detected in all (100%) of the children at 3 years of age. (From Lif Holgerson, P. et al., *PloS One*, 10, e0128534, 2015.)

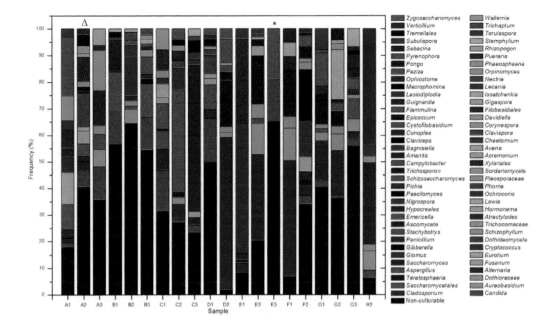

Figure 2.19 Overall distribution of fungi in oral rinse samples obtained from 20 healthy individuals. The triangle and asterisk indicate samples containing 16 and 3 fungal genera, respectively. (From Ghannoum, M.A. et al., *PLoS Pathog.*, 6, e1000713, 2010.)

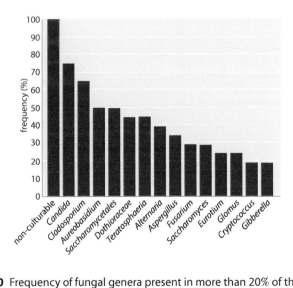

Figure 2.20 Frequency of fungal genera present in more than 20% of the tested samples. (From Ghannoum, M.A. et al., *PLoS Pathog.*, 6, e1000713, 2010.)

Oesophagus

The upper gastrointestinal tract contains low numbers of microorganisms ($<10^2$/mL to 10^4–10^5/mL) compared to the mouth and large intestine. As one moves down the alimentary tract beyond the mouth, it is no longer possible to obtain samples non-invasively and an endoscope or colonoscope is necessary. One advantage of using such instruments is the ability to biopsy the mucosa and thus examine adherent (biofilm) microorganisms rather than planktonic bacteria that are predominant in stool. As might be expected, the microbiota of the squamous oesophagus largely resembles that of the oral cavity, with the genera *Streptococcus*, *Prevotella* and *Veillonella* most prevalent. A survey of the microbiota of the normal and Barrett's oesophagus is shown in **Figure 2.21**.

Stomach

The stomach is the most acidic environment in the human body with a pH in the range of 1.5 to 3.5. The major component of gastric fluid is hydrochloric acid (0.05–1.0 M) which is produced by parietal cells in the body and fundus of the stomach. Beyond breaking down food and denaturing proteins, gastric acidity provides an important barrier to the ingress of microbial pathogens to the small and large intestines. Furthermore, there is evidence that the stomach provides an ecological filter that dictates the composition of the microbiota of the small and large intestines. The pH of the stomach is higher in neonates (> pH 4.0 in premature neonates) and above pH 6.0 in the elderly. Both of these groups are more susceptible to enteric infections as a result of the reduced acidity of the stomach. Unlike the oesophagus, the stomach appears to support its own ecosystem, comprising more than 100 phylotypes. *Helicobacter pylori*, when present, is the dominant organism followed by *Streptococcus* species and *Prevotella* species. The gastric microbiota also includes the genera *Caulobacter, Actinobacillus, Corynebacterium, Rothia, Gemella, Leptotrichia, Porphyromonas, Capnocytophaga, Flexistipes* and *Deinococcus* (**Figure 2.22**). However, there are strong similarities between the microbiota of the stomach and those of the mouth and respiratory tree (**Figure 2.23**).

Small intestine

The biomass of the intestines increases as one moves from the duodenum (10^4–10^5 cells/g) through the jejunum (10^6–10^7 cells/g) to the ileum (10^7–10^8 cells/g). Like the stomach, the small intestine is a formidable environment for microorganisms because of the secretion of bile and digestive enzymes and the rapid transit of nutrients. It is likely that the duodenum, jejunum and ileum support different microbiotas, with the ileum more similar to the colon, but the difficulty in sampling the small intestine makes confirming such differences difficult. The dominant phylogenetic

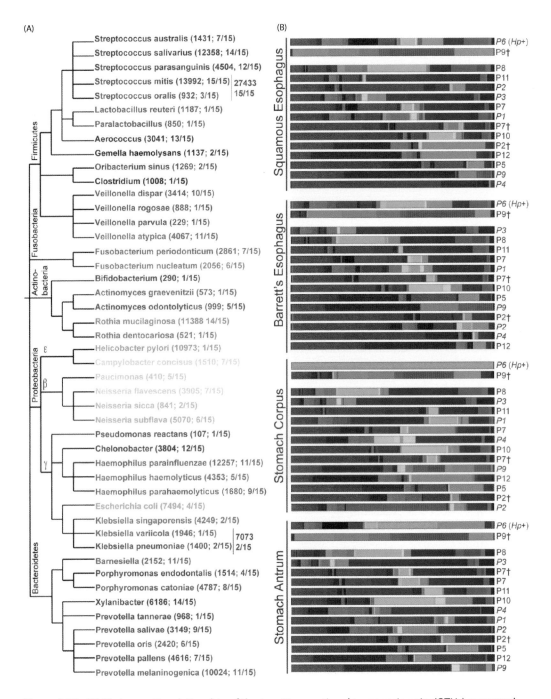

Figure 2.21 (A) Phylogenetic relationship of the top 45 operational taxonomic units (OTUs) recovered from each of four sites sampled in individual participants. Respective phyla are noted above main branches of the phylogenetic tree. Numbers in parentheses represent total number of pyrosequencing reads recovered for a given species or genera across all samples followed by the fraction of participants in whom a relative abundance of ≥1.3% of a given species or genera were detected. (B) Species/genera-level profiles of top 45 OTUs detected by 454 sequencing in squamous esophagus, Barrett's esophagus, stomach corpus and stomach antrum of indicated participants. Data arranged in order of increasing Firmicutes dominance. Individual species/genera are colour-coded according to scheme presented in (A). (From Gall, A. et al., *PLoS One*, 10, e0129055, 2015.)

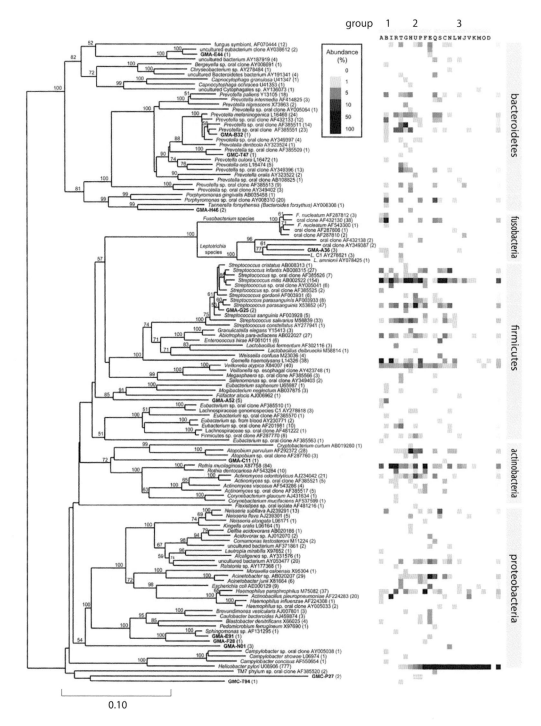

Figure 2.22 Phylogenetic tree with the 128 gastric 16S rDNA phylotype representatives from 23 human subjects. GenBank entries are shown in normal font; names of previously uncharacterised phylotype representatives are shown in bold. Numbers of clones within each phylotype are shown in parentheses. The scale bar represents evolutionary distance (10 substitutions per 100 nucleotides). The right side of the figure shows the relative abundance of phylotypes per gastric specimen in grey values (white, 0% present; black, 100% of clone library); letters above the abundance graph correspond to subjects A–W. Subjects are grouped according to *H. pylori* status as determined by conventional and molecular tests, in increasing order of percentage of *H. pylori* clones. (From Bik, E.M. et al., *PNAS*, 103, 732–737, 2006.)

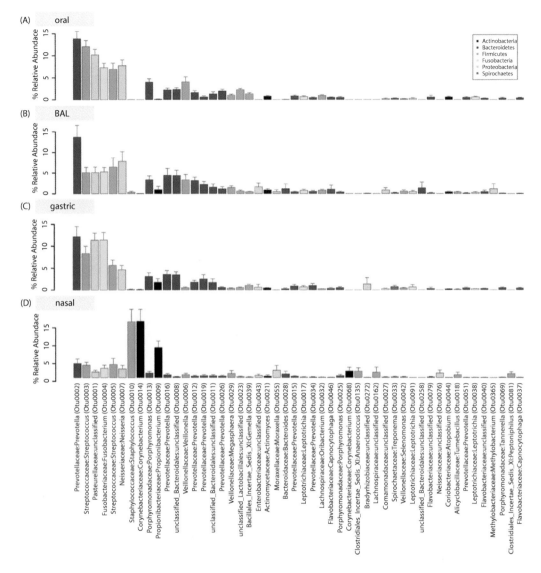

Figure 2.23 Similarities between the microbiotas of the oral cavity, nasal cavity, respirator tree and stomach. (A–D) Rank abundance plots for each of the sampling locations based on the top 50 operational taxonomic units (OTUs) from the overall order (greatest to smallest) taken from all of the samples combined. The bars depict the mean plus or minus the standard error of the mean. Bars are coloured according to their phyla. The family, genus, and OTU identification of the bacterial community members are displayed along the x-axis of the bottom panel. (From Bassis, C.M. et al., *MBio*, 6, e00037–15, 2015.)

groups of the small intestine are *Clostridium* species, *Streptococcus* species and coliforms, but the composition of the microbiota fluctuates over time. Metagenomic studies have revealed that the small intestine microbiota is enriched for pathways and functions related to carbohydrate uptake and metabolism and amino acid metabolism (**Figure 2.24**). Genes related to the synthesis of biotin and cobalamin are also enriched. Consistent with the high level of carbohydrate metabolism are high concentrations of

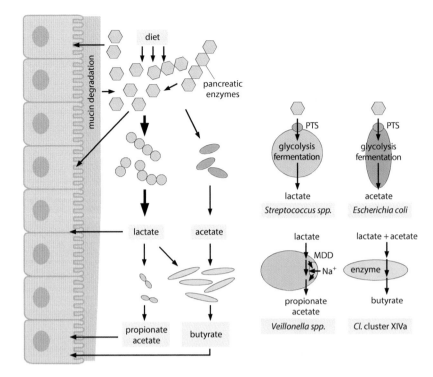

Figure 2.24 Small intestinal microbiome model based on the 16S rRNA, metabolite, metagenome and metatranscriptome data. (From Zoetendal, E.G. et al., *ISME J.*, 6, 1415–1426, 2012.)

volatile and non-volatile fatty acids. An ecological perspective of the small intestinal microbiota is shown in **Figure 2.25**.

Large intestine

The majority of studies of the microbiota of the large bowel are based on the study of microorganisms in faeces. However, faeces represents the consortium of planktonic bacteria of the alimentary canal to this point. It does not reflect the mucosa-adherent bacteria and planktonic bacteria of the colon, alone. Colonoscopy is required to sample mucosa-adherent microorganisms. In fact, studies that have utilised biopsies of the mucosae of the ascending, transverse and descending colon have shown that the mucosa-adherent bacterial populations are distinct from the microbiota of faeces (**Figures 2.26** and **2.27**). The biomass of the mucosal biofilm is far smaller than the biomass of faeces.

As is the case at other barrier epithelia, microbial colonisation begins at birth. The composition of the neonatal large bowel microbiota as determined from faeces samples is a function of whether the neonate is delivered vaginally or by caesarean delivery, and this is likely true

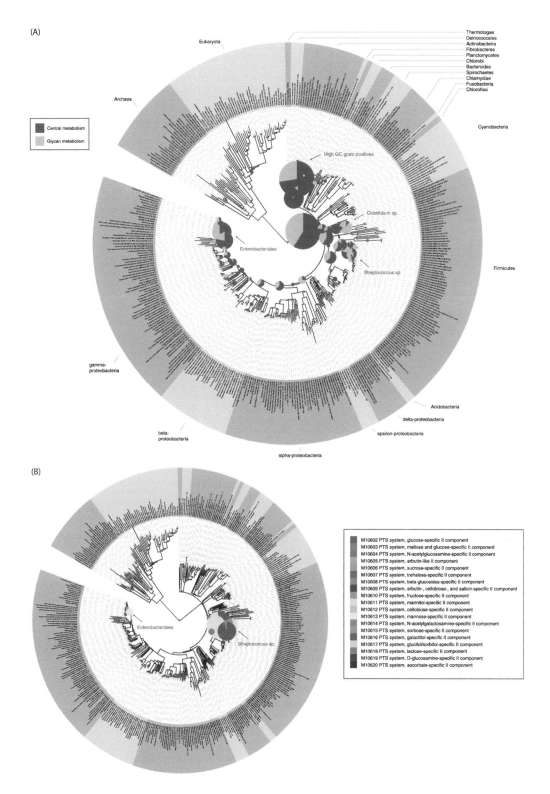

Figure 2.25 (A) Bacteria in the small intestine involved in central (purple) and glycan (green) metabolism. (B) Expression of simple carbohydrate transport phosphotransferase systems (PTS) and carbohydrate metabolic genes by components of the small intestine microbiota. The colours indicate the different types of PTS. (From Zoetendal, E.G. et al., *ISME J.*, 6, 1415–1426, 2012.)

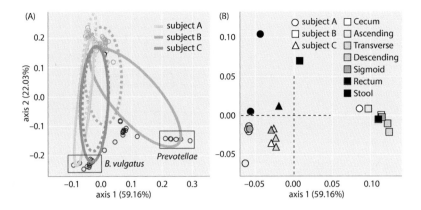

Figure 2.26 Differences in the microbiota between anatomical sites, stool and mucosa in three subjects. Double principal coordinate analysis (DPCoA) for (A) colonic mucosa (solid lines) and stool (dashed lines). (B) is an enlarged view of (A), depicting the centroids of each site-specific ellipse.

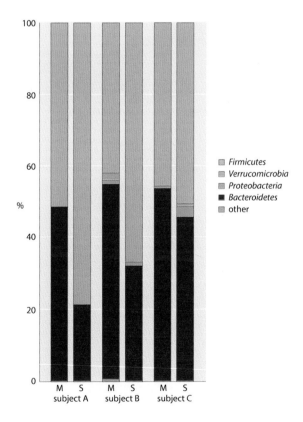

Figure 2.27 Relative abundance of sequences from stool and pooled mucosal samples per subject. The sequence frequencies are grouped according to phylum. 'Other' represents the *Fusobacteria*, *Actinobacteria*, and unclassified near *Cyanobacteria* phyla, each containing less than 0.2% of the total sequences. 'M' denotes pooled mucosal sequences per subject and 'S' refers to stool sample. (From Eckburg, P.B. et al., *Science*, 308, 1635–1638, 2005.)

for the microbiotas at other barrier epithelia. The faecal microbiota of neonates born vaginally resembles that of the mothers', with a dominance of the genera *Enterococcus*, *Escherichia/Shigella*, *Streptococcus*, and *Rothia* (**Figure 2.28**). By 4 months *postpartum*, the signature genera of the faecal microbiota are, *Bifidobacterium*, *Lactobacillus*, *Collinsella*, *Granulicatella* and *Veillonella*. By 1 year of age, *Anaerostipes*, *Anaerotruncus*, *Clostridiales* and *Eikenella* emerge (**Figure 2.29**). In contrast, the faecal microbiota of neonates born by caesarean delivery (Figure 2.28) is more heterogeneous, reflecting contributions from the skin, the oral cavity and the environment. However, differences the faecal microbiota between vaginally and caesarean delivery disappear by 1 year *postpartum* except for the genus *Bacteroides*, which remain more prevalent in faeces from vaginally-delivered infants (**Figure 2.30**). The disappearance of bacteria that derive from other barrier epithelia may reflect their lack of fitness to compete with other members of the

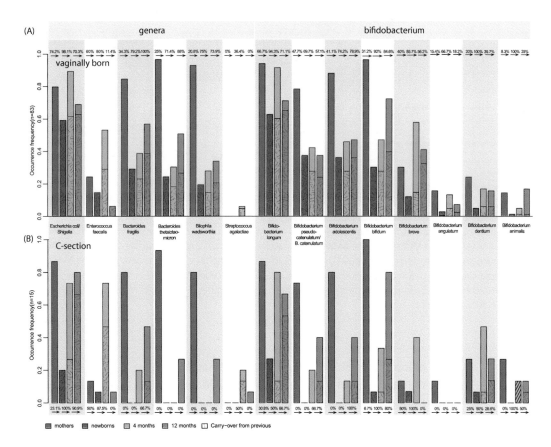

Figure 2.28 Comparison of the mother's gut microbiome in caesarean (C)-section infants. Occurrence frequency of selected operational taxonomic units (OTUs) in the different stages for (A) vaginally born and (B) C-section infants. Within each bar, the hatches mark the part shared with the previous stage, i.e., newborns with their own mothers, 4-month samples with their corresponding newborn samples, 12-month samples with their corresponding 4-month samples. The percentage of such possible carry-overs is indicated on the arrows. (From Backhed, F. et al., *Cell Host Microbe*, 17, 690–703, 2015.)

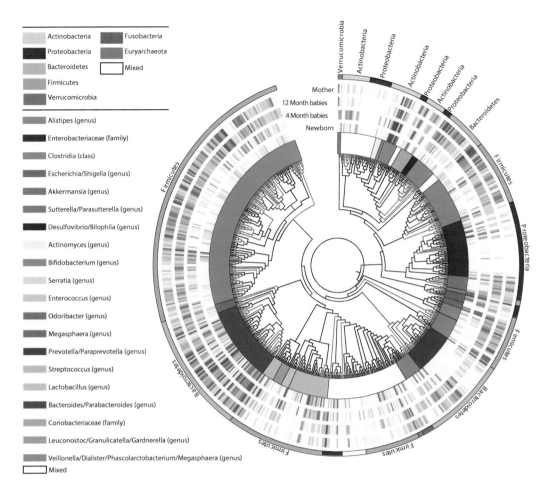

Figure 2.29 Faecal samples from 98 full-term Swedish infants, 15 of whom were delivered by C-section, were collected during the first days of life and at 4 and 12 months of age. The cohort included infants who were exclusively breast-fed, mixed fed (formula + breast-feeding) or exclusively formula-fed. Averaged genome-genome MUMi distance of MetaOTU pairs was used to construct the tree according to the neighbour-joining method. Novel MetaOTUs are shown as red branches. Colored blocks indicate phyla (outermost circle). The heatmap circles show relative abundance of each MetaOTU in the newborns, 4-month-old infants, 12-month-old infants and mothers. (From Backhed, F. et al., *Cell Host Microbe*, 17, 690–703, 2015.)

colonic microbiota. Because mother and infant share the same genera and species of bacteria, it is concluded, perhaps correctly, that the mother is the source of her infants' microbiota. However, for this assertion to be confirmed, it must be shown that mother and infant share the same clones. As is the case for the oral microbiota, the microbiotas of the other parts of the alimentary canal are influenced by breast- or formula-feeding. By 4 months *postpartum*, there are clear differences in

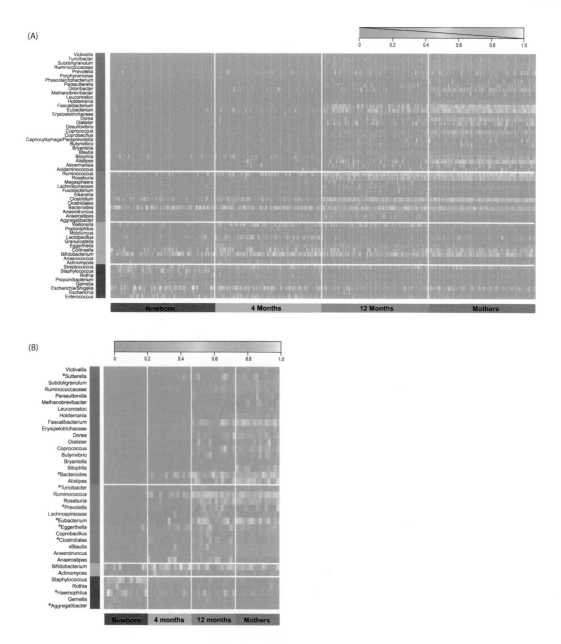

Figure 2.30 Heat map of the relative abundance of the signature genera in (A) vaginally or (B) caesarean (C)-section-born infants at birth, after 4 months, after 12 months, and in their mothers. Each vertical lane corresponds to one sample. Signature genera only seen in C-section or shifted in different stages compared with vaginally born infants are highlighted with an asterisk. (From Backhed, F. et al., *Cell Host Microbe*, 17, 690–703, 2015; See also Table S5.)

the microbiota of exclusively breast-fed infants and exclusively formula-fed infants. The faecal microbiota of the former has increased levels of *Lactobacillus* species and *Bifidobacterium longum*, whereas the latter has elevated levels of *Clostridium difficile, Granulicatella adiacens, Citrobacter species, Enterobacter cloacae, Bisophila wadsworthia* and *Bifidobacterium adolescentis* (**Figure 2.31**).

In addition to breast- or formula-feeding, there are many other physiological variables, such as the introduction of solid food, that play important roles in determining the composition of the faecal microbiota. Despite the many variables that contribute to the development of the gut microbiotas of adults, the inter-individual composition of the microbiotas at the phylum level are quite stable, but there is considerable variability at the species level. The high level of species variability may reflect the functional redundancy of the gut and other microbiota with different microbial groups being able to carry out the same functional processes (**Figure 2.32**).

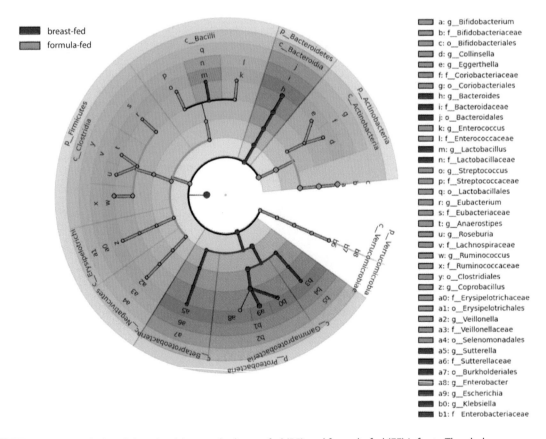

Figure 2.31 Clades of the microbiota under breast-fed (BF) and formula-fed (FF) infants. The cladograms show the taxa (highlighted by *small circles* and by *shading*) showing different abundance values. Colours of circle and shading indicate the microbial lineages that are enriched within corresponding samples. The class *Bacilli* is of low abundance in BF samples with high abundant *Lactobacillus* lineage (indicated with a *red shade over green* for indices *m* and *n* [see adjacent key]). (From Praveen, P. et al., *Microbiome*, 3, 41, 2015.)

(A)

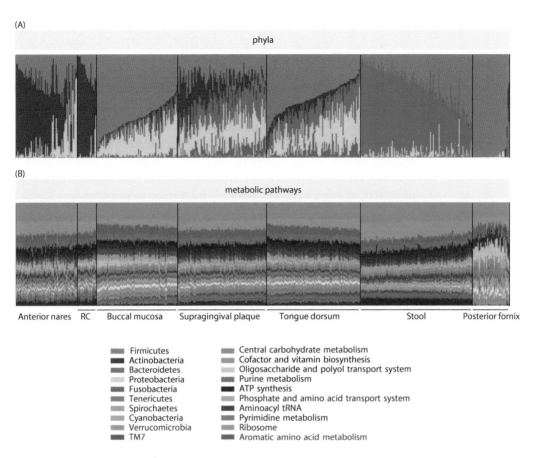

Figure 2.32 Carriage of microbial taxa varies, whereas metabolic pathways remain stable within a healthy population. Vertical bars represent microbiome samples by body habitat in the seven locations bars indicate relative abundances coloured by (A) microbial phyla and (B) metabolic pathways. Key indicates most abundant phyla/pathways by average within one or more body habitats. RC, retroauricular crease. A plurality of most communities' memberships consists of a single dominant phylum and often genus; but this is universal neither to all body habitats nor to all individuals. Conversely, most metabolic pathways are evenly distributed and prevalent across both individuals and body habitats. (From Huttenhower, C. et al., *Nature*, 486, 209–214, 2012.)

As is the case for other resident microbiotas of the human body, the faecal microbiota is simple immediately *postpartum* and increases in complexity and functional similarity over time to approach the adult microbiota. Similar to that observed in the small intestine, the faecal microbiota of infants is rich in bacteria such as *Bacteroides thetaiotaomicron* that can uptake and metabolise complex sugars and starch. As infants enter their first year of life, methanogens, such as *Desulfovibrio* species and *Methanobrevibacter* species, appear to dispose of hydrogen generated by the increased fermentative capacity of the increasingly complex microbiota. Consumption of an increasingly diverse diet also drives increasing diversity of the microbiota. The faecal microbiota of the neonate and infant is rich in bacteria such as *Bacteroides* and *Escherichia/Shigella* that are capable of producing vitamins. As might be expected, the faecal microbiota of infants becomes increasingly anaerobic with increasing age (Figures 2.28 through 2.30). There is evidence that the faecal microbiota changes over the lifetime of an individual, with the microbiota of centenarians showing a rearrangement of the firmicutes population and an enrichment in proteobacteria containing many bacteria that are considered to be pathobionts (**Figure 2.33**). Examination of faecal metagenomes from subjects from Europe, Scandinavia, Japan and the United States has led to the identification of a core faecal microbiome consisting of 57 species (**Figure 2.34**) and of 3 robust enterotypes in human faeces. **Table 2.3** shows the genera over-represented in each enterotype, and **Figure 2.35** shows the proportions of the genera from which these enterotypes were derived.

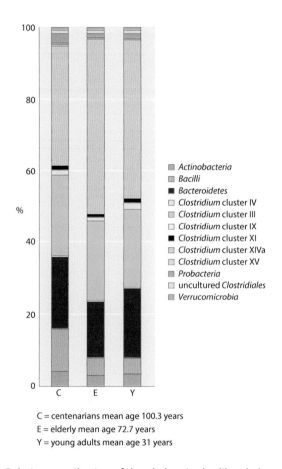

C = centenarians mean age 100.3 years
E = elderly mean age 72.7 years
Y = young adults mean age 31 years

Figure 2.33 Relative contribution of the phylum/order-like phylogroups to the microbiota of centenarians, elderly and young adults. In the key, phylum/order-like phylogroups that contribute for at least 0.5% to one of the profiles are indicated. C, centenarians mean age of 100.3 years; E, elderly mean age of 72.7 years; Y, young adults mean age of 31 years. (From Biagi, E. et al., *PloS One*, 5, e10667, 2010.)

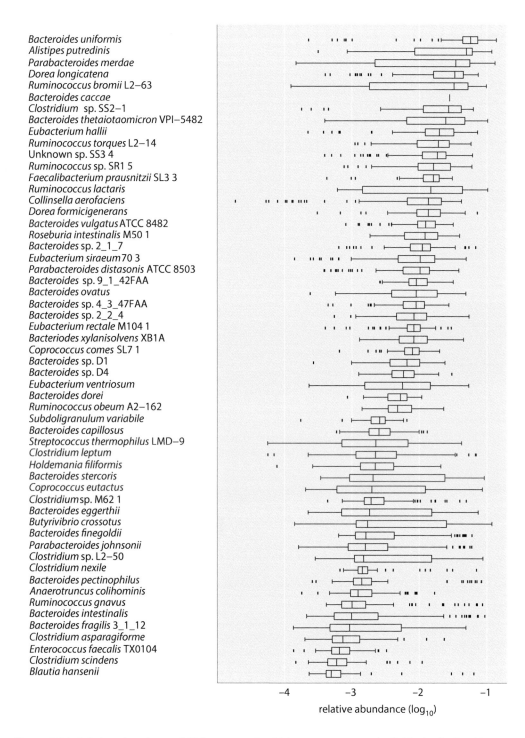

Figure 2.34 Relative abundance of 57 frequent microbial genomes among individuals obtained by deep sequencing DNA from faeces samples of 124 European adults. (From Qin, J. et al., *Nature*, 464, 59–65, 2010.)

enterotype	genus
1	*Acidaminococcus*
	Bacteroides
	Roseburia
	Faecalibacterium
	Anaerostipes
	Parabacteroides
	Clostridiales
2	*Prevotella*
	Streptococcus
	Enterococcus
	Desulfovibrio
	Lachnospiraceae
3	*Akkermansia*
	Alistipes
	Klebsiella
	Ruminococcus
	Escherichia/Shigella
	Dialister
	Mitsuokella
	Methanobrevibacter
	Eggerthella
	Ruminococcaceae
	Subdoligranulum
	Coprococcus
	Collinsella
	Blautia
	Eubacterium
	Dorea

Table 2.3 Genera over-represented in faced enterotypes (clusters that are not nation or continent specific) (From Arumugam, M. et al., *Nature*, 473, 174–180, 2011.)

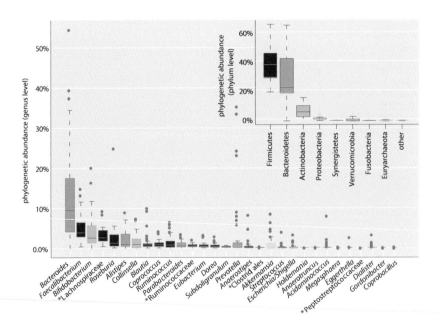

Figure 2.35 Genus abundance for the 30 most abundant genera. Genera are coloured by their respective phylum (see inset for colour key). Inset shows phylum abundance. (From Arumugam, M. et al., *Nature*, 473, 174–180, 2011.)

The difficulty in defining the contribution of fungi to the microbiota of the alimentary canal is the inability to determine which fungi are residents and which are transients, derived from food and the environment, that are incapable of colonising and persisting in the gut. Culture-independent studies indicate that fungal genes account for less than 0.1% of the faecal microbiome. Nevertheless, 66 genera and an additional 13 lineages that could not be classified to the genus level have been detected in faeces. However, the most prevalent genera were but three, *Saccharomyces*, *Candida* and *Cladosporium* (**Figure 2.36**). **Table 2.4** shows fungi proposed to inhabit the human alimentary tract based on a compilation of data from 36 studies dating from 1917 to 2015.

The archaea *Methanobrevibacter* species and *Nitrososphaera* species inhabit the gastrointestinal tract. There is a large number of virus-like particles in the human bowel. These are lysogenic or temperate phages that infect the resident faecal bacteria and modify their function in the ecosystem.

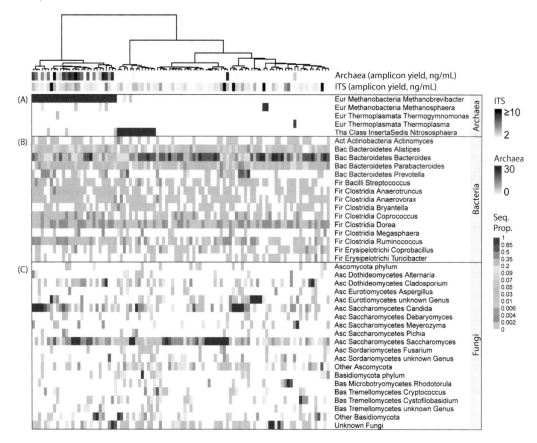

Figure 2.36 The archaeal and fungal components of the human gut microbiome. The heat maps show the relative proportions of fungal lineages. The lineages are marked on the right, with phylum (abbreviated), class, and genus. Archaeal genera are shown in (A), representative bacterial genera in (B), and fungal genera in (C). Asc, Ascomycota; Bas, Basidiomycota. Other Ascomycota and Other Basidiomycota are composed of genera that were detected in only one sample. (From Hoffmann, C. et al., *PloS One*, 8, e66019, 2013.)

Candida albicans
Candida tropicalis
Candida parapsilosis
Candida glabrata
Candida krusei
Candida lusitaniae
Candida dubliniensis
Candida rugosa
Candida ortopsilosis
Candida intermedia
Malassezia species
Cladosporium species
Galactomyces geotrichum
Saccharomyces cerevisiae

Table 2.4 Fungi whose habitat is likely the human gastrointestinal tract. (From Suhr, M.J. and Hallen-Adams, H.E., *Mycologia*, 107, 1057–1073, 2015.)

KEY CONCEPTS

- Resident microbiotas have co-evolved with hominids for millions of years.
- Resident microbiotas exist in a state of dynamic equilibrium with their human host.
- Each ecosystem contains a core group of microorganisms that are common to all human beings.
- Pertubations of resident microbiotas reduce the resistance of the host to infection and disease.
- Resident microbiotas protect the barrier epithelial surfaces of the human body from colonisation by pathogens.

BIBLIOGRAPHY

Aagaard K et al. 2012. A metagenomic approach to characterization of the vaginal microbiome signature in pregnancy. *PloS One*, 7(6): e36466.

Achtman M. 2004. Population structure of pathogenic bacteria revisited. *International Journal of Medical Microbiology*, 294: 67–73.

Al-Shehri SS et al. 2016. Deep sequencing of the 16S ribosomal RNA of the neonatal oral microbiome: A comparison of breast-fed and formula-fed infants. *Scientific Reports*, 6: Article number: 38309.

Arumugam M et al. 2011. Enterotypes of the human gut microbiome. *Nature*, 473(7346): 174–180.

Avila M et al. 2009. The oral microbiota: Living with a permanent guest. *DNA and Cell Biology*, 28(8): 405–411.

Backhed F et al. 2015. Dynamics and stabilization of the human gut microbiome during the first year of life. *Cell Host & Microbe*, 17: 690–703.

Banerjee S et al. 2018. Keystone taxa as drivers of microbiome structure and functioning. *Nature Reviews Microbiology*, 16: 567–576.

Bassis CM et al. 2015. Analysis of the upper respiratory tract microbiotas as the source of the lung and gastric microbiotas in Healthy Individuals. *MBio*, 6(2): e00037-15.

Biagi E et al. 2010. Through ageing, and beyond: Gut microbiota and inflammatory status in seniors and centenarians. *PloS One*, 5(5): e10667.

Bik EM et al. 2006. Molecular analysis of the bacterial microbiota in the human stomach. *PNAS*, 103(3): 732-737.

Byrd Al et al. 2018. The human skin microbiome. *Nature Reviews Microbiology*, 16: 143-155.

Cao B et al. 2014. Placental microbiome and its role in preterm birth. *NeoReviews*, 15(12): e537.

Charlson ES et al. 2011. Topographical continuity of bacterial populations in the healthy human respiratory tract. *American Journal of Respiratory and Critical Care Medicine*, 184: 957-963.

Cogen AL et al. 2008. Skin microbiota: A source of disease or defence? *British Journal of Dermatology*, 158(3): 442-455.

Dong Q et al. 2011. Diversity of bacteria at healthy human conjunctiva. *Investigative Ophthalmology & Visual Science*, 52(8): 5408-5413.

Eckburg PB et al. 2005. Diversity of the human intestinal microbial flora. *Science*, 308(5728): 1635-1638.

El Kaoutari A et al. 2013. The abundance and variety of carbohydrate-active enzymes in the human gut microbiota. *Nature Reviews Microbiology*, 11(7): 497-504.

Feil EJ, Spratt BG. 2001. Recombination and the population structures of bacterial pathogens. *Annual Review of Microbiology*, 55: 561-590.

Fettweis JM et al. 2012. A new era of the vaginal microbiome: Advances using next-generation sequencing. *Chemistry and Biodiversity*, 9(5): 965-976.

Findley K et al. 2013. Topographic diversity of fungal and bacterial communities in human skin. *Nature*, 498: 367-370.

Gall A et al. 2015. Bacterial composition of the human upper gastrointestinal tract microbiome is dynamic and associated with genomic instability in a Barrett's esophagus cohort. *PloS One*. doi:10.1371/journal.pone.0129055.

Ghannoum MA et al. 2010. Characterization of the oral fungal microbiome (mycobiome) in healthy individuals. *PloS Pathogens*, 6(1): e1000713.

Grice EA et al. 2008. A diversity profile of the human skin microbiota. *Genome Research*, 18(7): 1043-1050.

Grice EA, Segre JA. 2011. The skin microbiome. *Nature Reviews Microbiology*, 9(4): 244-253.

Haque SZ, Haque M. 2017. The ecological community of commensal, symbiotic, and pathogenic gastrointestinal microorganisms: An appraisal. *Clinical and Experimental Gastroenterology*, 10: 91-103.

Hoffmann C et al. 2013. Archaea and fungi of the human gut microbiome: Correlations with diet and bacterial residents. *PloS One*, 8(6): e66019.

Huttenhower C et al. (The Human Microbiome Project Consortium). 2012. Structure, function and diversity of the healthy human microbiome. *Nature*, 486: 207-214.

Lif Holgerson P et al. 2015. Maturation of oral microbiota in children with or without dental caries. *PloS One*, 10(5): e0128534. doi:10.1371/journal.pone.0128534.

MacIntyre DA et al. 2015. The vaginal microbiome during pregnancy and the postpartum period in a European population. *Scientific Reports*, 5: 8988. doi:10.1038/srep08988.

Manzoor MAP, Rekha P-D. 2017. Microbiome: The 'unforeseen organ'. *Nature Reviews Urology*, 14: 521-522.

Martin DH et al. 2012. The microbiota of the human genitourinary tract: Trying to see the forest through the trees. *Transactions of the American Clinical and Climatological Association*, 123: 242-256.

Nelson DE et al. 2012. Bacterial communities of the coronal sulcus and distal urethra of adolescent males. *PloS One*, 7(5): e36298.

Nguyen LDN et al. 2015. The lung microbiome: An emerging field of the human respiratory microbiome. *Frontiers in Microbiology*, 6: article 89.

Pearce MM et al. 2014. The female urinary microbiome: A comparison of women with and without urgency urinary incontinence. *MBio*, 5(4): e01283-14.

Praveen P et al. 2015. The role of breast-feeding in infant immune system: A systems perspective on the intestinal microbiome. *Microbiome*, 3: 41. doi:10.1186/s40168-015-0104-7.

Qin J et al. 2010. A human gut microbial gene catalogue established by metagenomic sequencing. *Nature*, 464: 59–67.

Romero R et al. 2014. The composition and stability of the vaginal microbiota of normal pregnant women is different from that of non-pregnant women. *Microbiome*, 2(1): 4.

Spor A et al. 2011. Unravelling the effects of the environment and host genotype on the gut microbiome. *Nature Reviews Microbiology*, 9(4): 279–290.

Suhr MJ, Hallen-Adams HE. 2015. The human gut mycobiome: Pitfalls and potentials – a mycologist's perspective. *Mycologia*, 107(6): 1057–1073.

The Human Oral Microbiome Database [HOMD], http://www.homd.org.

Wang J, Jia H. 2016. Metagenome-wide association studies: Fine-mining the microbiome. *Nature Reviews Microbiology*, 14(8): 508–522.

Whiteside SA et al. 2015. The microbiome of the urinary tract: A role beyond infection. *Nature Reviews Urology*, 12: 81–90.

Willing BP et al. Shifting the balance: Antibiotic effects on host–microbiota mutualism. *Nature Reviews Microbiology*, 9(4): 233–243.

Wilson M. 2008. *Bacteriology of Humans: An Ecological Perspective*. 1st ed. Blackwell Publishing, Malden, MA.

Wilson M. 2019. *The Human Microbiota in Health and Disease: An Ecological and Community-based Approach*. 1st ed. CRC Press, Boca Raton, FL.

Zmora N et al. 2018. Personalized gut mucosal colonization resistance to empiric probiotics is associated with unique host and microbiome features. *Cell*, 174: 1388–1405.

Zoetendal EG et al. 2012. The human small intestinal microbiota is driven by rapid uptake and conversion of simple carbohydrates. *The ISME Journal*, 6: 1415–1426.

Chapter 3: Biofilms

INTRODUCTION

In textbooks of medical microbiology, each pathogenic microorganism is considered in isolation as if it existed in a vacuum. Indeed, the central tenet of clinical microbiology is to isolate etiologic agents of infections from clinical specimens in pure culture in order to identify and determine their susceptibility to antimicrobial agents. The clinical specimens may be from usually sterile sites such as blood and cerebrospinal fluid (CSF) but they may be from external and internal body surfaces that are associated with complex microbiotas (see Chapter 2). In nature, almost all microorganisms grow as part of communities attached to surfaces that are bathed by liquid, generally water. These organised surface-adherent communities are termed biofilms and are the way that bacteria and fungi grow on external and internal surfaces of the human body, or on catheters and prostheses. In nature, and in the human body, biofilms are generally multi-species consortia; however, single-species biofilms can occur on catheters and indwelling prostheses in the human body. Biofilms are important in the pathogenesis of infectious diseases for several reasons. They protect the microbial community from host defences, enable persistence of microbes in flowing systems such as blood and urine, protect microbes against desiccation, and, importantly, allow efficient horizontal transfer of genes encoding virulence factors and resistance to antibiotics.

The discovery of biofilms is not new. Van Leeuwenhoek in 1676 was the first to observe a biofilm when he examined material that he had obtained from between his teeth and from the teeth of two ladies and two old men and which he described as *a little white matter, which is as thick as if 'twere batter*. Van Leeuwenhoek wrote, *I then most always saw, with great wonder, that in the said matter there were many very little living animalcules, very prettily a-moving*. The term 'gelatinous plaque' was first used by G. V. Black in 1898 to describe the 'felt-like mass of microorganisms' which had been observed by Leeuwenhoek in 1676 and in 1897 by J. L. Williams covering the surface of carious human enamel (**Figure 3.1**). Between the end of the nineteenth century and beginning of the twentieth century, research by the dental scientists, G. V. Black, J. L. Williams and W. D. Miller established that dental plaque consisted of a consortium of metabolising microorganisms invested in a gelatinous (polysaccharide) diffusion-limiting matrix. Moreover, it was realised that the extracellular matrix (ECM) of dental plaque could become mineralised and give rise to dental calculus. By the end of the 1960s, dental scientists understood the fundamentals of biofilm development and organisation through the use of electron microscopy. However, artefacts produced by dehydration during specimen preparation made the dental plaque biofilm microcolonies appear more compacted

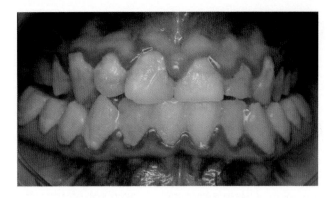

Figure 3.1 Dental plaque on the facial surfaces of the upper and lower dentition. Note the inflammation of the gingival margins and interdental papillae. (From http://medscoopdaily.com/is-dental-plaque-bothering-you-heres-how-to-brush-it-off/.)

than they are in nature. Importantly, such studies showed that the dental plaque biofilm was essential to the development of dental caries and periodontal disease, the two most prevalent diseases that affect humankind. It would be several decades before the significance of biofilms in infectious disease was apparent to medical microbiologists and infectious disease physicians. It is regrettable that, despite the significant contributions of dental scientists and oral biologists to the understanding of the biology of biofilms, their work is rarely cited in the published literature. It is now appreciated that biofilms play a role in the majority of bacterial and fungal infections of humans, including those associated with indwelling medical devices and hospital equipment (**Table 3.1**). This chapter will focus on bacterial biofilms.

BIOFILMS STRUCTURE AND PROPERTIES

Mucosae versus skin

There are several reasons why bacteria and fungi adhere to surfaces and form biofilms in flowing systems such as the mucosal surfaces of the body that are bathed by secretions that flow across their surfaces (**Table 3.2**). For example, nutrients are concentrated on surfaces and, in flowing systems, attachment enables microorganisms to persist in their environment rather than being washed away. However, the skin is quite different from mucosal surfaces because it is keratinised and, for the most part, dry. The bacteria that colonise the skin are largely gram-positive and their cell envelopes contain extensively cross-linked peptidoglycans to resist desiccation; thus, one might ask whether the skin has a biofilm. The answer is, yes. The skin is moistened by sweat consisting mainly of water containing small amounts of minerals, lactic acid, and urea secreted by eccrine glands. In addition, sebum composed primarily of triglycerides, wax esters, squalene and free fatty acids is secreted by sebaceous glands associated with hair shafts. Skin biofilms are

infection or disease	common bacterial species involved
dental caries	acidogenic gram-positive cocci (*Streptococcus* sp.)
periodontitis	gram-negative anaerobic oral bacteria
otitis media	non-typeable *Haemophilus influenzae*
chronic tonsillitis	various species
cystic fibrosis pneumonia	*Pseudomonas aeruginosa, Burkhoideria cepacia*
endocarditis	viridans group streptococci, staphylococci
necrotising fasciitis	group A streptococci
musculoskeletal infections	gram-positive cocci
osteomyelitis	various species
biliary tract infection	enteric bacteria
infectious kidney stones	gram-negative rods
bacterial prostatitis	*Escherichia coli* and other gram-negative bacteria
infections associated with foreign body material	
contact lens	*P. aeruginosa*, gram-positive cocci
sutures	staphylococci
ventilation-associated pneumonia	gram-negative rods
mechanical heart valves	staphylococci
vascular grafts	gram-positive cocci
arteriovenous shunts	staphylococci
endovascular catheter infections	staphylococci
cerebral spinal fluid-shunts	staphylococci
peritoneal dialysis (CAPD) peritonitis	various species
urinary catheter infections	*E. coli*, gram-negative rods
intrauterine devices	*Actinomyces israelii* and others
penile prostheses	staphylococci
orthopaedic prostheses	staphylococci

Table 3.1 Partial list of human infections involving biofilms. (From Fux, C.A. et al., *Trends Microbiol.*, 13, 34–40, 2005.)

located in the orifices of the ducts of these glands rather than on the skin surface, although biofilms may form below the most superficial layers of the squames.

Initial steps in biofilm formation

The mechanisms employed by pathogenic microorganisms to adhere to host surfaces is the topic of Chapter 4, and adhesion is the essential first step in biofilm formation. Microorganisms forming biofilms do not attach directly to a surface, termed a substratum; instead, they attach to a conditioning film that selectively adsorbs to the substratum. Conditioning films consists largely of proteins from the fluid phase such as saliva, respiratory secretions, plasma and urine that bathe the mucosal surface or endothelium. In the case of dental plaque, the conditioning film is known as the acquired pellicle. Therefore, initial colonisers adhere by interaction between molecules on the surface of the microorganisms, termed adhesins, and motifs of molecules that make up the conditioning film. These

general property	dental plaque example
open architecture	presence of channels and voids
protection from host defences, desiccation, etc.	production of extracellular polymers to form a functional matrix; physical protection from phagocytosis
enhanced tolerance to antimicrobials[a]	reduced sensitivity to chlorhexidine and antibiotics; gene transfer of resistance genes
neutralisation of inhibitors	β-lactamase production by neighbouring cells to protect sensitive organisms
novel gene expression[a]	synthesis of novel proteins on attachment; upregulation of glucosyltransferases in mature biofilms
coordinated gene responses	production of cell–cell signalling molecules (e.g., CSP, AI-2)
spatial and environmental heterogeneity	pH and O_2 gradients; coadhesion
broader habitat range	obligate anaerobes in an overtly aerobic environment
more efficient metabolism	complete catabolism of complex host macromolecules (e.g., mucins) by microbial consortia
enhanced virulence	pathogenic synergism in abscesses and periodontal diseases

Table 3.2 General properties of biofilms and microbial communities. AI-2, autoinducer-2; *CSP*, competence-stimulating peptide. [a]One consequence of altered gene expression can also be an increased tolerance of antimicrobial agents. (From Marsh, P.D. (Ed.), Chapter 5, Dental plaque, in *Marsh and Martin's Oral Microbiology*, 6th ed., Elsevier, New York, 2016.)

interactions are primarily protein–carbohydrate (lectin) but can also be protein–protein. In natural multi-species biofilms – that is, those that form in nature and on the mucosal epithelia – the initial colonising bacteria that adhere to the substratum comprise but a few species and they are termed pioneers. The nature of the pioneer species differs depending on the particular barrier epithelium. In the mouth, they comprise a few species of α-haemolytic streptococci. In contrast, biofilms that form on indwelling catheters and prostheses, such as artificial hip and knee joints, are usually composed of a single species of bacterium, frequently *Staphylococcus epidermidis, Staphylococcus aureus*, alpha-haemolytic streptococci, *Enterococcus faecalis, Escherichia coli, Klebsiella pneumoniae, Proteus mirabilis*, and *Pseudomonas aeruginosa*. However, over time single-species biofilms may progress to multi-species biofilms. The pioneer microorganisms adhere to the conditioning film-coated substratum and begin to divide, forming microcolonies. The pioneer bacteria synthesise and invest themselves in a gel of extracellular polysaccharide (EPS) that forms the biofilm matrix. The matrix also contains molecules continuously adsorbed from the fluid (planktonic) phase and extracellular bacterial DNA. The phenotypes of bacteria growing on surfaces in biofilms are quite distinct from the phenotypes of the same bacteria growing in the planktonic phase. Nutrients for bacterial growth are provided by host

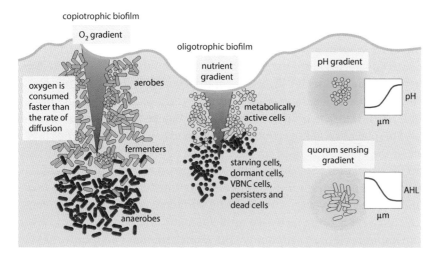

Figure 3.2 The formation of the extracellular polysaccharide matrix leads to the establishment of stable gradients that provide different localised habitats. In an aerobic copiotrophic (nutrient rich) biofilm, organisms are stratified according to oxygen availability, which becomes depleted in the lower layers of the biofilm because the consumption of oxygen by aerobic organisms in the higher layers of the biofilm is faster than the rate of diffusion. Similarly, in aerobic oligotrophic (nutrient rich) biofilms, nutrient consumption by organisms in the upper layers results in the starvation of organisms in the lower layers, which may lead to slow growth states, such as those found in dormant cells, or even in cell death. Other gradients that are present in biofilms include pH gradients, which are produced by heterotrophic metabolism, and gradients of signalling molecules, in which the concentration of quorum sensing molecules varies according to the distance from producing cells. (From Flemming, H.-C. et al., *Nat. Rev. Microbiol.*, 14, 563–575, 2016.)

proteins, including glycoproteins such as mucins, glycolipids and carbo-hydrates in the secretions that bathe the mucosae or sweat and sebum on the skin. The microcolonies continue to expand until they coalesce encased in the polysaccharide matrix. In fact, the majority of the biofilm is composed of heterogeneous hydrated EPS that is responsible for the mechanical and physiological properties of the biofilm. The EPS under-goes constant remodelling due to bacterial glycosidases and, thus differ-ent areas of the biofilm can have different viscosities that may allow the movement of bacteria within the biofilm. The metabolism of pioneer bac-teria changes the environment of the biofilm by, for example, reducing the oxygen tension, generating products of metabolism and enzymatic activity, and providing new ligands to which other bacteria can attach (**Figure 3.2**).

Biofilm development and the climax community

The adhesion of one bacterial species to another is termed coadhesion or intergeneric coaggregation and is the mechanism by which additional species attach to already resident bacteria. The resulting increase in diversity of the biofilm community is termed succession. Co-adhesion is mediated by receptor–ligand interactions between bacteria and is, therefore, a specific interaction that may have evolved to bring different bacterial species close together based on the nutritional relationships between them. Species diversity continues until the biofilm reaches a steady state termed the climax community. The stages of biofilm formation on the surface of teeth are shown in **Figure 3.3**. Whether or not a biofilm reaches a climax community depends on two factors: the stability of the substratum and the rate of flow of the planktonic phase that applies shear force to the biofilm. Therefore, it is unlikely that biofilms attain a climax community on shedding epithelial surfaces such as skin and mucosa. However, non-shedding surfaces such as teeth, catheters, and indwelling prostheses like artificial joints and perhaps heart valves can support climax community. The climax biofilm community of certain types of biofilm has an open architecture with water channels that run both perpendicular and parallel to the substratum (**Figure 3.4**). However, this may not be the case for all biofilms such as those that form in chronic wounds. In the open architecture biofilms, the water channels function like a primitive circulatory system. This together with the close spatial relationships of microcolonies of different species

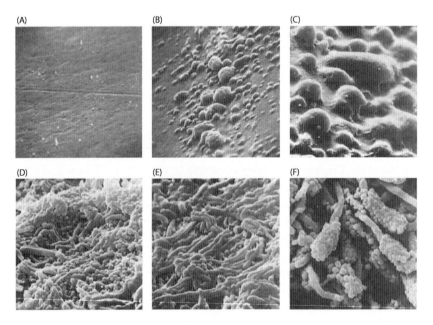

Figure 3.3 Development of dental plaque on a clean enamel surface: (A) coccal bacteria attach to the conditioning film (acquired pellicle); (B) the cocci multiply to form microcolonies; (C) the microcolonies expand and coalesce, resulting in confluent growth embedded in a matrix of extracellular polymers of bacterial and salivary origin; (D, E) the diversity of the community increases and rod and filamentous bacteria colonise; (F) interactions between different morphotypes, suggesting perhaps nutritional dependence. (From Marsh, P.D. (Ed.), Chapter 5, Dental plaque, in *Marsh and Martin's Oral Microbiology*, 6th ed., Elsevier, New York, 2016.)

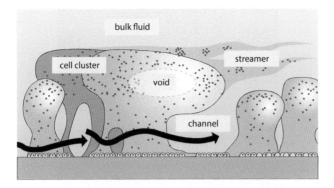

Figure 3.4 Biofilm structure. (From Fux, C. A. et al., *Expert Rev. Anti Infect. Ther.*, 1, 667–683, 2003.)

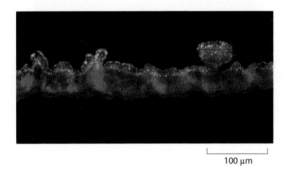

100 μm

Figure 3.5 Biofilm-embedded bacteria show non-uniform distributions of physiological activity. In this *Pseudomonas aeruginosa* biofilm, green indicates cells capable of synthesising alkaline phosphatase in response to phosphate starvation. Red indicates all cells independent of their activity. (Courtesy of Ruifang Xu; Fux, C. A. et al., *Trends Microbiol.*, 13, 34–40, 2005.)

of bacteria resulting from co-adhesion allows creation of numerous microenvironments that, although close together, can be quite different physiologically. This physiological variation is seen in **Figure 3.5** that shows the non-uniform capacity of *P. aeruginosa* cells in a biofilm to synthesise alkaline phosphatase in response to phosphate starvation and in **Figure 3.6** that shows variability in biocide susceptibility in a two-species biofilm.

Quorum sensing in biofilms

The apposition of different species of bacteria allows communication between them by a mechanism known as quorum sensing. Quorum sensing in gram-positive bacteria is mediated by oligopeptides and in gram-negative bacteria by *N*-acyl homoserine lactones. In addition, a family of autoinducers known as autoinducer-2 function in quorum sensing for both gram-negative and gram-positive bacteria. Quorum sensing detects population density and controls multicellular behaviour such as

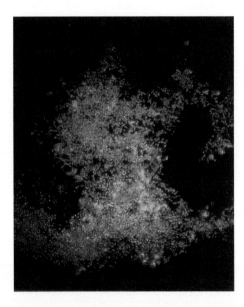

Figure 3.6 Susceptibility to biocide treatment. A two-species biofilm treated with the oxidatively active biocide monochloramine. This figure shows that there is heterogeneity within the biofilm in terms of the response of individual cells to biocide treatment. Areas of red-orange staining correspond to respiratory activity. Green cells have no respiratory activity. Yellow regions represent a mixture of respiring and non-respiring bacteria. The bottom of the image is the portion of the biofilm attached to the substratum and the top of the image is the portion of the biofilm exposed to the bulk medium. (From Huang, C.-T. et al., *Appl. Environ. Microbiol.*, 61, 2252–2256, 1995.)

physiological activities, virulence, competence, conjugation, antibiotic production, motility, sporulation and, indeed, biofilm formation itself.

The range of interactions between biofilm bacteria can run the gamut between cooperation and competition. Host proteins and glycoproteins in the planktonic phase are the principal source of nutrients for bacteria in biofilms forming on surfaces in humans, as has been shown in individuals undergoing tube feeding or intravenous (i.v.) feeding. In order to acquire nutrients, biofilm bacteria and fungi cooperate in food webs or chains in which hydrolytic enzymes produced by consortia of organisms result in the breakdown of macromolecules and provision of amino acids and simple sugars. In addition, the products of metabolism of one microorganism may serve as nutrients for others. Alternatively, microorganisms interact antagonistically to seek a competitive advantage by producing antimicrobial molecules called bacteriocins. Bacteriocins are a diverse group of proteins that vary considerably in size and that usually show antimicrobial activity to strains closely related to the producer strain but, in some cases, have a much broader spectrum of activity. In addition to bacteriocins, the metabolic end products of some bacteria such as volatile and non-volatile fatty acids, hydrogen peroxide and hydrolytic enzymes may be toxic to other bacteria. Furthermore, some bacteria inject toxins into competitors using type VI secretion systems (see Chapter 5).

Biofilm dispersal

Once a biofilm can no longer resist the shear force of the planktonic phase, pieces of the biofilm at the biofilm–planktonic interface break off and are carried downstream where they may reattach and form new bio-film. In some cases, the whole biofilm moves across a surface through shear-mediated transport, sometimes called 'rippling' TRANSPORT. This mechanism may be important in the extension of biofilms in endotra-cheal tubes. Some bacteria release themselves from the biofilm matrix by chemical means. For example, *Bacillus subtilis* secretes D-amino acids and norspermidine, which breaks down exopolysaccharides. Swarming dis-persal is a type of dispersal seen in non-mucoid *Pseudomonas aeruginosa* biofilms. Microcolonies differentiate to form an outer wall of stationary bacteria while the inner region liquefies, allowing motile cells (planktonic phenotype) to swim out leaving a hollow mound.

BIOFILMS IN HUMAN INFECTIONS

Two features of biofilm growth are pertinent to biofilms involved in infec-tious disease. The first is the extent of horizontal gene transfer that can occur in the biofilm when bacteria are close together. Biofilm communi-ties may act as a reservoir of antibiotic resistance genes that can be trans-ferred to extrinsic pathogens. The second property of bacteria growing in a biofilm is their increased resistance to antibiotics (Table 3.2). There are several reasons for increased resistance to antibiotics, including the diffu-sion-limiting nature of the biofilm EPS matrix, the slow growth rate of the bacteria as a result of nutrient limitation, thick cell walls, and high hydro-lytic activity that may inactivate or destroy antibiotics (**Figure 3.7**). Bacte-ria deep in biofilms exist in a stationary phase or in a state termed 'viable but nonculturable' (VBNC) in which they are resistant to antibiotics.

Peripheral and central i.v. catheters

The microbial composition of biofilms that form on indwelling devices reflects, primarily, microbes that are members of the endogenous microbiota at the site of entry of the catheter through the skin. However, microorganisms from the environment, from healthcare personnel or, rarely, from non-sterile i.v. fluids may be the source. Both the exterior and interior surfaces of catheters may be contaminated during passage through the skin. The nutritional composition of the i.v. fluid may favour the growth of some microorganisms over others. The most common microorganism associated with i.v. catheter biofilms is *Staphylococcus epidermidis*, followed by *S. aureus*, *Candida albicans*, *Pseudomonas aeruginosa*, *Klebsiella pneumoniae* and *Enterococcus faecalis*. **Figure 3.8** shows an i.v. catheter biofilm formed by methicillin-resistant *Staphylo-coccus aureus* (MRSA).

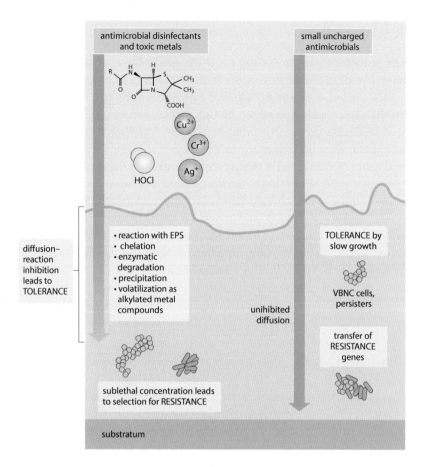

Figure 3.7 An important property of biofilms is an increased ability to survive exposure to antimicrobial compounds including disinfectants, toxic metals and small-molecule antibiotics. Tolerance, can arise when extracellular polysaccharides in the matrix quench the activity of antimicrobials using diffusion–reaction inhibition or as a consequence of the slow growth states that are adopted by many biofilm cells, which enables tolerance of the numerous antimicrobial drugs that target metabolic (or other) processes that occur during growth. Furthermore, the diffusion–reaction inhibition that decreases the concentration of antimicrobials to sublethal concentrations can lead to the survival of exposed cells and to the development of antimicrobial resistance. Resistance to antimicrobials can also increase in biofilms as a result of the dissemination of resistance genes between cells by horizontal gene transfer, which is facilitated by the close proximity of biofilm cells to one another and, by the presence of extracellular DNA in the matrix. VBNC cells, viable-but-nonculturable cells. (From Flemming, H.-C. et al., *Nat. Rev. Microbiol.*, 14, 563–575, 2016.)

Urinary catheters

Urinary catheter biofilms are composed of skin bacteria (male) and faecal bacteria (female). Microorganisms that commonly form biofilm on or in urinary catheters are uropathogenic *Escherichia coli* (UPEC), *S. epidermidis*, *E. faecalis*, *Proteus mirabilis*, *P. aeruginosa* and *K. pneumoniae*. A mixed-species biofilm growing on a urinary catheter is shown in **Figure 3.9**.

Figure 3.8 Methicillin-resistant *Staphylococcus aureus* (MRSA) biofilm on an i.v. catheter. (http://www.ucalgary.ca/biofilm/multimedia.)

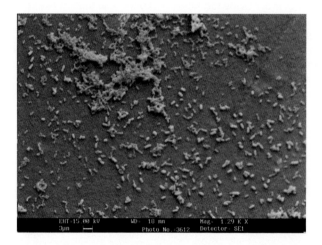

Figure 3.9 Scanning electron microscopy micrograph of a urinary catheter showing a mixed bacterial biofilm. (From Singhai, M. et al., *J. Glob. Infect. Dis.*, 4, 193–198, 2012.)

Bladder biofilms

It is thought that recurrent and chronic urinary tract infections are caused by strains of UPEC which are able to form not only extracellular biofilms on the surface of the bladder epithelium but also intracellular biofilm-like communities termed intracellular bacterial communities (IBC). An IBC in bladder epithelium is shown in **Figure 3.10**.

Endotracheal tubes

Almost all endotracheal tubes acquire biofilms (**Figure 3.11**). The most frequent microorganisms isolated are *Acinetobacter baumannii*, *P. aeruginosa*, coagulase-negative staphylococci and *Candida* species.

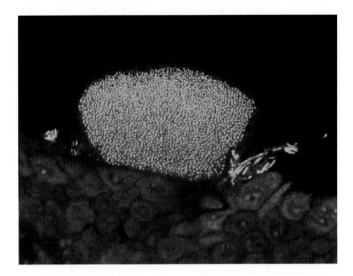

Figure 3.10 *E. coli* progresses through a series of distinct stages, from association with the bladder to the formation of intracellular biofilm-like communities, and release from this habitat into the bladder lumen to allow reinfection of the epithelium. Uropathogenic *E. coli*: (green) replicate within bladder cells to form large intracellular bacterial communities which are protected from many host defences and antibiotics. (From http://cwidr.wustl.edu/research/urinary-tract-infections/.)

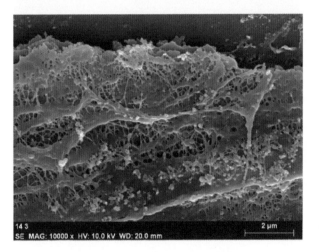

Figure 3.11 Scanning electron microscopy micrographs of biofilm in an endotracheal tube (ETT). Biofilm at low magnification is composed of a matrix that attaches onto the surface of the ETT. Scale bar: 2 μm. (From Gil-Perotin, S. et al., *Crit. Care*, 16, R93, 2012.)

Peritoneal cavity dialysis catheters

Staphylococci and *Pseudomonas* species (**Figure 3.12**) are the most commonly isolated bacteria from peritoneal dialysis catheters.

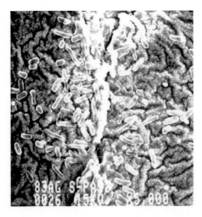

Figure 3.12 *Pseudomonas aeruginosa* biofilm on a peritoneal dialysis catheter. (From http://www.qmed.com/mpmn/medtechpulse/modifying-device-surfaces-key-future-infection-control-strategies.)

Prosthetic joints

Prosthetic joint biofilms may form on the joint components, the cement securing the prosthesis, and bone and fibrous tissues associated with the prosthesis. In addition, pieces of biofilm can be found in the joint space fluid. Generally skin microorganisms contaminate the joint during placement. Accordingly, staphylococci that include both coagulase-negative staphylococci and *Staphylococcus aureus* dominate. However, *Propionibacterium acnes* has been found to be an important cause of infection after shoulder joint surgery. Other organisms isolated from infected prosthetic joints are streptococci, enterococci, gram-negative rods and anaerobes. **Figure 3.13** shows various configurations of bacteria and biofilms in three different orthopaedic patients.

Heart valves

Mechanical heart valves, congenitally malformed valves and valves scarred following acute rheumatic fever are particularly susceptible to colonisation by bacteria in the blood. However, native valves may also be colonised. These microorganisms adhere to the valves and form a biofilm. The sources of the microorganisms are skin during placement surgery, the oral mucosa during tooth extraction or scaling, and existing indwelling devices such as catheters. The route of infection determines the microbial aetiology. Staphylococci such as *S. epidermidis* and *S. aureus* and viridans streptococci predominate followed by *Candida* species, enterococci, and gram-negative rods.

Chronic wounds

Chronic wounds harbour biofilms consisting of complex microbial consortia (**Table 3.3**). The fact that only small areas of chronic wounds

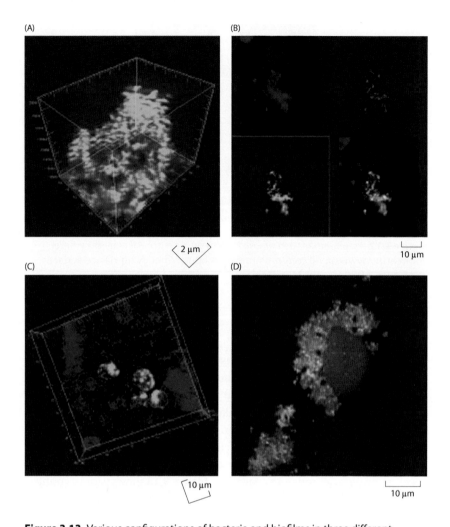

Figure 3.13 Various configurations of bacteria and biofilms in three different orthopaedic patients. (A) Biofilm of live cocci (green) attached to a screw removed from a fixation device in a non-union; (B) patch of biofilm attached to periprosthetic tissue from a failed ankle arthroplasty. The upper left panel shows reflected light, demonstrating the surface of the tissue (blue); the upper right panel shows a fluorescent *in situ* hybridisation (FISH) 'sau' probe, demonstrating *Staphylococcus aureus* bacteria (red); the lower left panel shows a FISH 'Eub' probe, demonstrating all stained bacteria (green); and the lower right panel shows an overlay, demonstrating the *S. aureus* biofilm cluster attached to the tissue. (C) Periprosthetic tissue from the same patient as (B), showing bacteria that appear to be intracellular. (D) Intraoperative fluid from a patient with a failed elbow showing clumps of live cocci (green). The large red object is a nucleolus from a host cell that appears to have been 'attacked' and damaged by the cocci. (From McConoughey, S. J. et al., *Future Microbiol.*, 9, 987–1007, 2014.)

can be examined at a time means that some biopsies may reveal biofilm, whereas others may reveal isolated cells (**Figure 3.14**). Unlike catheter biofilms, the composition of wound biofilms is significantly different from the resident microbiota of healthy skin: the predominant microorganisms of wound biofilms being members of the genera *Bacteroides, Fusobacterium, Pseudomonas, Proteus, Escherichia, Klebsiella,*

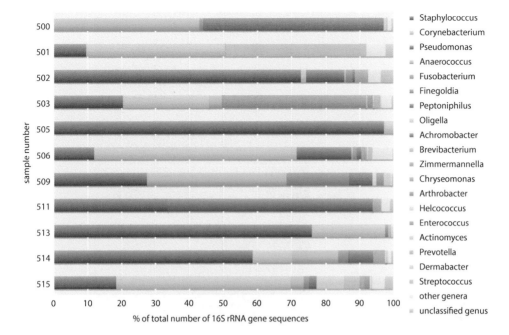

Table 3.3 Culture-independent 16S rRNA-based identification of bacteria from curette samples from chronic wounds. (From Han, A. et al., *Wound Repair Regen.*, 19, 532–541, 2011.)

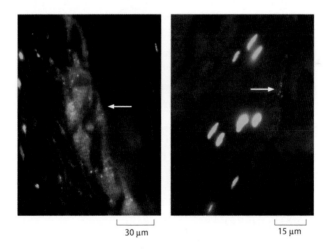

Figure 3.14 Wound biofilms in the left panel shows extensive biofilm formation. By contrast, the sample in the right panel shows scattered individual cells. Left-hand panel scale bar: 30 µm. Right-hand panel scale bar: 15 µm. (From Han, A. et al., *Wound Repair Regen.*, 19, 532–541, 2011.)

Staphylococcus and *Streptococcus*. **Figure 3.15** shows a multi-species biofilm from a diabetic foot ulcer.

Otitis media

Chronic otitis media is caused largely by *Streptococcus pneumoniae*, non-typeable *Haemophilus influenzae*, and *Moraxella catarrhalis*, and

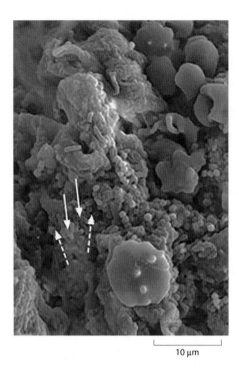

10 μm

Figure 3.15 Rod-shaped bacteria (solid arrows) and cocci-shaped bacteria (dashed arrows) in close proximity in diabetic foot wound biofilm. (From Sun, Y. et al., *Wound Repair Regen.*, 16, 805–813, 2008.)

these bacteria can be detected at high frequency using sensitive nucleic acid–based techniques when cultures are negative. In many cases, chronic otitis media is a mono-infection of one of these three bacteria, but in some cases, two or even all three of these bacteria may be present. Furthermore, these bacteria have been shown to form biofilms on the tympanic membrane that have been implicated in the chronicity of this infection (**Figure 3.16**).

Cystic fibrosis

Cystic fibrosis (CF) is a multi-organ genetic disorder that primarily affects the lungs but also involves the pancreas, liver, kidneys, and intestines. It is caused by a mutation in both alleles of the gene that encodes the cystic fibrosis transmembrane conductance regulator (CFTR) protein. CFTR controls the passage of water and chloride ions in and out of cells. In the lung, this defect results in the production of thick mucus that cannot be effectively removed by the ciliary escalator. Thus, the elimination of microbes from the lungs is significantly impaired, leading to, first, intermittent and then chronic infection. Microbial infection of the lung of individuals with CF is an example of microbial succession in a biofilm where different microorganisms replace one another over time. This is shown in **Figure 3.17** where *Staphylococcus aureus* and other microbes

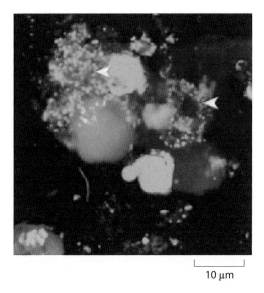

Figure 3.16 Bio films (arrowheads) on an middle ear membrane specimen from an ear with effusion that was PCR-positive for *H. influenzae*.

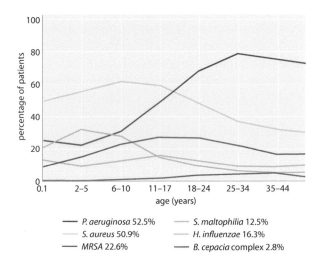

Figure 3.17 Prevalence of several common respiratory pathogens in cystic fibrosis (CF) as a function of age. (From Hauser, A. R. et al., *Clin. Microbiol. Rev.*, 24, 29–70, 2011; Adapted from the 2008 Annual Data Report of the Cystic Fibrosis Foundation Patient Registry, Bethesda, MD, 2009, http://www.cff.org/UploadedFiles/research/ ClinicalResearch/2008-Patient-Registry-Report.pdf. With permission from Cystic Fibrosis Foundation.)

are replaced as the dominant bacteria by *Pseudomonas aeruginosa*. As a result of better management of individuals with CF, the dominant pathogens that are shown in Figure 3.17 have been largely supplanted by opportunists such as *Stenotrophomonas maltophilia*, *Achromobacter xylosoxidansi*, *Alcaligens* species, *Ralstonia* species, *Pandoraea* species, and several fugal genera such as *Aspergillus*, *Penicillium*, *Candida* and *Exophiala*. However, *P. aeruginosa* and *B. cepacia* remain significant

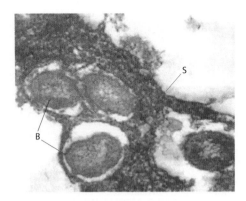

Figure 3.18 Electron micrograph of a thin section of ruthenium red–stained alveolar material obtained by postmortem excision from a *P. aeruginosa*-infected cystic fibrosis (CF) patient. Note the very thick fibrous mass of bacterial exopolysaccharide (S) surrounding the gram-negative bacterial cells (B) in the alveolar spaces. Scale bar: 0.1 μm. (From Lam, J. et al., *Infect. Immun.*, 28, 546–556, 1980.)

pathogens in CF. Although many of these opportunists can form biofilm, *P. aeruginosa* is foremost among biofilm-forming microorganisms in this setting. The inflammatory response to lung infection results in the release of reactive oxygen and nitrogen intermediates, cytokines and chemokines from neutrophils as well as the release of granule contents from these cells by exocytosis. In addition, chronic infection leads to mucus metaplasia in which increased amounts of thick mucus are secreted in response to inflammation and flatten cilia. In addition, squamous metaplasia may occur in which ciliated epithelial cells are replaced by non-ciliated squamous epithelium. It is clear that *P. aeruginosa* that have colonised the CF lung are physiologically distinct. They overproduce exopolysaccharide termed alginate (**Figure 3.18**), are non-flagellate and antibiotic resistant. These properties may result from mutations induced by reactive oxygen and nitrogen intermediates and reduced oxygen tension in the airway mucus. **Figure 3.19** shows a schematic of the process of colonisation of *P. aeruginosa* in the CF lung.

BIOFILM FORMATION BY FILAMENTOUS FUNGI

The formation of biofilms has been considered to be limited to bacteria and yeast; however, there are suggestions that filamentous fungi may be capable of forming biofilms based on the criteria that they produce an extracellular polysaccharide matrix, exhibit differential gene expression and are more resistant to antifungal agents than planktonically growing fungi. *Aspergillus* mycetoma, wound and toenail infections have been suggested to reflect the formation of biofilms.

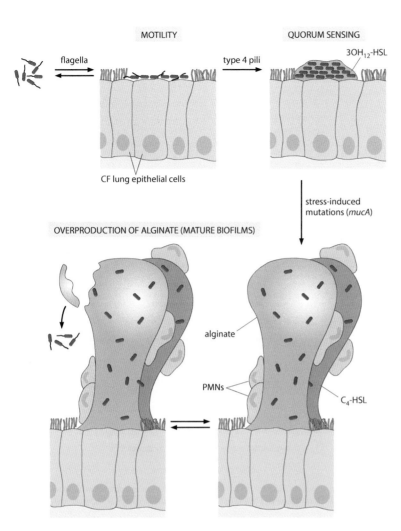

MOTILITY

flagella

type 4 pili

QUORUM SENSING

$3OH_{12}$-HSL

CF lung epithelial cells

stress-induced
mutations (*mucA*)

OVERPRODUCTION OF ALGINATE (MATURE BIOFILMS)

alginate

PMNs

C_4-HSL

Figure 3.19 A hypothetic process of *P. aeruginosa* colonisation leading to biofilm formation in cystic fibrosis (CF). Flagella-mediated motility is involved in the
initial attachment. Type 4 pili and a quorum sensing signal ($3OC_{12}$-HSL) participate in microcolony differentiation. Overproduction of alginate, which results in
the formation of mature biofilms, can be caused by the *mucA* mutations induced
through stresses such as oxidants released by PMNs. The level of C_4-HSL is
increased in mature biofilms. Sessile cells can be sloughed off as planktonic cells
due to mechanical factors (e.g., coughing). Biofilms formed as a result of medi-
ation through motility, quorum sensing and overproduction of alginate may
be different from each other physiologically, and may represent various stages
of a bacterial developmental process. C_4-HSL, *N*-butyryl-ʟ-homoserine lactone;
$3OC_{12}$-HSL, *N*-3-oxodedecanoyl homoserine lactone; PMN, polymorphonuclear
leucocyte. (From Yu, H. and Head, N.E., *Front. Biosci.*, 7, d442–d457, 2002.)

BIOFILM FORMATION BY VIRUSES

The proposal that viruses may be capable of forming biofilms has come
from observations of the human T-cell leukaemia virus type 1 (HTLV-1).
This virus has been found on the surface of infected T cells embedded a

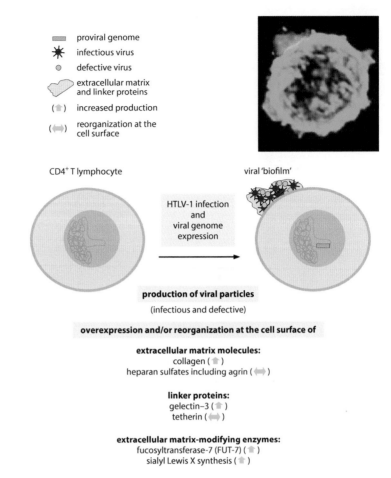

Figure 3.20 Viral components are clustered and tightly attached to the surface of infected cells. Confocal microscopy of primary CD4+ T cells from HTLV-1-infected individuals showing cell-surface glycoproteins stained with Con A (green) and viral proteins stained with HTLV-1–specific serum (red) and with Gag antibody (blue) (arrow). Model of 'viral biofilm' development on HTLV-1–infected lymphocytes. Viral genome expression drives both the production of viral particles and of a matrix enriched in certain carbohydrate moieties (e.g., sialyl Lewis X), ECM components (e.g., agrin and collagen), and some linker proteins (e.g., galectin-3 and tetherin). Some components of this matrix are reorganised at the surface of infected cell (⇔), whereas the production of others is increased upon infection. Together, infectious and defective viral particles embedded in the carbohydrate-rich matrix form the infectious structure termed a 'virus biofilm'. (From Thoulouze, M.I. and Alcover, A., *Trends Microbiol.*, 19, 257–262, 2011.)

glycoprotein matrix produced by the infected cell. This feature has lead to referring to this infectious structure as a virus biofilm (**Figure 3.20**).

HOW BIOFILMS ARE STUDIED

A detailed account of approaches to biofilm research is beyond the scope of this book; however, we will examine general approaches to studying biofilm development and their limitations. As stated at the beginning of this chapter, the earliest studies of biofilms by Van Leeuwenhoek used a primitive light microscope to examine dental plaque scraped from tooth surfaces, and dental plaque has remained the focus of much of biofilm research. However, the application of the electron microscope was required to observe the detailed structure of the dental plaque biofilm, albeit with artefacts introduced by dehydrating the specimens. Study of the development of dental plaque biofilm required placing thin wafers (coupons) of human dental enamel in a dental appliance in the mouth. These coupons were removed over time and the accrued biofilm examined with an electron microscope (see Figure 3.3). These studies took advantage of the non-invasive access of teeth as abiotic, non-shedding surfaces on which biofilm forms. Study of biofilm development on epithelial surfaces or on indwelling prostheses or catheters *in situ* has not yet proven possible for technical and ethical reasons. For this reason, *in vitro* model systems were developed. There are two fundamental requirements to model biofilm formation: a relevant surface (substratum) and a relevant flowing planktonic phase that is the source of the conditioning film. However, these conditions are rarely met. The simplest and most primitive method, known as the microplate static biofilm model, involves measuring the extent to which a microbe attaches to the abiotic plastic surface of wells of the microplate. It is arguable whether this method measures biofilm formation or simply adherence. The plastic substratum is appropriate only if the object is to examine adhesion/biofilm formation to the plastic from which the microplate is made. Generally, the microbes are suspended in physiological saline or, perhaps, nutrient broth, neither of which are appropriate planktonic phases. There is no flow and, thus, no shear force applied to the adherent microbes. Adherent bacteria/biofilm are detected and quantified by emptying and washing the wells and staining the adherent microbes with a dye the density of which is measured using a spectrophotometer (**Figure 3.21**).

Chemostats (continuous culture) allow a closer approximation of the way that biofilms form in nature. A chemostat consists of a vessel that has an inlet through which fresh nutrient medium is introduced with a pump and an overflow outlet that maintains a constant volume of the medium in the vessel. The vessel has a sealed lid with several ports that allow placement of a pH electrode, a tube to introduce a gas mixture, and several coupons. The medium in the vessel, for example hog gastric mucin, is inoculated with a single species of bacterium or with several species or genera of bacteria. By changing the rate at which fresh medium is added to the vessel, the specific growth rate of the microorganism(s) can be regulated. The rate of medium exchange in the vessel is termed the dilution rate D. An important property of chemostats is that microorganisms grow under constant environmental conditions known as steady state. In steady state, microbes grow at a constant specific growth rate and all culture variables such as culture volume, dissolved gas concentration, nutrient and waste product concentrations,

(A)

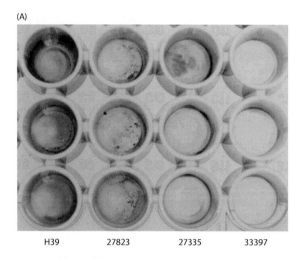

H39   27823   27335   33397

(B)

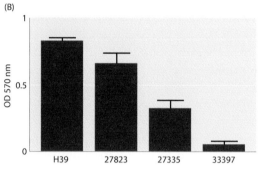

Figure 3.21 (A) Biofilm formation on microtitre plates. Biofilm production of four bacterial strains H39, 2782, 27355 and 33397. A polystyrene microtitre plate: a set of microtitre plate wells (three wells for each strain) stained with 0.1% crystal violet solution after 24 h of incubation. (B) Optical density of sets of three wells. Bars indicate standard deviations. (From Yamanaka, T. et al., *J. Bacteriol. Parasitol.*, 4, 160, 2013.)

pH and cell density remain constant. Coupons can be removed from the vessel at different time points and under different growth rates or growth conditions. Biofilm on the coupons can be analysed by electron microscopy or other methods. An example of such a system is shown in **Figure 3.22**. One of the disadvantages of the system described above using a vessel (bioreactor) is that the ratio of the volume of the planktonic phase to the surface area of the substratum is extremely large. This may be appropriate for modelling biofilm formation in marine and fresh water environments, but it does not replicate the conditions in the human body, where surfaces are generally bathed by thin films of secretions (the planktonic phase). However, this limitation can be ameliorated by placing the substrata downstream of the culture vessel (bioreactor) in flow cells. Generally, microscope slide coverslips serve as the substratum and sometimes these are coated with biomolecules such as collagen to simulate adhesion and biofilm formation on

(A)   (B)

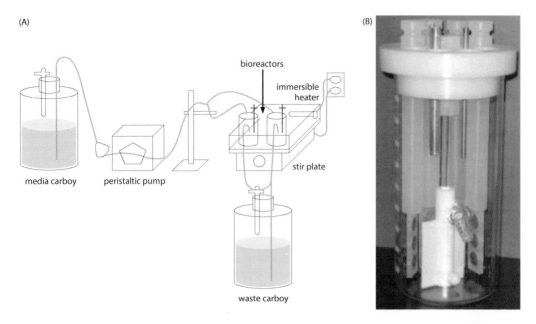

Figure 3.22 (A) Schematic of a bioreactor system. Medium is pumped from the media carboy (reservoir) to the reactors using a peristaltic pump. The bioreactors are contained in a water bath maintained at the temperature selected for the experiment. The medium in the bioreactors is stirred at a constant rpm over the course of the experiment. Liquid exits the bioreactors by gravity into a waste carboy. (B) Close-up view of the CDC biofilm reactor. The bioreactor has openings for eight polypropylene rods. Each rod contains three spaces for coupons which can be removed to assay for bacterial growth. At the centre of the bioreactor is a stirring baffle that maintains constant shear stress. There is a spout located about one-third of the length of the vessel from the bottom to allow for the exit of medium. (From Stewart, C.R. et al., *PLoS One*, 7, e50560, 2012.)

an ECM. Another advantage of flow cells is the ability to use confocal laser microscopy to 'look through' the biofilm. Utilisation of techniques such as polymerase chain reaction (PCR)–based bacterial detection and identification, species-specific peptide–nucleic acid fluorescent *in situ* hybridisation (PNA-FISH) allow detailed examination of the biofilm by confocal laser scanning microscopy. Modelling biofilm formation on epithelial surfaces *in vitro* is difficult because of the technical complexity in maintaining the viability of polarised epithelium over the period required to establish a biofilm. **Figure 3.23** shows a flow cell designed to examine biofilm formation on an oral mucosa analogue. The upper chamber of the flow cell is fed with a batch culture of the microorganism(s) to be tested grown in brain-heart infusion (BHI) broth supplemented with human whole saliva. The lower chamber is fed with tissue culture medium. The upper and lower chambers are separated by a sheet of immortalised human oral keratinocytes (OKF6/TERT-2) seeded on collagen type I–embedded fibroblasts (3T3 fibroblasts) and fed with tissue culture fluid. Advantages and disadvantages of this system are listed in **Table 3.4**.

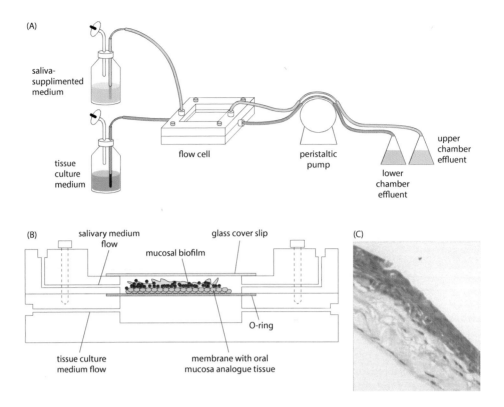

Figure 3.23 Schematic representation of flow cell device that supports mucosal biofilm formation in the presence of salivary flow. (A) Components of the assembled flow cell system. (B) Cross-sectional view of the flow cell device, which consists of two pieces of polytetrafluoroethylene juxtaposed to each other and held together by screws. A membrane containing a pre-grown oral mucosa analogue tissue is placed in the middle, supported by an O-ring, forming two separate chambers for independent saliva-supplemented medium and tissue culture medium flow. (C) Example of an H&E-stained section of an oral mucosa analogue tissue over which microorganisms are inoculated to form mucosal biofilms in the device. (From Diaz, P.I. et al., *Infect. Immun.*, 80, 620–632, 2012.)

strengths	weaknesses
biofilm development in a flowing system	attachment under stop flow condition
microbes grown in batch culture – not reproducible	microbes not grown in continuous culture under steady-state conditions
microbes grown in complex bacteriological medium with saliva supplementation	microbes not grown in saliva alone
ability to examine the ability of combinations of microbes to form biofilms	conditioning film not derived entirely from saliva
ability to examine biofilm on epithelium	
planktonic phase is a thin film	
ability to employ confocal laser microscopy and other techniques to image throughput the thickness of the biofilm	

Table 3.4 Strengths and weakness of the flow cell system described in the text.

KEY CONCEPTS

- In nature, biofilms form on all surfaces bathed by a fluid phase.

- In the human body, biofilms form on all mucosal surfaces and perhaps the skin.

- Biofilm formation is involved in almost all bacterial and fungal infections in humans.

- Microbes in biofilms are physiologically different from planktonic microbes.

- Microbes in biofilms are more resistant to antimicrobial agents than are planktonic microbes.

- Biofilms protect microbes from the immune system.

- Biofilms are reservoirs of antibiotic resistance genes and other genes involved in pathogenesis.

BIBLIOGRAPHY

Achermann Y et al. 2014. *Propionibacterium acnes*: From commensal to opportunistic biofilm-associated implant pathogen. *Clinical Microbiology Reviews*, 27(3): 419–440.

Allen HB et al. 2014. The presence and impact of biofilm-producing staphylococci in atopic dermatitis. *JAMA Dermatology*, 150(3): 260–265.

Bjarnsholt T. 2013. The role of bacterial biofilms in chronic infections. *APMIS*, 121(136): 1–58.

Bjarnsholt, T, Jensen PØ, Moser C, Høiby N (Eds.). 2011. *Biofilm Infections*. Springer, New York.

Burmølle M et al. 2014. Interactions in multispecies biofilms: Do they actually matter? *Trends in Microbiology*, 22(2): 84–91.

Cheng Y et al. 2019. Population dynamics and transcriptomic responses of *Pseudomonas aeruginosa* in a complex laboratory microbial community. *NPJ Biofilms and Microbiomes*, 5 Article number: 1.

Chew SC, Yang L. 2017. Biofilms: Microbial cities wherein flow shapes competition. *Trends in Microbiology*, 25(5): P331–P332.

Costerton JW et al. 1995. Microbial biofilms. *Annual Review of Microbiology*, 49: 711–745.

Costerton JW et al. 1999. Bacterial biofilms: A common cause of persistent infections. *Science*, 84: 1318–1322.

Costerton JW. 2001. Cystic fibrosis pathogenesis and the role of biofilms in persistent infection. *Trends in Microbiology*, 9(2): 50–52.

De Vos WM. 2015. Microbial biofilms and the human intestinal microbiome. *NPJ Biofilms and Microbiomes*, 1: Article number 15005.

Diaz PI et al. 2012. Synergistic Interaction between *Candida albicans* and commensal oral streptococci in a novel in vitro mucosal model. *Infection and Immunity*, 80(2): 620–632.

Donlan RM. 2001. Biofilms and device-associated infections. *Emerging Infectious Diseases*, 7(2): 277–281.

Donlan RM. 2002. Biofilms: Microbial life on surfaces. *Emerging Infectious Diseases*, 8(9): 881–890.

Donlan RM, Costerton JW. 2002. Biofilms: Survival mechanisms of clinically relevant microorganisms. *Clinical Microbiology Reviews*, 15(2): 167–193.

Flemming H-C et al. 2016. Biofilms: An emergent form of bacterial life. *Nature Reviews Microbiology*, 14(9): 563–575.

Folkesson A et al. 2012. Adaptation of *Pseudomonas aeruginosa* to the cystic fibrosis airway: An evolutionary perspective. *Nature Reviews Microbiology*, 10(12): 841–851.

Fux CA et al. 2003. Bacterial biofilms: A diagnostic and therapeutic challenge. *Expert Reviews of Anti-infectious Therapy*, 1(4): 667–683.

Fux CA et al. 2005. Survival strategies of infectious biofilms. *Trends in Microbiology*, 13(1): 34–40.

Gil-Perotin S et al. 2012. Implications of endotracheal tube biofilm in ventilator-associated pneumonia response: A state of concept. *Critical Care*, 16: R93.

Hall-Stoodley L et al. 2006. Direct detection of bacterial biofilms on the middle-ear mucosa of children with chronic otitis media. *JAMA*, 296(2): 202–211.

Han A et al. 2011. The importance of a multi-faceted approach to characterizing the microbial flora of chronic wounds. *Wound Repair and Regeneration*, 19(5): 532–541.

Harding MW et al. 2009. Can filamentous fungi form biofilms? *Trends in Microbiology*, 17(11): 475–480.

Harriott MM, Noverr MC. 2011. Importance of *Candida*–bacterial polymicrobial biofilms in disease. *Trends in Microbiology*, 19(11): 557–563.

Hassett DJ et al. 2008. *Pseudomonas aeruginosa* hypoxic or anaerobic biofilm infections within cystic fibrosis airways. *Trends in Microbiology*, 17(3): 130–138.

Hauser AR et al. 2011. Clinical significance of microbial infection and adaptation in cystic fibrosis. *Clinical Microbiology Reviews*, 24(1): 29–70.

Huang C-T et al. 1995. Nonuniform spatial patterns of respiratory activity within biofilms during disinfection. *Applied and Environmental Microbiology*, 61(6): 2252–2256.

Jenkinson HF, Lappin-Scott HM. 2001. Biofilms adhere to stay. *Trends in Microbiology*, 9(1): 9–10.

Kanaparthy A, Kanaparthy R. 2012. Biofilms—The unforgiving film in dentistry (clinical endodontic biofilms). *Dentistry*, 2: 145.

Kanematsu H, Barry DM (Eds.). 2015. *Biofilm and Materials Science*. Springer International Publishing, Cham, Switzerland.

Lam J et al. 1980. Production of mucoid microcolonies by *Pseudomonas aeruginosa* within infected lungs in cystic fibrosis. *Infection and Immunity*, 28(2): 546–556.

Marsh PD (Ed.). 2016. Chapter 5, Dental plaque. In: *Marsh and Martin's Oral Microbiology*, 6th ed., Elsevier, Edinburgh, UK.

McConoughey SJ et al. 2014. Biofilms in periprosthetic orthopedic infections. *Future Microbiology*, 9(8): 987–1007. doi:10.2217/fmb.14.64.

Nadell CD et al. 2016. Spatial structure, cooperation and competition in biofilms. *Nature Reviews Microbiology*, 14: 589–600.

Peters BM et al. 2012. Polymicrobial interactions: Impact on pathogenesis and human disease. *Clinical Microbiology Reviews*, 25(1): 193–213.

Rickard AH et al. 2003. Bacterial coaggregation: An integral process in the development of multi-species biofilms. *Trends in Microbiology*, 11(2): 94–100.

Roder HL et al. 2016. Studying bacterial multispecies biofilms: Where to start? *Trends in Microbiology*, 24(6): 503–513.

Singhai M et al. 2012. A study on device-related infections with special reference to biofilm production and antibiotic resistance. *Journal of Global Infectious Diseases*, 4(4): 193–198.

Soto SM. 2014. Importance of biofilms in urinary tract infections: New therapeutic approaches. *Advances in Biology*, 2014, Article ID 543974.

Stewart CR et al. 2012. *Legionella pneumophila* persists within biofilms formed by *Klebsiella pneumoniae*, *Flavobacterium* sp., and *Pseudomonas fluorescens* under dynamic flow conditions. *PLoS One*, 7(11): e50560.

Sun Y et al. 2008. In vitro multispecies Lubbock chronic wound biofilm model. *Wound Repair and Regeneration*, 16: 805–813.

Tan CH et al. 2017. All together now: Experimental multispecies biofilm model systems. *Environmental Microbiology*, 19(1): 42–53.

Thoulouze M-I, Alcover A. 2011. Can viruses form biofilms? *Trends in Microbiology*, 19(6): 257–262.

Wilson M. 2001. Bacterial biofilms and human disease. *Science Progress*, 84(3): 235–254.

Yamanaka T et al. 2013. Biofilm-forming capacity on clinically isolated *Streptococcus constellatus* from an odontogenic subperiosteal abscess lesion. *Journal of Bacteriology and Parasitology*, 4(1): 1000160.

Chapter 4: Adhesion to Host Surfaces

INTRODUCTION

In this chapter, we examine mechanisms by which the various classes of pathogens adhere to host cells, molecules and inanimate objects such as catheters and prostheses. We will see that there are two sequential steps in adhesion to surfaces, initial non-specific ionic and hydrophobic interactions followed by specific receptor–ligand interactions. The specific interactions are predominantly protein–carbohydrate (lectin) or protein–protein in nature. These receptor–ligand interactions occur between molecules on the surface of the microbe termed adhesins and molecules on the surfaces of the host cell, intercellular matrix or inanimate objects termed receptors. Microbes do not rely on a single adhesin; rather, they express several, each of which may target a different receptor. Some adhesins bind a single receptor, whereas others bind several. Some adhesins are dedicated attachment organelles, whereas others are multifunctional. In addition, individual adhesins may bind both sugar motifs and peptides. Furthermore, microbes can display adhesins on their cell surface at different stages of their infectious cycle. While pathogens can adhere to many different host surfaces during their infectious cycle, most pathogens enter the human body by first attaching to the mucous membranes (mucosae) or skin. These surfaces are the physical barriers that protect the inner and outer surfaces of the body from the environment. Collectively, they are referred to as the barrier epithelia. For all but a few pathogens, adhesion to mucous membranes or skin is an essential first step to entering host cells and extending into deeper tissues and structures of the human body. However, pathogens transmitted by vectors such as mosquitoes or ticks whose biting mouthparts penetrate the skin and those that enter *via* damage to the barrier epithelia are exempt from the requirement to adhere to epithelium. In addition, some pathogens that secrete exotoxins (proteins that can affect the function of cells or destroy them) that are absorbed through mucous membranes can cause tissue injury in the absence of the microorganism. One such example is food-borne botulism, which results from the production of botulinum toxin by the bacterium *Clostridium botulinum* in improperly sterilised canned food. In this case, the consumption of the preformed toxin causes the disease. In addition to binding to host surfaces and abiotic surfaces such as catheters, bacteria and fungi utilise adhesins to bind to other microorganisms during biofilm formation (see Chapter 3). This process is termed co-aggregation or inter-generic aggregation. We will begin by considering the barrier epithelia and the molecules on their surfaces to which microbes adhere. The host cell surface receptors bound by microbial adhesins are, essentially, any molecule that is exposed on the host cell membrane. These include transport molecules, hormone receptors and intercellular matrix molecules. Because host cells display a finite number of different types of molecule, it is not surprising that different classes of pathogen bind to the same complement of receptors. The surfaces to which microbes adhere are called substrata (singular = substratum).

BARRIER EPITHELIA

Skin

The skin is composed of keratinised stratified squamous epithelium (**Figure 4.1**), and this type of epithelium is also found on the dorsum of the tongue, hard palate and the free and attached gingivae. The most superficial cells are termed corneocytes and are flattened tile-like cells. The cornified cell envelope consists of involucrin, loricrin, small proline-rich proteins, elafin, keratin filaments, filaggrin, cystatin-A and desmosomal proteins. The corneocytes are sealed together by specialised lipids. These molecules are likely those to which microbes attach. Although the skin environment is much less diverse than the mucous membranes, different areas of the skin are exposed to differences in temperature, humidity and anatomical location. These physiological variables dictate the composition and density of the resident microbiota which colonise the different areas of the skin (see Chapter 2). The physiological and anatomical properties of the skin also influence the areas that are targeted by exogenous pathogens. The hair shafts, hair follicles and their associated sebaceous glands, and eccrine and apocrine glands all serve as sites for adhesion to, and subsequent penetration of, the skin. A histological section of skin showing anatomical features of the skin is shown in Figure 4.1.

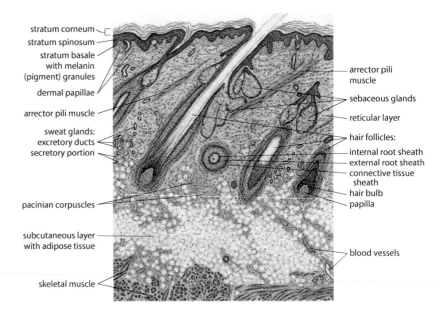

Figure 4.1 Anatomy of the skin. (From Summer Ekelund, https://www.pinterest.com/pin/369928556885000125/?lp=true.)

Mucous membranes (mucosae)

The mucosae vary in their structure according to anatomic location. Different mucosae are composed of different types of epithelial cells, including squamous, cuboidal, columnar, transitional and glandular, each of which may be simple or stratified. In addition, columnar epithelial cells may be ciliated, as in the cases of the tracheal and bronchial regions of the pulmonary tree and in the fallopian tubes of the female reproductive system. **Figure 4.2** shows the mucosal epithelium of the small intestine, with a Peyer's patch located in the centre of the field.

Mucosal epithelia are covered by a mucus layer the thickness of which varies with its anatomical location (**Figure 4.3**). The mucus layer is chemically

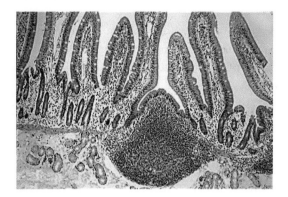

Figure 4.2 Histology of small intestine showing single layer of columnar epithelial cells and a Peyer's patch. (From http://www.kumc.edu/instruction/medicine/anatomy/histoweb/gitract/large/Gi14.jpg.)

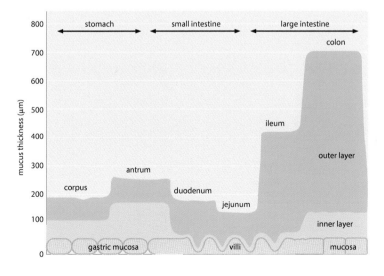

Figure 4.3 Difference in thickness of the mucus layers along the rat gastrointestinal tract. The intestinal epithelial surface is covered by two mucus layers (an inner, firmly adherent layer and an outer, loosely adherent layer). Microbes are mostly associated with the outer layer. (From Juge, N., *Trends Microbiol.*, 20, 30–39, 2012.)

complex and is composed principally of the glycoprotein mucin. The mucus layer can be divided into an outer semi-fluid layer (gel-forming or secreted mucins) and an inner layer termed the glycocalyx, which is firmly attached to the apical surface of the epithelial cells (Figure 4.3). The outer mucus layer is continuously moving across the surface of the epithelium, driven either by muscular action such as chewing, swallowing and peristalsis or by the beating of cilia as in the respiratory tract and fallopian tubes. Mucins are large, heavily *O*-glycosylated glycoproteins that form homo-oligomers. The gel-forming mucins In the human alimentary canal consist of five oligomerising secreted mucins and one non-oligomerising secreted mucin which are produced in different regions of the gastrointestinal tract. Mucins that make up the glycocalyx in the gastrointestinal tract include MUC1, MUC3A, MUC3B, MUC4, MUC12, MUC13, MUC15, MUC16 and MUC17. The distribution of mucins varies between different regions of the gastrointestinal tract and between different cell types (**Table 4.1**).

Mucus also contain proteoglycans, immunoglobulins, antimicrobial peptides, and lipids. In order to reach receptors such as glycolipids, phospholipids, proteins and glycoproteins on the apical surface of epithelial cells, pathogens must cross the mucin barrier. However, cells known as

tissue	gel-forming mucins	cell surface mucins
oral cavity	MUC5B	MUC1
	MUC7[§]	MUC4
	MUC19	MUC16
stomach	MUC5AC	MUC1
	MUC6	MUC16
small intestine	MUC2	MUC1
		MUC3A
		MUC3B
		MUC4
		MUC12
		MUC13
		MUC15
		MUC16
		MUC17
large intestine	MUC2	MUC1
	MUC5AC	MUC3A
	MUC6	MUC3B
		MUC4
		MUC12
		MUC13
		MUC15
		MUC16
		MUC17

Table 4.1 Types of mucins (MUC) throughout the gastrointestinal tract. (From McGuckin, M.A. et al., *Nat. Rev. Microbiol.*, 9, 265, 2011.)

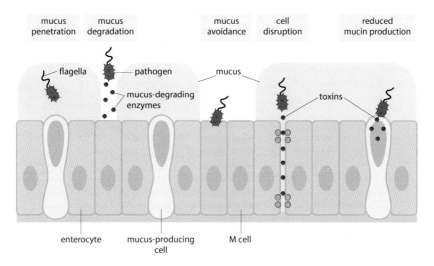

Figure 4.4 Pathogenic strategies to evade the mucus barrier. (1) Pathogenic microbes can penetrate the mucus barrier using flagella-mediated motility. (2) Pathogens can degrade membrane-bound and secreted mucins by the use of degrading enzymes. (3) Pathogenic microbes can avoid the mucus barrier by entering *via* the M cells found in the epithelium covering the mucosa-associated lymphoid structures, which is deprived of a mucus layer. (4) Pathogens can secrete toxins that diffuse in the mucus layer and disrupt the tight junctions between epithelial cells. (5) Pathogenic microbes can secrete or inject toxins that interfere with cell signalling pathways to impact the mucin production. The reduction in mucus levels will allow pathogens to reach the epithelial surface and initiate their infectious process. (From Sperandio, B. et al., *Semin. Immunol.*, 27, 111–118, 2015.)

microfold (M) cells, that overly mucosa-associated lymphoid tissues such as Peyer's patches, isolated intestinal lymphoid follicles, the appendix, tonsils and other mucosa-associated lymphoid tissues do not produce mucins. Furthermore, M cells are specialised antigen-uptake cells that sample the secretions bathing mucosal surfaces. M cells engulf pathogens from the lumen and transport them across the epithelium. Thus, this route circumvents the need to penetrate the mucin layer but delivers pathogens to phagocytes and lymphocytes immediately beneath M cells. The mechanisms used by pathogens to circumvent the mucin barrier are shown in **Figure 4.4**.

Blood and lymphatic vessels

Blood and lymphatic vessels are structurally similar tubes (**Figure 4.5a and b**) that have two surfaces to which microorganisms can adhere: the inner surface or lumenal surface and the outer surface or ablumenal surface. The lumenal surface of the vessels is covered by a single layer of various types of squamous epithelial cells and is called the endothelium. The total area of the endothelial layer of all of the blood and lymphatic vessels in the human body is estimated to be between 4,000 and 7,000 m^2. Like the barrier epithelia, the endothelium of arteries, arterioles, capillaries, venules and veins differ structurally and functionally, depending on anatomical and physiological demands. The smallest capillaries are

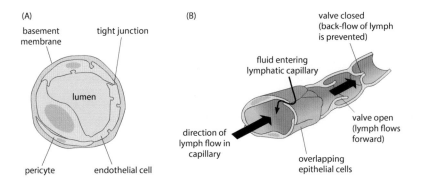

Figure 4.5 (A) Capillary. Pericyte is a contractile cell. (https://www.histology.leeds. ac.uk/circulatory/capillaries.php.) (B) Lymphatic vessel. (http://www.assignment-point.com/science/biology/lymphatic-vessel.html.)

formed by single endothelial cells that wrap themselves into a tube to produce the lumen of the vessel. The endothelium is defined as being continuous or discontinuous. Continuous endothelium can be divided into two types, non-fenestrated and fenestrated. Non-fenestrated continuous endothelium is found in arteries, veins, and capillaries of the brain, skin, heart, and lung. Continuous fenestrated endothelium has transcellular pores (fenestrae) that extend through the cell from cell membrane to cell membrane. The fenestrae have a non-membranous diaphragm across their opening. Continuous fenestrated endothelium is found in the capillaries of exocrine and endocrine glands, gastric and intestinal mucosa, the choroid plexus, glomeruli, and the kidneys. Discontinuous endothelium has large fenestrae without diaphragms which form gaps between one cell and the next. Integrins, proteoglycans, and the hyaluronic acid receptor are present on the ablumenal surface, while several inducible cell adhesion molecules such as selectins and intercellular adhesion molecules (ICAMs) are expressed on the lumenal surface. The density and types of adhesion molecules on the lumenal surface are increased during inflammation. All of these molecules on the ablumenal and lumenal surfaces serve as receptors for microbial adhesins.

Blood-brain barrier

The blood-brain barrier is made up of specialised continuous non-fenestrated endothelial cells which separate the blood from the cerebrospinal fluid. The tight junctions between endothelial cells and the very low rate of transcytosis enable control of the passage of molecules from the blood to the brain. Brain microvascular endothelial cells rest on a basement membrane consisting of type IV collagen, fibronectin, laminin, and proteoglycans. Selectins and ICAMs are found on the lumenal surface of the endothelium and glycosylphosphatidylinositol (GPI)-anchored proteins may also serve as receptors for microbial adhesion. A diagram comparing a general capillary to a capillary in the brain is shown in **Figure 4.6**.

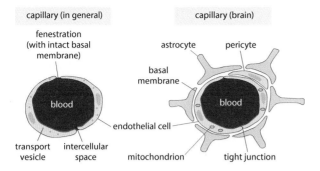

Figure 4.6 Comparison between a general capillary and a capillary in the brain. (From https://web.stanford.edu/group/hopes/cgi-bin/hopes_test/cerebrovascular-and-blood-brain-barrier-impairments-in-huntingtons-disease-potential-implications-for-its-pathophysiology/, Figure 1.)

Foeto-placental interface

The maternal–foetal interface begins with the invasion of blastocyst trophoblasts into the endometrium, termed the decidua, where they rupture uterine capillaries (**Figure 4.7**). The outer trophoblasts form into a syncytiotrophoblast that forms finger-like projections, termed villi, in the endometrium (panel A). These villi support a network of foetal capillaries and are surrounded by maternal blood. Trophoblasts residing outside the villi are known as extra-villus trophoblasts and invade the uterine wall (panel B). Figure 4.7 panel C shows the relationship between the fetal and maternal blood at the blood-trophoblast interface.

The resistance of the placenta and foetus to infection may in part be due to a paucity of receptors displayed on the surface of syncytiotrophoblasts to which pathogens can adhere. Unfortunately, little is known about

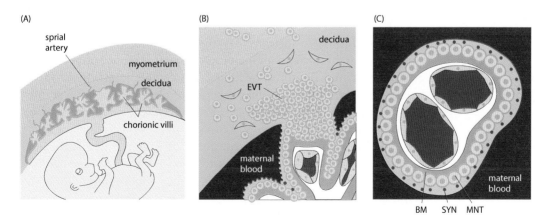

Figure 4.7 The maternal-fetal interface. (A) The villus structure of the placenta anchored in the endometrium; (B) the extravillus trophoblasts (EVT) extending into the endometrium; and (C) the blood–trophoblast interface showing the relationship between the foetal and maternal blood. BM, basement membrane; MNT, mononuclear trophoblast; SYN, syncytiotrophoblast. (Adapted from Robbins, J.R. and Bakardjiev, A.I., *Curr. Opin. Microbiol.*, 15, 36–43, 2012, Figure 1.)

molecules on the cell membrane of the extravillus trophoblast (EVT) or syncytiotrophoblast (SYN). It should be noted that pathogens may colonise the placenta by ascending from the vagina into the uterus. Recently it has been proposed that the placenta has a microbiome. Bacterial colonisation of the human placenta has been found not only in patients with clinical infections or in preterm births, but also in normal pregnancy and term placentas.

THE EXTRACELLULAR MATRIX AND INTERCELLULAR ADHESION MOLECULES

The extracellular matrix (ECM) consists of a variety of extracellular molecules secreted by fibroblasts and provides structural and biochemical support to the surrounding cells. Its organisation, biochemistry and spatial relations are tissue specific. The ECM is composed of two classes of macromolecules: fibrous proteins and proteoglycans. The fibrous proteins are mainly collagens, elastins, fibronectins and laminins. Proteoglycans have a range of functions that include buffering, hydration, and force resistance.

Cell adhesion molecules (CAM) are used by cells to communicate with each other and with the ECM. Motifs on these molecules serve as receptors for adhesins on the surface of microbes to allow them to attach to host cells or ECM molecules. CAMs comprise the ICAMs, cadherins, integrins, selectins, receptor protein tyrosine phosphatases and hyaluronate receptors. Cellular adhesins, such as integrins, discoidin domain receptors and syndecans that tether cells one to another *via* the ECM also serve as receptors for microbial adhesins.

Abiotic surfaces

As well as adhering to surfaces of living cells (biotic surfaces) microbes can adhere to the surfaces of inanimate objects (abiotic surfaces) such as contact lenses, dentures, catheters, prosthethic joints, heart valves and the enamel of teeth. Within seconds of placing an abiotic surface into an aqueous medium such as saliva and other secretions or blood and other body fluids, macromolecules from these secretions or fluids selectively adsorb onto the substratum. This acquired layer is termed a conditioning film and its deposition changes the physicochemical properties of the substratum. An example of a conditioning film is the acquired pellicle that forms on the enamel and exposed cementum surfaces of teeth and is derived from saliva. Therefore, it is to the conditioning film that microbes adhere and not to the substratum. It is unclear whether the skin has a conditioning film. As described earlier, the surface layer of skin is composed of dead anucleate stratified squamous, keratinised epithelial cells termed corneocytes. The corneocytes are surrounded by material containing the proteins loricrin, involucrin and cytokeratin. It is conceivable that these proteins together with components of the secretions of sebaceous glands and/or sweat glands may constitute a conditioning film. The apical surfaces of mucosal epithelial cells, as we have seen, are coated with mucins that constitute a conditioning film.

Initial adhesion events

Most of the studies of microbial adhesion in humans have focussed on abiotic surfaces such as human dental enamel and have employed oral bacteria. The concentration of research on adhesion to abiotic surfaces results from the simplicity of the experimental design. For example, wafers of enamel can be suspended in saliva to allow the formation of the conditioning film and the conditioned substratum can be incubated with different strains and species of bacteria suspended in physiological saline or saliva. These wafers can be removed at various time points and examined using different microscopic and other imaging techniques. This experimental design is amenable to the study of adhesion of microbes to prosthetic devices such as heart valves, prosthetic joints and catheters. Studying adhesion to epithelium and other biotic surfaces is a much more difficult proposition. The study of adhesion and facilitated microbial uptake by viable host cells are considered in Chapter 5, *Facilitated Cell Entry*.

Based mostly on studies of bacterial adhesion to abiotic surfaces, the process of adhesion begins with the microorganism making random transient contact with the conditioning film *via* Brownian motion. The contact is transient because the both the bacteria and the conditioning film substatum are negatively charged, so the bacteria are repulsed. The ionic forces involved in bringing bacteria close enough to the substratum to allow receptor–ligand (adhesin–receptor) interactions, have been described by the DLVO theory, named for Derjagiun, Landau, Verwey and Overbeek and extended by van Oss. The extended DLVO theory describes the force between charged surfaces interacting through a liquid medium. The extended DLVO theory has been applied to the adhesion of planktonic bacteria, viruses and yeast to substrata, and the extended theory likely applies to the adhesion of other unicellular microorganisms. The extended DLVO theory allows the binding strength between colloidal particles such as microorganisms and substrata, to be determined as the summation of attractive van der Waals forces and repulsive electrostatic forces as the distance between the microorganism and the substratum decreases. As mentioned earlier, most cells and substrata are negatively charged, but these negatively charged surfaces attract positively charged ions such as Ca^{2+} from saliva, tissue fluid and blood, for example, forming what is termed a mobile electrical double layer. Thus, when two charged particles approach one another or approach a surface, the electrical double layers overlap and they are repelled. The size of the electrical double layer is inversely proportional to the ionic strength of the fluid phase; therefore, with increasing ionic strength, cells can get closer to each other and to surfaces. Counteracting the repulsive electrical double layer are attractive van der Waals oscillating dipole forces, which are much more powerful than the repulsive forces of the electrical double layer, but act only over short distances. At a distance of approximately 10 nm from the substratum, cells experience a mild attraction for the surface, termed the secondary minimum (**Figure 4.8**), mediated by van der Waal forces, but closer to the substratum, they encounter the repulsive electrical double layer that acts as a barrier. If they can traverse this barrier, then, at a distance of about 1–2 nm from the substratum, they are subject to powerful attractive van der Waals forces. This is termed the primary minimum (**Figure 4.9**). At submicrometer distances between the microbial cell and substratum, hydrophobic interactions become important. Thus, for bacteria and for fungi, the more hydrophobic their cell wall, the stronger they adhere. How do microorganisms overcome the electrostatic, repulsive forces to approach close enough for the strongly attractive van der Waal attractive forces to come into play? The ionic strength of the fluid phase may be sufficient to reduce or overcome repulsion. Additionally, divalent cations, such as calcium in the fluid phase, can bridge the negative charges on the surface of the microorganism and the substratum, or surface

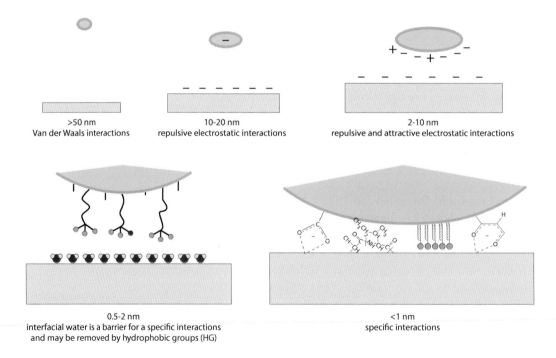

Figure 4.8 The forces mediating initial attachment to substrata. (From Fletcher, M. (Ed.), *Bacterial Adhesion: Molecular and Ecological Diversity* (Wiley Series in Ecological and Applied Microbiology), 1st ed., Wiley-Liss, New York, 1996.)

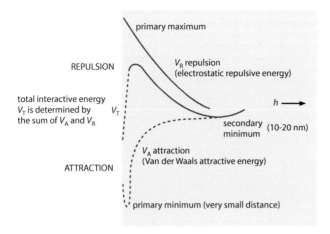

Figure 4.9 The extended DLVO theory describes the interaction between colloidal particles such as microorganisms and substrata. (From Marsh, P. and Martin, M.V., *Oral Microbiology*, 4th ed., Elsevier, 1999.)

appendages on the microorganism can traverse the electrostatic barrier. In addition, motile microorganisms can use their flagella to overcome repulsive forces because they can swim to surfaces.

Although electromagnetic interactions bring microorganisms close to surfaces, the forces that retain them on the surface and allow them to induce uptake by host cells are mediated by specific 'lock-and-key' receptor–ligand interactions between microbial adhesins and host receptors. In the following sections, we will consider the various types of adhesin–receptor interactions used by the major classes of pathogens. It is fair to say that microbes can bind to almost any macromolecule displayed on cell surfaces, the ECM or the conditioning film. Surprisingly, the identity of the host cell molecules (receptors) to which pathogen adhesins attach is known or characterised for only a relatively small number of microbes.

ADHESIN–RECEPTOR INTERACTIONS

Protein–carbohydrate (lectin) interactions

Adhesins binding to sugars (receptors) that are part of glycoproteins, glycolipids, proteoglycans and glycoaminoglycans displayed on the membrane of host cells or in conditioning films are the most common way by which microbes bind to surfaces. All classes of pathogens employ this type of adhesion. The lectin binds to a mono- or disaccharide by way of a motif that is complementary in shape to the surface of the mono- or disaccharide to which it binds (**Figure 4.10**). Binding is mediated by multiple non-covalent interactions.

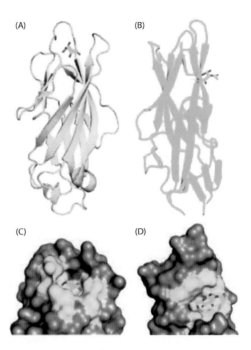

Figure 4.10 Crystal structures of F17-G and FimH fimbrial adhesins. Side-by-side comparisons of (A) FimH and (B) F17-G (PDB 1O9Z) lectin domains with bound D-mannose and GlcNAc ligands, respectively. Both domains share the immunoglobulin-like fold of the pilins, despite lack of sequence identity. Close-up views of (C) FimH and (D) F17-G carbohydrate-binding sites (surface representation). The mannoside-binding pocket of FimH is deep, with a large surface for interactions with the D-mannose monosaccharide, whereas the GlcNAc ligand does not fill the whole F17-G-binding site, providing structural insights into differences of ligand affinity and specificity. (From Juge, N., *Trends Microbiol.*, 20, 30–39, 2012.)

Bacteria

Pili/fimbriae are bacterial adhesins that bind sugars of glycoconjugates on cell membranes of the ECM. Pili/fimbriae are a diverse group of hair-like, tubular organelles that are anchored in the cytoplasmic membrane and extend from the surface of both gram-negative and gram-positive bacteria. They are dedicated adhesins. Pili/fimbriae are composed of thousands of 15- to 25-kDa protein subunits called pilins/fimbrillins. The lectin motif which is specific for particular sugars on host macromolecules is located at the tip of the pilus/fimbria. Based on their assembly pathways, pili of gram-negative bacteria are divided into four groups. Group 1 pili are those assembled *via* the chaperone–usher pathway; group 2 pili known as type IV pili; group 3 pili are called curli and are assembled *via* the extracellular nucleation/precipitation pathway; and group 4 pili are called CS1 pili and are assembled *via* the alternative chaperone–usher pathway. A single bacterial cell can express more than one type of pilus. In group 1 pili, the best characterised are the type 1 pili, which are found in the Enterobacteriaceae and many other gram-negative bacteria such as *Haemophilus influenzae, Yersinia pestis* and *Proteus mirabilis.* Type 1 pili are composed of a pilus rod comprising polymerised FimA subunits surmounted by a tip fibrillum consisting of Fim F, G and the tip adhesin FimH. Type 1 pili are prominent on the surface of uropathogenic *Escherichia coli* (UPEC). FimH has specificity for mannose groups found on uroplakins which are integral membrane glycoproteins on the lumenal surface of bladder epithelium. P pili are a family of pili that are also members of group 1. P pili are found on the surface of UPEC that infect the kidney. These pili bind to sugar moieties that are abundant on uroepithelial and kidney cells *via* the PapG adhesin. There are three variants of P pili that all bind to a common Galactose (α-1-4) Gal moiety linked to a ceramide group by an α-glucose residue. Afa/Dr adhesins are pillus-like structures found in UPEC and diffusely adhering *E. coli* (DAEC) that recognise the decay accelerating factor (DAF), a complement regulatory protein which is found on erythrocytes and urinary epithelium and carcinoembryonic antigen-related cell adhesion molecules (CEACAMs). Other adhesins in group 1 include the Hif pili of *Haemophilus influenzae*, the *Yersinia pestis* F1 antigen and the colonisation factor antigen I (CFA/I) of enterotoxigenic *E. coli* (ETEC). Chaperone-usher pili (Type 1) pili are shown in **Figure 4.11** and listed in **Table 4.2**.

Group 2 pili are otherwise known as type IV pili (**Table 4.3**) and are the most common pili observed in bacteria. They are found in a wide range of gram-negative bacteria but also in several genera of gram-positive bacteria such as *Clostridium*. Type IV pili are typically thin homopolymers

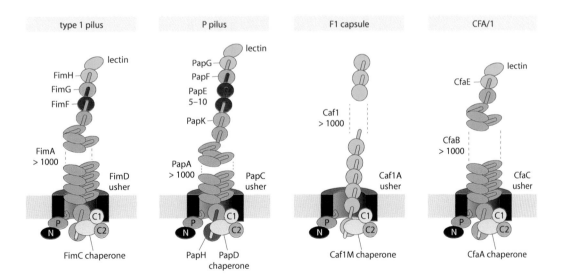

Figure 4.11 Schematic of four different chaperone–usher pili systems. The type 1 pili and the P pili from uropathogenic *Escherichia coli*, the F1 capsule from *Yersina pestis* and the CFA/I pili from the enterotoxigenic *E. coli*. Each pilus is built up with a different combination of subunits (Fim, Pap, Caf or Cfa), represented by different colours. The type 1 and P pili are made of two distinct regions: a flexible end tip and a rigid long rod. The pili expose their binding sites (lectin) at the cell surface and are anchored at the bacterial outer membrane by the usher. The usher is composed of a transmembrane pore domain and four soluble domains: the *N*-terminal domain (N), the plug domain (P), and the *C*-terminal domains (C1 and C2). (From Busch, A. et al., *Philos. Trans. Royal Soc. A*, 2015.)

pili	organism	major pilin	chaperone/ usher		adhesin	host cells	receptors	disease
Type 1	E. coli	FimA	FimC	FimD	FimH	bladder and kidney epithelial cells, buccal cells, erythrocytes, mast cells, neutrophils, macrophages	uroplakin UP1a, β1α3 integrins, laminin, CD48, collagen (type I and IV)	cystitis, sepsis, meningitis
P	UPEC	PapA	PapD	PapC	PapG	kidney epithelial cells, erythrocytes	GbO3, GbO4, GbO5	polynephritis
S	E. coli	SfaA	SfaE	SfaF	SfaS	bladder and kidney epithelial cells, erythrocytes, endothelial cells	sialic acid residues plasminogen	UTT, meningitis, sepsis
Hif	H. influenzae	HifA	HifB	HifC	HifE	nasopharyngeal cells	unkonwn	oitis media
PMF	P. mirabilis	FmfA	FmfC	PmfD	PmfF	bladder and kidney epithelial cells	unknown	UTI
Dr	UPEC	DraA	DraB	DraC	DraE	bladder and kidney epithelial cells	CD55/DAF, CEACAMs	polynephritis
	DAEC					neutrophils, erythrocytes	collagen (type IV)	cystitis
Afa	UPEC	AfaA	AfaB	AfaC	AfaE	uroepithelium, erythrocytes	CD55/DAFm CEACAMs α5β1 integrin	cystitis, diarrhoea
F1	Y. pestis	Caf	Caf1M	Caf1A		respiratory tract epithelial cells	human IL-1β	plague

Table 4.2 Chaperone usher pathway assembled pili. (From Proft, T. and Baker, E.N., *Cell. Mol. Life Sci.*, 66, 613–635, 2009.)

pili	organisms	major pilin	putative adhesin	host cells	putative receptors	disease
a) Type IVa pili						
GCP	N. gonorrhoaeae	PilE	PilC	epithelial and endothelial cells	MCP (CD46), C4BP	gonorrhoea
MCP	N. meningitis	PilE	PilC	epithelial and endothelial cells	MCP (CD46), C4BP	sepsis, meningitis
Pa pilus	P. aeruginosa	PilA		epithelial cells	asialo-GMl and GM2	pneumonia, sepsis
FT pilus	F. tularensis	PilE		unknown	unknown	tularemia
b) Type IVb pili						
BFP	EPEC	BfpA		epithelial cells	conflicting results	diarrhoea
TCP	V. cholerae	TcpA		CTXφ phage		cholera
CFA/III	ETEC	CofA		enterocytes	unknown	diarrhoea
Longus	ETEC	LngA		unknown	unknown	diarrhoea

Table 4.3 Type IV pili. (From Proft, T. and Baker, E.N., *Cell. Mol Life Sci.*, 66, 613–635, 2009.)

consisting of 15- to -20-kDa pilin subunits called pilE, and many have an adhesion subunit at the tip of the pilus termed pilC. They have a tendency to aggregate and form bundles (bundle-forming pili). An interesting feature of type IV pili is their ability to retract, generating significant mechanical force that allows the bacterium to exhibit a form of movement called twitching motility. In addition, it is suggested that the extension and retraction of type IV pili allows the bacterium to drag itself along a solid surface or that the type IV pili can act as a catapult to propel the bacterium over a cellular surface. These functions are thought to play a role in biofilm formation. Type IV pili have been shown to mediate adhesion to many different substrates.

Group 3 pili, also known as curli, are coiled pili consisting of repeating subunits of the protein curlin (CsgA) and have been found in *Escherichia coli* and *Salmonella* species, among others. Curlin is unlike any other pilus protein. Curli share some biochemical and structural properties with amyloid. Its receptor appears to be fibronectin and perhaps plasminogen.

Group 4 pili, or CS1 pili, are found on the surface of ETEC. CS1 pili are composed almost entirely of the major pilin, CooA. A minor pilin, CooD, is found only at the pilus tip. The ligand for CS1 pili is currently unknown. Pili (fimbriae) have been identified in an increasing number of genera of gram-positive bacteria. So far, two types of pili have been found in these bacteria. The two types are short thin rods 1–2 nm in diameter that extend 70–500 nm from the cell surface, sometimes termed fibrils, and long, thicker, flexible rods 3–10 nm in diameter that extend up to 4 μm from the bacterial surface. Gram-positive flexible pili (**Table 4.4**) comprise pilin subunits that form the pilus shaft with a pilus tip that determines receptor specificity. The receptor specificity of both types of pili in

pili	organism	major pilin	minor pilins	host cells	receptors	disease
Spa	C. diphtheria	SpaA	SpaB, SpaC	pharyngeal epithelial cells	unknown	diphtheria
PI-1 pilus	S. agalactiae	GBS80	GBS52, GBS104	pulmonary cells	unknown	neonatal sepsis and
PI-2 pilus	S. agalactiae	GBS1477	GBS1474, GBS1478	lung and cervical epithelial cells	unknown	meningitis
GAS M1 pilus	S. pyogenes	Spy0128	Cpa, Spy0130	pharyngeal epithelial cells, tonsil epithelium cells, skin keratinocytes	unknown	pharyngitis, impetigo, sepsis, toxic shock
Rrg pilus	S. pneumoniae	RrgB	RrgA, RrgC	lung epithelial cells	unknown	sinusitis, pneumonia
Type I	A. naeslundii	FimP	FimQ	tooth enamel	proline-rich salivary proteins	dental caries, peridontitis
Type II	A. naeslundii	FimA	FimB	various host cells	glycoproteins/ glycolipids	dental caries, peridontitis

Table 4.4 Pili of gram-positive bacteria. (From Proft, T. and Baker, E.N., *Cell. Mol. Life Sci.*, 66, 613–635, 2009.)

gram-positive bacteria are largely unknown. The receptor specify of some gram-negative and gram-positive pili/fimbriae is shown in **Table 4.5**.

Fungi

The most important fungal pathogen in humans is *Candida albicans*, and so most information is known about adhesion of this organism. Less is known about the adhesion of other species in this genus and other genera of human fungal pathogens. Three gene families, *ALS, HWP* and *IFF/HYR* contain adhesins that bind to a variety of receptors. Als1 binds fucose-containing glycans. Other Als family members, HWP and IFF/HYR bind to several unrelated proteins and will be considered later. The adhesin families of *C. albicans* exist in other *Candida* species. However, the largest adhesin family in *C. glabrata* is the Epa family. The receptor (ligand) binding domains of Epa adhesins are lectins that bind to oligosaccharides that contain terminal galactose residues such as the mucin-type *O*-glycans. The Pwp family of *C. glabrata* is also believed to be composed of lectins, but their ligands have not been determined. A 32-kDa lectin specific for the terminal sialic acid residues of glycoconjugates on human epithelial cells has been detected on the surface of *Aspergillus fumigatus* conidia.

Viruses

Many viruses bind to carbohydrate ligands (receptors) (**Table 4.6**) on the host cell surface, likely because the ligands are prevalent and the sugar chains extend some distance from the cell membrane, perhaps allowing the virus to dock on the host cell membrane receptor without needing

organism	target tissue	carbohydrate	form*
E. coli Type 1	urinary	Man(α1-3)[Man (α1-6)] Man	GP
P	urinary	Gal(α1-4)Gal	GSL
S	neural	NeuAc(α2-3)Gal(β1-3)GalNAc	GSL
CFA/1	intestinal	NeuAc(α2-8)-	GP
K1	endothelial	GlcNAc(β1-4)GlcNAc	GP
K99	intestinal	NeuGc(α2-3)Gal(β1-4)Glc**	GSL
H. influenza	respiratory	[NeuAc(α2-3)]$_{0,1}$ Gal(β1-4)GlcNAc-(β1-3) Gal(β1-4)GlcNAc**	GSL
H. pylori	stomach	NeuAc(α2-3)Gal(β1-4)Glc(NAc)**	GP
		Fuc(α1-2)Gal(β1-3)[Fuc(α1-4)] Gal**	GP
K. pneumoniae	respiratory	Man	GP
M. pneumoniae	respiratory	NeuAc(α2-3)Gal(β1-4)Glc(Nac)**	GP
N. gonorrhoea	genital	Gal(β1-4)Glc(NAc)**	GSL
N. meningitidis	respiratory	[NeuAc(α2-3)]$_{0,1}$ Gal(β1-4)GlcNAc-(β1-3) Gal(β1-4)GlcNAc**	GSL
P. aeruginosa	respiratory	Ga(β1-3)Glc(NAc)(β1-3)Gal(β1-4)-Glc**	GSL
S. typhimurium	intestinal	Man	GP
S. pneumoniae	respiratory	[NeuAc(α2-3)]$_{0,1}$ Gal(β1-4)GlcNAc-(β1-3) Galβ(1-4)GlcNAc**	GSL
S. suis	respiratory	Gal(α1-4)Gal(β1-4)Glc	GSL

Table 4.5 Carbohydrates serving as receptors for pili/fimbriae. *Predominant form in tissue: GP, glycoproteins; GSL, glycolipids. **Structures present in human milk oligosaccharides. (From Sharon, N. and Ofek, I., *Glycoconj. J.*, 17, 659–664, 2000.)

to overcome repulsive ionic interactions between virus and cell. Mucins are examples of such ligands (receptors). Sialic acid (*N*-acetylneuraminic acid) is the most frequent terminal monosaccharide on glycoproteins and glycolipids on the surface of eukaryotic cells. Therefore, it is not surprising that a large number of viruses utilise this sugar for binding. Viruses that use sialic acid for adhesion tend to have a broad host range because of the ubiquity of this sugar. The influenza virus is an example of a virus that binds sialated oligosaccharides that are displayed on the respiratory epithelium using the molecule haemaglutinin (HA) as an adhesin.

JC virus, Sendai virus and certain strains of rotavirus bind to sialyloligosaccharides that are part of proteoglycans or glycosphingolipids. Sugars other than sialic acid may serve as the ligands for attachment of other viruses. Glycosphingolipids are a common component of the apical surface of the cytoplasmic membrane of enterocytes, urinary epithelium, neuroepithelium and myelin and serve as receptors for viral adhesins. The human immunodeficiency virus 1 (HIV-1) envelope proteins gp120 and gp41 both bind galactoceramide. Glycosphingolipids are also receptors for binding Ebola, Marburg and measles viruses to epithelial cells. The glycosaminoglycan moiety of the proteoglycan heparan sulphate (HS) serves as a receptor for the attachment of both enveloped and non-enveloped viruses such as echovirus, herpes simplex virus (HSV)-1, human herpes virus (HHV)-8, respiratory syncytial virus (RSV), human papillomavirus (HPV)-16, HPV-33 and foot and mouth disease virus (FMDV).

human

virus	family	characteristics	epithelial tropism	attachment carbohydrate*	protein*
herpes simplex virus (HSV-1, -2)	herpesviridae-α	enveloped dsDNA	retinal pigment epithelial cell, cornea	HSPG	nectin 1, HVEM
varicella-zoster Virus	herpesviridae-α	enveloped dsDNA	gastrointestinal tract, retinal pigment epithelial cell	HSPG	Man6-P/IGFII-R, nectin 1
human cytomegalovirus (HCMV)	herpesviridae-β	enveloped dsDNA	retinal pigment epithelial cell	HSPG	–
epstein-barr virus (EBV)	herpesviridae-γ	enveloped dsDNA	–	–	CR2 (CD21), poly Ig-receptor
vaccinia virus	poxviridae	enveloped dsDNA	rhinopharynx, skin	–	–
human immunodeficiency virus 1 (HIV-1)	retroviridae	enveloped ssRNA	gastrointestinal and genital tracts	galactosylceramide	CCR5
respiratory syncytial virus (RSV)	paramixoviridae	enveloped ssDNA	pulmonary and respiratory tracts	HSPG	ICAM1, VUDLR
sendai virus	paramixoviridae	enveloped ssRNA	bronchial tract, upper airway	sialyloligosaccharide	–
measles virus	paramixoviridae	enveloped ssRNA	respiratory tracts	–	CD46 CD46/moesin
black creek canalvirus	bunyaviridae	enveloped ssRNA	pulmonary tract	–	Integrin-β_3
influenza virus	orthomyxoviridae	enveloped ssRNA	bronchial epithelium	sialyloligosaccharide	–
vesicular stomatitis virus (VSV)	rhabdoviridae	enveloped ssRNA	bronchial epithelium	GlcNAc	–
rotavirus	reoviridae	naked dsRNA	intestinal tract	–	Integrins-$\alpha_2\beta_3$, -$\alpha_4\beta_1$, -$\alpha_2\beta_3$
reovirus-1 and -3	reoviridae	naked dsRNA	intestinal tract including M cells	sialyloligosaccharide	tight-junction-associated protein
human papilloma virus (HPV)	papillomaviridae	naked dsDNA	mucosa, oesophagus, skin	HSPG	–
adeno-associated virus (AAV)	parvoviridae	naked ssDNA	airway	HSPG	–

(Continued)

virus	family	characteristics	epithelial tropism	attachment carbohydrate*	protein*
Jamestown Canyon (JC) virus	polyomaviridae	naked dsDNA	colorectal tract, neuroepithelial cells	sialyloligosaccharide	–
adenovirus	adenoviridae	naked dsDNA	airway, ocular and gastrointestinal tracts	HSPG	Integrin-$\alpha_2\beta_5$, CAR
coxsackievirus	picornaviridae	naked ssRNA	airway	–	CAR
poliovirus	picornaviridae	naked ssRNA	gastrointestinal tract	–	Ab D171, PVR, PRR 1,2
rhinovirus major group	picornaviridae	naked ssRNA	respiratory tract	–	ICAM1
rhinovirus minor group	picornaviridae	naked ssRNA	respiratory tract	–	LDLR family
echovirus and human parechovirus (HPeV)	picornaviridae	naked ssRNA	intestinal tract	DAF	Integrin-$\alpha_2\beta_1$

Table 4.6 Virus receptors on epithelial cells. *Together the attachment carbohydrate and the protein form the epithelial receptor. CAR, coxsackie adenovirus receptor; CEA, carcino-embryonic antigen; DAF, decay acceleration factor; ds, double stranded; HLA-1; human leukocyte antigen 1; HSPG, heparan sulphate proteoglycan; ICAM1, intracellular adhesion molecule 1; Ig, immunoglobulin; LDLR, low density lipoprotein receptor; Man6-P/IGFI1-R, mannose 6-phosphate/insulin-like growth factor receptor; PRR, polio virus related receptor; PVR, polio virus protein receptor; ss, single stranded; VLDLR, very low density lipoprotein receptor. (From Bomsel, M. and Alfsen, A., *Nat. Rev. Mol. Biol.*, 4, 57–68, 2003.)

Parasites

The principle adhesins of *Plasmodium* species are lectins that recognise host cell glycans. In addition to adhering to hepatocytes and erythrocytes, once inside red blood cells, the parasite places adhesin molecules on the surface of erythrocytes that cause them to aggregate, a process termed cytoadhesion. *Plasmodium* has a secretory organelle termed the microneme, which produces several adhesins among which is a family of erythrocyte-binding-like (EBL) and reticulocyte binding-like (RBL) proteins that have multiple domains. The best described EBL adhesin, EBA-175, contains two Duffy blood group antigen binding-like domains that bind to sialic acid on glycophorin A, an erythrocyte membrane glycoprotein. Inside the red blood cell, *P. falciparum* expresses an adhesin called *P. falciparum* erythrocyte membrane protein 1 (PfEMP1) of which there are some 60 types. These molecules are involved in cytoadhesion. The best studies of these, VAR2CSA, interacts with heparin-like carbohydrates. Interestingly, the PfEMP1 adhesins can also bind proteins. *Toxoplasma gondii* adhesion is mediated by the micronemal protein M1C1 that binds various oligosaccharides with terminal sialic acid. In addition, *T. gondii* coccidia express a large family of surface-antigen glycoprotein 1 (SAG)–related sequences (SRSs). SAG1, -2 and -3 recognise host glycans. In the case of *Giardia lamblia*, the cyst form of the parasite breaks open in the small intestine, releasing transient excyzoites that immediately divide into four trophozoites, which are the active, motile feeding stage of the parasite. Adhesion of the trophozoites to the microvilli of the enterocytes is accomplished by a ventral sucking disc but also by specific receptor-ligand interactions involving *Giardia* α-giardins. By comparing wild-type with adhesion-deficient *Giardia*, a 200-kDa protein has been shown to be important in adhesion. The ligand(s) for this protein is/are currently unknown. A surface-associated trypsin-activated *Giardia* lectin (taglin) may be important for *G. lamblia* attachment. *Cryptosporium* sporozoites secrete a lectin, termed p30, from their apical organelles that is thought to mediate binding to mucin. Using gliding motility, a type of smooth movement that occurs without the aid of propulsive organelles, the parasite penetrates the mucus layer and attaches to the apical surface of the epithelium. Lectins have also been implicated in epithelial adhesion. After the rupture of the cysts, infectious sporozoites of both *Toxoplasma gondii* and *Cryptosporidium* species target the mucosa of the ilium, jejunum and duodenum. *T. gondii* sporozoites orient themselves perpendicular to the apical epithelial surface. Initially, they concentrate around intercellular junctions and probably use this route to cross the epithelial barrier. Adhesion is mediated by parasite surface antigen-1 (SAG1) binding to host laminin *via* the β1 integrin receptor on the epithelial surface. *Entamoeba histolytica*, the causative agent of amoebiasis in humans, adheres to the intestinal epithelium using lectins that include the 220-kDa lectin, which is specific for GlcNAc, and the Gal/GalNAc lectin, which binds galactose. Among the best-characterised adhesins of *Trichomonas vaginalis (TV)* are the lipoglycans. Lipoglycans are the principal molecules on the parasite

parasite	stage	protein	specificity
Plasmodium falciparum	merozoite	EBA-175	Neu5Acα2-3Gal/glycophorin A
	merozoite	EBA-140	sialic acid/glycophorin B?
	merozoite	EBA-180	sialic acid
	sporozoite	circumsporozoite protein	HS
	parasitized erythrocytes	the anchor protein VAR2CSA	chondroitin sulfate A
Trypanosoma cruzi	trypomastigote	*trans*-sialidase	Neu5Acα2-3Gat
	trypomastigote	penetrin	HS
Entamoeba histolytica	trophozoite	Gal/GalNAc lectin	Gal/GalNAc
Entamoeba invadens (areptilian pathogen)	cyst	cyst wall protein (Jacob lectin)	chitin
Giardia lamblia	trophozoite	taglin (α-1 giardin)	Man-6-PO$_4$-, HS
Cryptosporidium parvum	sporozoite	Gal/GalNAc lectin	Gal/GalNAc
	sporozoite	Cpal35 protein	?
Acanthamoeha keratitis	trophozoite	136 kDa mannose-binding protein	mannose
Toxocara canis	larval	TES-32	?
Haemonchus contortus	gut-localized	galectin (Hco-gal)	β-galactosides

Abbreviation: HS, heparan sulfate.

Table 4.7 Some major parasites and their glycan-binding proteins. (From Cummings, R. and Turco, S., Parasitic infections, in *Essentials of Glycobiology*, 2nd ed., Ch. 40, Cold Spring Harbor Laboratory Press, Cold Spring Harbor, NY, 2009.)

surface. The ligand for the lipoglycans on epithelial cells appears to be galectin-1. Galectins are a family of carbohydrate-binding proteins with an affinity for β-galactosides. Microsporidia are fungus-like unicellular parasites that are emerging as human pathogens. They are obligatory intracellular and invade host cells by extruding an organelle termed the polar tube that introduces infectious spores into the cell. Microsporidia parasitise a wide range of hosts from insects to human beings. They survive outside their hosts as environmentally resistant spores. The spores are believed to adhere to host cells using a spore wall protein termed EnP1, which recognises glycosaminoglycans on host cell surfaces. In addition, the *O*-mannosylated moiety of one of the polar tube proteins (PTP1) can interact with host cell mannose receptors. Some major parasites and their glycan-binding proteins are listed in **Table 4.7**.

PROTEIN–PROTEIN INTERACTIONS

Microbial surface components recognising adhesive matrix molecules

Many pathogenic microorganisms bind peptide or sugar motifs on ECM molecules such as fibronectin, fibrinogen, collagen and glycosaminoglycans such as heparin sulphate and the chondroitin sulphates. Major types of ECM molecules are listed in **Table 4.8**.

ECM components	characteristics	distribution	function
major proteins			
collagens	fibrous proteins, 29 different types, >90% of total ECM	interstitial spaces, basement membranes, in all organs and tissues	structural scaffolds, tissue integrity, cellular and tissue repair and growth
elastin	fibrous protein builds elastic fibers	connective tissues, skin, blood vessels, cartilage, and lungs	providing elasticity to the tissues
fibrillin	fibrous proteins, four different types, associated with elastin	connective tissues, skin, blood vessels, cartilage, and lungs	associated with elastin and involved in cell-matrix interaction
laminin	heterotrimeric, 15 different types	basement membranes, connective tissues, cell surface, skin, blood vessels	structural scaffold, cell migration, signaling
fibronectin	440 kDa binds to integrins, collagen, fibrin and heparan sulfate	insoluble incorporated to connective tissues and soluble in plasma	cell adhesion, migration, growth, differentiation
vitronectin	75-65 kDa binds to integrins, uPA, uPAR and PAI-1	plasma, connective tissues, cell surface	fibrinolytic system regulator, Cell migration, growth, differentiation, regulation of complement system
proteoglycans			
heparan sulfate (HS), Chodroitin sulfate (CS), Keratan sulfate (KS)	polysaccharide chain attaches to ECM protein moities	HS; all tissues. CF and KS; cartilage, bone joints	HS; angiogenesis, blood coagulation and tumour cell migration. CF and KS; Tissue integrity, provide resistance to compression, and flexibility of joints
non-proteoglycans			
hyaluronic acid	consisting of alternative residues of D-glucuronic acid and N-acetylglucosamine	connective, epithelial, and neural tissues	embryonic development, healing processes, inflammation and tumor development

Table 4.8 Major ECM components and their various properties. (From Singh, B. et al., *FEMS Microbiol. Rev.*, 36, 1122–1180, 2012.)

Microbial surface components recognising adhesive matrix molecules (MSCRAMMs) are a family of adhesins that have a similar modular design and an IgG-like folded domain organisation. As with other types of adhesin, an individual MSCRAMM can bind several different ECM molecules, and microorganisms can have several distinct MSCRAMMs that bind the same ECM molecule. The majority of the ligands of MSCRAMMs are amino acid sequence motifs, but a few are sugar motifs. Although MSCRAMMs are adhesins, it is believed that they may have other functions as well.

Fibronectin-binding MSCRAMMs in bacterial adhesion: Over 100 bacterial proteins have been identified that bind soluble and/or insoluble fibronectin. Much research has focussed on the MSCRAMMs of *Staphylococcus aureus* and other staphylococcal species, *Streptococcus pyogenes* and other streptococci and entero-cocci. An *S. aureus* fibronectin-binding protein was the first MSCRAMM identified, and fibronectin is a common ligand (recep-tor) for all classes of pathogen. The fibronectin-binding domain of *S. aureus* fibronectin-binding protein is a 38-amino acid–repeated domain located at the *N*-terminus and this characteristic appears to be similar for other fibronectin-binding MSCRAMMs from other gram-positive bacteria. Other fibronectin binding sites may also be located in the *C*-terminal region of the molecule. The ligand-binding mechanism of several *S. aureus* MSCRAMMs has been termed 'dock, lock and latch' (DLL). DLL involves the ligand (receptor) docking in a binding groove of the adhesin. The adhesin then undergoes con-formational change to 'lock' the receptor in place. This mechanism is a common mode of ligand binding for structurally related cell wall–anchored MSCRAMMs of gram-positive bacteria (**Figure 4.12**). How-ever, it is unlikely that this mechanism is used by all MSCRAMMs.

In gram-positive bacteria, a common way of anchoring MSCRAMMs in the bacterial envelope is by the use of a Leucine-Proline-Any amino acid-Threonine-Glycine (LPXTG) motif at the *C*-terminus of the adhesin. A list of gram-positive bacteria with the ability to bind fibronec-tin is shown in **Table 4.9**.

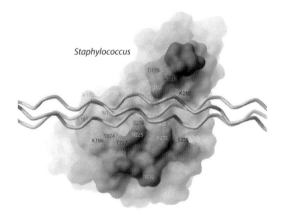

DOCK an open conformation allows access of the ligand to
a binding cleft

LOCK following binding, a structural rearrangement is induced
at the C-terminus of the protein such that access to and from
the binding cleft is blocked

LATCH to stabilize the structure, the rearranged C-terminal beta-sheet
inserts between two beta sheets in an adjacent domain,
'latching' the protein-ligand complex together

Figure 4.12 *Dock, lock and latch steps* in binding an ECM molecule to a MSCRAMM. The example shown is for a *S. aureus* collagen-binding MSCRAMM. View of the molecular surface of the 19-kDa subdomain of the collagen-binding MSCRAMM (microbial surface components recognizing adhesive matrix molecules) looking into the trench on the β-sheet. The figure shows the surface area of residues within 6 Å from docked collagen. The orange lines represent the collagen triple helix. (From Foster, T.J. and Höök, M., *Trends Microbiol.*, 6, 484–488, 1998.)

bacterium	fibronectin-binding protein
A. pyogenes	CbpA
Bac. fragilis	Omp
C. difficile	Fbp68(FbpA)
C. perfringens	FbpA
	FbpB
E. faecium	SagA
E. rhusiopathiae	RpsA
	RpsB
L. brevis	SlpA
L. plantarum	enolase
L. monocytogenes	FbpA
	five undefined proteins
Mycobacteria spp.	Antigen85A
	Antigen85B
	Antigen85C
	MPT51/Antigen85D
	Apa/FAP
	Wag22 antigen
	malate synthase
Mycoplasma spp.	EfTu
	PDH-β
	Hlp3
	PlpA
P. acnes	undefined
S. aureus	FnBPA
	FnBPB
	Eap
	Emp
	Ebh
	Aaa/Sle 1
S. caprae	AtlC
S. epidermidis	Aaa/Aae
	Embp/Ebh
S. saprophyticus	Aas (AtlC)
	UaFA
Strep. agalactiae	ScpB
Strep. dysgalactiae	FnBA
	FnBB
Strep. equi ssp. equi	FnE
	FnEB
	SFS
Strep. equi ssp. zooepidemicus	FNZ
	FNZ2
Strep. gordonii	CshA
	CshB
	FbpA
Strep. intermedius	antigen I/II
Strep. mutans	antigen I/II
	FbpA/SmFnB
	FBP-130

(Continued)

bacterium	fibronectin-binding protein
Strep. pneumonia	PavA (FbpA)
	SP-0082
	PbfA
Strep. pyogenes	PrtF1/SfB1
	PrtF2
	FbaA
	FbaB/PrtF2-like
	serum opacity factor
	SfbX
	Fbp54 (FbpA)
	GAPDH
	protein H
	M1 protein

Table 4.9 Gram-positive bacteria that bind fibronectin. (From Henderson, B. et al., *FEMS Microbiol. Rev.*, 35, 147–200, 2011.)

The PapE and PapF proteins that make up the P fimbriae of gram-negative uropathogenic *Escherichia coli* bind fibronectin as do the type 1 pili and curli of *E. coli* and the type 1 pili of *Salmonella enteritidis*.

Multivalent adhesin molecule 7 (MAM7) is found in many gram-negative bacteria, including *Vibrio*, *Yersinia*, *Salmonella* and enteropathogenic *E. coli*. This adhesin binds fibronectin and the host cell membrane phospholipid, phosphatidic acid. It appears that MAM7 forms a tripartite complex with fibronectin and phosphatidic acid on the host cell surface. In some gram-negative bacteria, porins are fibronectin-binding proteins. Such is the case for *Neiserria gonorrhoeae*, *Campylobacter jejuni*, *Borrelia burgdorferi* and *Vibrio vulnificus*. Some of the fibronectin-binding proteins in gram-negative bacteria are autotransporters. Autotransporters are proteins that promote their own transport across the bacterial cytoplasmic membrane to the cell surface. This class of molecules forms a trimer with a head-stalk-anchor structure. In addition to adhesion, autotransporter proteins have been implicated in aggregation, invasion, biofilm formation and toxicity. Gram-negative bacteria that bind fibronectin are listed in **Table 4.10**.

Fibronectin-binding MSCRAMMs in fungal adhesion: In *Candida albicans*, many of the adhesins that have lectin activity – for example, members of the Als family (Als1, -3 and -5), and Hwp1 of the Hwp1 family – can also bind ECM molecules (**Table 4.11**). These adhesins

bacterium	fibronectin-binding protein
B. henselae	BadA (YadA-like)
	Pap31 (porin)
B. burgdorferi	BBK32
	RevA/B
C. jejuni	CadF (porin)
	Cj1279c (FlpA)
E. coli	Curli
	P-fimbriae
	Type I fimbriae
	aggregative adherence fimbriae
	haemorrhagic coli pili (HCP)
	H6/H7 pili
	Tsh (autotransporter)
	Upag (YadA-iike)
H. ducreyi	DrsA (YadA-like)
H. influenza	Pili
	Hap (autotr.)
Hel. Pylori	VacA toxin (autotr.)
L. interrogans	unidentified
	LigA
	LigB
	LigB
	LenB/C/D/E
	LipL32
N. gonorrhoeae	OpaA (porin)
Pasteurellaceae spp.	ComE1
Past. multocida	Omp87, Omp16, tran
Por. gingivalis	Fimbrillin/FimA
Prev. intermedia	AdpB (CadF-like)
Pseud. aeruginosa	OprD, OprE 1,
	OprE3, OprF
Sal. typhimurium	ShdA (autotr.)
	MisL
Tann. forsythia	BspA
Trep. denticola	OppA
	Msp
Trep. pallidum	several unidentified
V. vulnificus	OmpU (porin)
V pseudouberculosis	YadA (autotr.)

Table 4.10 Gram-negative bacteria that bind fibronectin. (From Henderson, B. et al., *FEMS Microbiol. Rev.*, 35, 147–200, 2011.)

abbreviation	protein name
fibronectin-binding proteins	
Als1	agglutinin-like sequence protein 1
Als3	agglutinin-like sequence protein 3
Als5	agglutinin-like sequence protein 5
Hwp1	hyphal cell wall protein 1
Tdh3	glyceraldedyhde-3-phosphate dehydrogenase
Adh1	alcohol dehydrogenase
vitronectin-binding proteins	
Adh1	alcohol dehydrogenase
Gpm1	phosphoglycerate mutase
laminin-binding proteins	
Als1	agglutinin-like sequence protein 1
Als3	agglutinin-like sequence protein 3
Als5	agglutinin-like sequence protein 5
Tdh3	glyceraldedyhde-3-phosphate dehydrogenase

Table 4.11 *Candida albicans* proteins that binding human fibronectin, vitronectin and laminin. (From Kozik, A. et al., *BMC Microbiol.*, 15, 197, 2015.)

probably exist in the other opportunistic human pathogenic *Candida* species: *C. dubliniensis, C. tropicalis* and *C. parapsilosis.*

Fibronectin-binding MSCRAMMs in virus adhesion: The viral envelope of influenza A, parainfluenza 1, and mumps viruses bind fibronectin *via* sialic acid moieties of the ECM molecule.

Fibronectin-binding MSCRAMMs in parasite adhesion: Adherence of the infective stage of *Trypanosoma cruzi* to ECM molecules such as fibronectin and laminin is an essential step in host cell invasion and is thought to be mediated by members of the *T. cruzi* gp85 glycoprotein superfamily. *Entamoeba histolytica* adheres to colonic epithelium using a variety of adhesins, one of which binds fibronectin. This fibronectin binding may be mediated by the intermediate subunit-2 of the Gal/GalNAc-specific lectin, which is a major determinant of pathogenesis in this organism. *Leishmania* species promastigotes and amastigotes bind to fibronectin and laminin by means of a 60-kDa surface protein.

Collagen- and laminin-binding MSCRAMMs in bacterial adhesion: There are five main types of collagen and both gram-negative and gram-positive bacteria can bind either one or several of the collagen types. *Staphylococcus aureus* binds collagen, as do various species of streptococci such as *S. pyogenes* and some viridans streptococci. The binding mechanism of the collagen-binding adhesin Cna of *S. aureus* is a modification of the DLL called the collagen hug. Collagen-binding gram-negative bacteria include many members of the Enterobacteriaceae such as *Klebsiella, Enterobacter, Proteus, Providencia,* and *Serratia* species. Collagen binding in the Enterobacteriaceae is mediated by type 3 fimbriae. **Table 4.12** lists the various classes of pathogen that bind the various ECM molecules.

pathogens	Ln/collagen-interacting protein/system	Ln	collagen	others
Burkholderia cepacia	BCAM0224 (trimeric autotransporter adhesion)		I	
Escherichia coli		Ln		
Haemophilus influenza	PE	Ln		Vn, Plg
H. influenzae nontypable (NTHi)	Haemophilus adhesion penetration (Hap)	Ln	IV	
			I	
			I, III	
Klebsiella pneumoniae	MrkD1P		V	
Legionella pneumophila	Mip		I, VI	
Moraxella catarrhalis	UspA1/UspA2	Ln		
	MID/Hag		IV	
Mycobacterium bovis	Ln-binding protein (LBP)	Ln		
M. leprae	Histone-like protein (Hlp)	Ln		
	Hlp/Lbp			
		Ln		
M. smegmatis	LBP	Ln		
M. tuberculosis	M. tuberculosis pili (Mtp)	Ln		
	Malate synthase	Ln		
Mycoplasma pneumoniae		Ln		
Neisseria meningitidis	PiLO, PorA	Ln		
			I, III, V	
Pseudomonas aeruginosa	Saccharides		I, II	
Streptococcus agalactiae	Ln-binding protein (Lmb)	Ln		
S. anginosus	Putative Ln-binding protein (Plbp)	Ln		
		Ln	I, IV	
S. pneumoniae	Pilus I (RgrA)	Ln		
	RrgA		I	
			VI	
			IV	
S. pyogenes	Ln-binding protein (Lbp)	Ln		
	SpeB	Ln		
	Lipoprotein of S. pyogenes (Lsp)	Ln		
	Streptococcal collagen-like protein (Scl-1)	Ln		
	ShrA	Ln		
	M1		VI	
	Cpa		I	
			VI	
Yersinia pestis	Plg activator (Pla)	Ln	I, IV, V	
		Ln		
Influenza virus type A	Envelope GP	Ln		Fn
Aspergillus fumigatus	AfCalAp	Ln		Murine pneumocytes
	Asp f 2	Ln		IgE
	72-kDa surface GP	Ln		
	37-kDa surface protein	Ln		
	extracellular serine protease	Ln		
	extracellular serine proteinase		I, III	elastin, Fn

(Continued)

pathogens	Ln/collagen-interacting protein/ system	Ln	collagen	others
	Aspergillopepsin F	Ln	collagen	elastin
Aspergillus spp.	33-kDa alkaline protease		collagen	elastin, fibrinogen
Coccidioides immitis	SOWgp-encoded cell surface GP	Ln	IV	Fn
Cryptococcus neoformans	75-kDa extracellular serine protease	Ln	IV	
	gelatinase		collagen	
Histoplasma capsulatum	50-kDa Ln-binding protein	Ln		
Paracoccidioides brasiliensis	GAPDH	Ln	I	Fn
	47- and 80-kDa surface components		I	
	30-kDa adhesin	Ln		
	Gp43	Ln		Fn
	Serine–thiol proteinase	Ln	IV	Fn, proteoglycans
	19-kDa and 32-kDa cell wall proteins	Ln		Fn, fibrinogen
Pneumocystis carinii	33-kDa adhesin	Ln		Fn, type 1 pneumocytes
Acanthamoeba castetianii	42-kDa extracellular serine proteinase		collagen	
	12-kDa extracellular serine proteinase		collagen	Fn, sIgA, IgG, Plg, fibrinogen, haemoglobin
	150-kDa metalloprotease and 130-kDa serine protease		I, III	Elastin, Plg, haemoglobin
Acanthamoeba culbertsoni	55-kDa Ln-binding protein	Ln-1		
Acanthamoeba healyi	33-kDa extracellular serine proteinase		I, IV	Fn, fibrinogen, IgG, IgA, albumin, haemoglobin
	A. healyi Ln-binding protein AhLBP	Ln		
Balamuthia mandrillaris	?	Ln-1	I	Fn
	40- to 50-kDa extracellular metalloproteases		I, III	elastin, Plg
Trypanosoma brucei	T. brucei prolyl oligopeptidase (POP Tb)		I, native collagen in rat mesentery	
Trypanosoma cruzi	T. cruzi prolyl oligopeptidase (POP Tc80)		collagen	Fn
	Gp 58/68 or HAG bp		collagen	Fn
	LAG bp		collagen	Fn
	Penetrin		collagen	heparin, heparin sulphate
	30-kDa cysteine protease		I	
	Ln-binding GP	Ln		

Table 4.12 Microorganisms with affinity for with collagen and various extracellular matrix (ECM) proteins. (From Singh, B. et al., *FEMS Microbiol. Rev.*, 36, 1122–1180, 2012.)

Fibrinogen-binding MSCRAMMs

Gram-positive bacteria, including staphylococci and streptococci, bind fibrinogen. Coagulase and clumping factor are examples of fibrinogen-binding molecules of *Staphylococcus aureus*. Lancefield groups A, C and G streptococci bind fibrinogen, as does the M protein of *Streptococcus pyogenes*. Certain anaerobic gram-negative rods and the fungi *Candida albicans* and *Aspergillus fumigatus* also bind fibrinogen.

Vitronectin-binding MSCRAMMs

Vitronectin (VN) is a high-affinity heparin-binding protein and binds to cell membrane–bound integrins. It is likely that VN functions in microbial adhesion as a bridge between pathogen adhesins and the host cell membrane. VN increases the efficiency of binding of *Escherichia coli, Staphylococcus aureus, Streptococcus pneumoniae* and *Clostridium difficile* to epithelial cells and increases adherence of *Staphylococcus epidermidis* to abiotic surfaces such as catheters. Adhesion of *Helicobacter pylori* to gastric epithelium is facilitated by its ability to bind VN with high affinity, and the adhesion of *Chlamydia trachomatis* is mediated, in part, by binding VN and other ECM molecules. A bridging role for VN has been demonstrated in the adhesion of *Neisseria meningitidis* and *N. gonorrhoeae via* binding to the opacity proteins, Opc and Opa, respectively, in the form of an Opa:heparin:VN complex (**Table 4.13**).

name of pathogen	protein/system interaction with vitronectin	role in adhesion
gram-negative bacteria		
Haemophilus influenzae type b (Hib)	*Haemophilus* surface fibrils (Hsf)	+
non-typable *H. influenza* (NTHi)	protein E (PE)	+
H. ducreyi	*Ducreyi* serum resistance protein A (DsrA)	(−)
Moraxella catarrhalis	ubiquitous surface protein A2 (UspA2)	+
Pseudomonas aeruginosa	complement regulator-acquiring surface proteins (CRASP)-2	+
Escherichia coli	unknown	+
Yersinia pseudotuberculosis	unknown	+
Neisseria meningitidis	opacity protein (Opc)	+
N. gonorrhoeae	opacity-associated outer membrane protein (Opa)	+
Chlamydia trachomatis	unknown	+
Clostridium difficile	Fibronectin-binding protein (Fbp), surface-layer protein (SLP) A	+
Helicobacter pylori	unknown	+
Porphyromonas gingivalis	fimbriae	+
	gingipain enzymes	
gram-positive bacteria		
Streptococcus pyogenes	unknown	+
S. pneumoniae	unknown	+
S. bovis	unknown	+
S. suis	unknown	+
S. dysgalactiae	unknown	+
Staphylococcus epidermidis	autolysin (Aae)	+
Staph. aureus	unknown	+
Enterococcus faecalis	unknown	+

Table 4.13 Vitronectin in adhesion of bacterial pathogens. (From Singh, B. et al., *Mol. Microbiol.*, 78, 545–560, 2010.)

Candida *albicans* and other species of *Candida* such as *C. tropicalis* and *C. parapsilosis* adhere to vitronectin, laminin and fibronectin. The asdhesin(s) that mediate binding to these ECM molecules may be integrin-like proteins in the cell wall.

Proteoglycan-binding adhesives

Proteoglycans consist of glycosaminoglycan (GAG) chains covalently attached to core proteins. GAGs are linear polysaccharides consisting of repeating disaccharide units. In most proteoglycans, GAGs comprise greater than half of the total molecular mass and are responsible for their biological functions. Proteoglycans are found on the cell surface, as part of the ECM, and intracellularly. Proteoglycans can be divided into heparan sulphate proteoglycans (HSPGs), chondroitin sulphate proteoglycans (CSPGs), dermatan sulphate proteoglycans (DSPGs), and keratan sulphate proteoglycans (KSPGs), depending on the types of GAG chains attached to the core proteins. Some proteoglycans are composed of both HS and chondroitin sulphate (CS) chains. Most pathogens appear to bind HSPGs.

Bacteria

The OpaA adhesin of the gram-negative bacterium *Neisseria gonorrhoeae* binds to syndecans, which are HS proteoglycans on urinary epithelium. The adhesion of *H. pylori* is mediated by the recognition of HS by the outer membrane proteins of the bacterium. Members of the genus *Chlamydia*, utilise HSPGs for adherence to and entry into human cells. GAGs are recognised by several *B. burgdorferi* surface proteins. Adhesion of *Mycobacterium tuberculosis* to pulmonary epithelial cells is mediated by a heparan-binding haemagglutinin adhesin. Several pathogenic streptococci bind proteoglycans. For example, the group A streptococcus *S. pyogenes* binds to DSPGs and HSPGs on host cell surfaces. The alpha-C protein of group B streptococcus *S. agalactiae* binds to HSPGs on host cell membranes and contributes to the internalisation of the bacterium. *S. pneumoniae* binds to heparin, HS and chondroitin 4-sulphate in the colonisation of respiratory mucosal epithelial cells. HSPGs are also involved in the adhesion of *Enterococcus faecalis* to different tissues. Bacteria that bind to proteoglycans are shown in **Table 4.14**.

Viruses

In the adhesion of viruses, proteoglycans function largely as co-receptors that concentrate virions on the plasma membrane of the host cell to facilitate binding to the virus's specific receptor(s). As described above

bacteria	glucosaminoglycans
Bordetella pertussis	glucosaminoglycans
Borrelia burgdorferi	heparan sulphate, dermatan sulfate
Chlamydia pneumoniae	heparan sulphate
Chlamydia trachomatis	heparan sulphate, dermatan sulfate
Enterococcus faecalis	heparan sulphate
Haemophilus influenzae	heparan sulphate
Helicobacter pylori	heparan sulphate
Listeria monocytogenes	heparan sulphate
Mycobacterium tuberculosis	heparan sulphate
Neisseria gonorrhoeae	heparan sulphate
Neisseria meningitidis	heparan sulphate, condroitin sulphate
Pseudomonas aeruginosa	heparan sulphate, glucosaminoglycans
Staphylococcus aureus	heparan sulphate
Streptococcus agalactiae	heparan sulphate
Streptococcus pneumoniae	heparan sulphate, condroitin sulphate
Streptococcus pyogenes	heparan sulphate, condroitin sulphate

Table 4.14 Bacteria that bind glucosaminoglycans.

for bacteria, most viruses bind to HS proteoglycans. For example, the envelope glycoproteins, gB and gC, of HSV serotypes 1 and 2 bind to HS proteoglycans *via* syndecan-1 on the membrane of skin, corneal, and urogenital epithelia. This interaction facilitates the binding of HSV gD to its specific receptors, which include the herpes virus entry mediator (HVEM) also known as tumour necrosis factor receptor superfamily member 14 (TNFRSF14) and nectin-1 and -2, which are cellular adhesion molecules. Envelope glycoprotein gp120 of HIV-1 binds to HS proteoglycans on the membrane of epithelial cells and other permissive cells such as macrophages and dendritic cells. In addition, HIV-1 envelope glycoprotein gp41 can bind to HS proteoglycans on the basement membrane. The major capsid protein, L1, of HPV types 11, 16, 33 and 39 binds with low affinity to cell membrane HS proteoglycans on skin and mucosal epithelial cells. This low-affinity interaction with HS proteoglycans, aided by cell membrane cyclophilin B, brings about a conformational change in the minor capsid protein L2, allowing high-affinity binding of L2 with host cell membrane internalisation receptors. The A27L protein of vaccinia virus binds to cell surface HS proteoglycans on various types of cells.

Parasites

Parasites such as *Plasmodium* species, trypanosomes, and *Leishmania* species bind cell membrane-associated HS as an initial event in cell entry.

Anchorless adhesins (Moonlighting proteins)

Another class of adhesins that bind ECMs and proteins of the coagulation cascade are termed anchorless adhesins because these proteins lack LPXTG motifs or choline-binding repeats to anchor them in the cell envelope, as is the norm for other molecules that function as adhesins. These proteins have been called moonlighting proteins because they perform activities in addition to their primary function. One such secondary function of these molecules is adhesion. The evidence for moonlighting enzymes functioning as adhesins varies depending on the class of pathogens. The evidence is strong that anchorless adhesins/moonlighting proteins such as enolase and glyceraldehyde 3-phosphate dehydrogenase (GAPDH) – enzymes in the later stages of the glycolytic pathway – function as plasminogen and ECM adhesins in bacteria. However, this is not the case for some protozoa and multicellular parasites.

Bacteria

Glycolytic pathway enzymes, functioning as adhesins, have been detected on the surface of several gram-positive bacteria, gram-negative bacteria, and cell wall-less bacteria such as *Mycoplasma pneumoniae* and *M. genitalium*. For example, *M. pneumoniae* has eight glycolytic enzymes on its surface, and all bind plasminogen. Examples of bacteria using the enzyme enolase as an adhesin are shown in **Table 4.15**. Other glycolytic enzymes functioning as adhesions are shown in **Table 4.16**.

bacterium[a]	moonlighting function of enolase
A. hydrophila	plasminogen binding
B. anthracis	plasminogen binding
Bifidobacterium spp.	plasminogen binding
B. burgdorferi	unknown
L. jensenii	inhibitor of *Neisseria* binding
L. plantarum	fibronectin binding protein/adhesin
M. fermentans	plasminogen binding
N. meningitidis	plasminogen binding
P. larvae	unknown
S. aureus	laminin binding protein
S. gordonii	MUC7 binding
S. mutans	salivary mucin MG2 and plasminogen binding
S. pneumoniae	plasminogen binding
S. suis	fibronectin binding protein/adhesin
T. vaginalis	plasminogen binding

Table 4.15 Bacteria that use enolase as an adhesion. [a] *B. anthracis, Bacillus anthracis; M. fermentans, Mycoplasma fermentans; S. mutans, Streptococcus mutans; T. vaginalis, Trichomonas vaginalis.* (From Henderson, B. and Martin, A., *Infect. Immun.,* 79, 3476–3491, 2011, Table 2.)

bacterium[b]	metabolic protein	moonlighting activity
N. meningitidis	aldolase	adhesin
S. pneumoniae	aldolase GAPDH	adhesion for atypical cadherin, Flamingo plasminogen binding protein
S. aureus	TPI	adhesion to fungal mannans
S. pyogenes	GAPDH GAPDH	fibronectin binding protein plasminogen binding protein
EHEC and EPEC	GAPDH	bind plasminogen and fibrinogen
M. pneumonia	GAPDH β-Subunit of pyruvate dehydrogenase	adhesin for mucin fibronectin binding protein
L. plantarum	GAPDH GAPDH	binds mucus and Caco-2 cells binds human ABO blood group antigens
oral streptococci	phosphoglycerate kinase phosphoglycerate mutase	plasminogen binding protein plasminogen binding protein
Group B streptococci	phosphoglycerate kinase	actin binding protein
L. monocytogenes	alcohol acetaldehyde dehydrogenase	binds to Human Hsp60
M. tuberculosis	superoxide dismutase malate synthase mycolyl transferases	adhesin binding host GAPD laminin/fibronectin binding protein fibronectin binding proteins
M. avium	superoxide dismutase	adhesin binding mucus-associated proteins

Table 4.16 Moonlighting actions of bacterial metabolic proteins.[a] [a] For enolase,[b] L. lactis, Lactococcus lactis. (From Henderson, B. and Martin, A., Infect. Immun., 79, 3476–3491, 2011, Table 3.)

Fungi

Candida albicans has a number of molecules on its surface – such as GAPDH, alcohol dehydrogenase, and phosphoglycerate mutase – that have activities different to their primary functions in the cytosol. *Paracoccidioides brasiliensis* displays several presumptive anchorless adhesins on its surface, including GAPDH, enolase, malate synthase, and triosephosphate isomerase.

Protozoa and multicellular parasites

Several putative moonlighting proteins have been identified on the surface of *Schistosoma mansoni*, including both GADPH and triosephosphate isomerase. *Entamoeba histolytica* has an alcohol dehydrogenase, and *Trichomonas vaginalis* GADPH and enolase, on their surfaces. *Plasmodium falciparum* displays enolase, and both *P. falciparum* and *Toxoplasma gondii* display aldolase on their surfaces, both of which are thought to play a role in the adhesion of these apicomplexan parasites. *Leishmania* displays enolase on its surface, where it functions as a plasminogen-binding adhesin. Glycolytic enzymes that function as adhesins in parasites are shown in **Table 4.17**.

glycolytic enzyme	moonlighting function
glucose-6-phosphate isomerase (PGI)	
aldolase (ALD)	plasminogen binding by various parasites
glyceraldehyde-3-phosphate dehydrogenase (GAPDH)	fibronectin binding by *Trichomonas vaginalis* plasminogen binding by various parasites
phosphoglycerate mutase (PGAM)	heparin binding motif in *Leishmania* spp.
enolase (ENO)	plasminogen binding by various parasites

Table 4.17 Moonlighting functions of extracellular glycolytic enzymes of parasites. (From Gómez-Arreaza, A. et al., *Mol. Biochem. Parasitol.*, 193, 75–81, 2014, Table 1.)

Cell wall glycopolymers

Teichoic and teichuronic acids are important cell wall glycopolymers in gram-positive bacteria. While both glycopolymers are found covalently linked to peptidoglycan, teichoic acid can also be anchored in the cell membrane, where it is called lipoteichoic acid. The teichoic acids of *S. aureus* and *S. pyogens* have been implicated in adhesion to epithelium. However, the ligands on host cells to which they bind are currently unknown. The teichoic acids are zwitterionic but have a net negative charge. It has been suggested that Ca^{2+} ions from the fluid phase, for example, nasal secretions or blood, can bridge lipoteichic acid and negatively charged molecules on the host cell surface. This mechanism is known as calcium bridging as described earlier in this chapter. The identity of molecules on the host cell membrane that act as ligands for teichoic and teichuronic acid remain unclear; however, several classes of receptors on phagocytes bind these glycopolymers, including C-type lectins, scavenger receptors, G-protein–coupled receptors and Toll-like receptor 2. These will be considered in Chapter 8.

Capsules

Capsules are located external to bacterial and fungal envelopes and they contribute to the pathogenicity of the organism in several ways, one of which is adhesion. It is likely that virtually all bacterial pathogens produce a capsule at least at some stage of their infectious cycle. The opportunistically pathogenic fungus *Cryptococcus neoformans* also produces a capsule. Generally, bacterial capsules are composed of complex carbohydrates, although the capsule of the anthrax agent *Bacillus anthracis* is composed of polyglutamic acid while other species of *Bacillus* have polysaccharide capsules. The capsule of *C. neoformans* consists of the polysaccharides glucuronoxylomannan and galactoxylomannan with a small proportion of mannoproteins. Although the principal roles of the capsule are in the evasion of antibodies, complement, and phagocytosis, they contribute to adhesion and biofilm formation.

GALECTINS AS BRIDGING MOLECULES IN MICROBIAL ADHESIONS

Galectins are a family of lectins with a characteristic domain organisation, affinity for β-galactosides, and a conserved carbohydrate recognition motif. There are 15 members of the galectin family in vertebrates, and they are distributed in different compartments of the cell (**Figure 4.13**). The vast majority of galectins are non-glycosylated soluble proteins; however, membrane-anchored galectins have been reported. Both galectins and calcium-dependent lectins (C-type lectins) are centrally involved in pathogen recognition and function as pattern recognition receptors. Galectins recognise both self and non-self glycans, whereas C-type lectins recognise non-self glycans only. Research on galectins has focussed on galectins-1, -3, and -9 because they predominate on epithelia and endothelia and on cells of the immune system. Galectins have been shown to bind glycans on the surface of microorganisms and worms. Although all galectin family members have an affinity for galactose, different members of the family vary in their preferences for complex glycan structures. Glycans that contain polylactosamine chains such as laminin, fibronectin, and mucins are the favoured endogenous ligands for galectins. In addition, galectins are either bivalent or polymeric. The multimeric nature of galectins allow them to bridge sugars on the surface of both the microorganism and the host cell membrane.

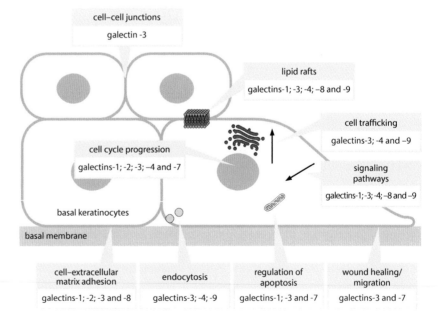

Figure 4.13 Physiological functions of galectins in epithelial cells. (From Viguier, M. et al., *Tissue Barriers*, 2, e29103, 2014, Figure 1.)

This bridging phenomenon has been observed for bacteria, viruses, and parasites. Galectin-3 can bind lipopolysaccharide (LPS) of *Pseudomonas aeruginosa* and act as a bridging molecule for adhesion of the bacterium to human corneal epithelial cells, perhaps facilitating corneal infection by this organism. In addition, galectin-3 facilitates adhesion of *Neisseria meningitidis* to human monocytes and macrophages by bridging lipooligosaccharides in the outer membrane of the bacterium (**Figure 4.14**). Galectins do not appear to be involved in adhesion to nasopharyngeal epithelial cells.

The influenza virus has a neuraminidase that produces a high density of terminal galactose residues on the envelope of its virions, and it has been shown that galectins promote binding to the respiratory epithelium. Lipophosphoglycans of *Leishmania major* can bind both galectin-3 and -9 and are involved in adhesion of the parasite to macrophages. *Trypanosoma cruzi* trypomastigotes bind to human coronary artery smooth muscle cells and ECM molecules such as laminin *via* galectin-3 bridging. *Trichomonas vaginalis* binding to cervical epithelial cells is mediated by galectin-1 bridging.

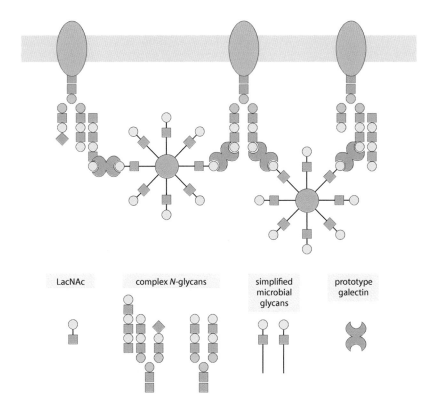

Figure 4.14 Galectins can promote microbial attachment to and infection of host cells by bridging pathogen glycans to host cell surface glycans. (From Baum, L.G. et al., *Front. Immunol.*, 5, 284, 2014.)

ADHESION TO OTHER BARRIERS

Endothelium of blood vessels and lymphatics

Integrins, proteoglycans, and the hyaluronic acid receptor are present on the basement membrane side of endothelial cells, termed the ablumenal surface, whereas several constitutive and inducible cell adhesion molecules such as selectins and ICAMs are displayed on the lumenal surfaces of these cells, particularly during inflammation. All of these molecules may serve as receptors for microbial adhesins as does the perivascular ECM.

KEY CONCEPTS

- Adhesion is a prerequisite for colonisation of the host by a pathogen except when the agent is transmitted by an intermediate vector or in some cases of toxin-mediated disease.

- Pathogens adhere predominantly to mucous membranes but also to the skin. These surfaces are called barrier epithelia.

- Adhesion involves two steps: non-specific, non-covalent interactions and specific receptor–ligand interactions.

- For the most part, the receptors are molecules on the surface of the pathogen termed adhesins, and the ligands termed receptors are glycoconjugates on the host cell surface or the ECM.

- Virtually any host cell membrane or ECM molecule can serve as a receptor.

- Adhesion mechanisms are either lectin or protein–protein in nature.

- A single pathogen has several adhesins that allows them to bind to different receptors.

REFERENCES

Aird WC. 2007. Phenotypic heterogeneity of the endothelium I. Structure, function, and mechanisms. *Circulation Research*, 100: 158–173.

Aird WC. 2012. Endothelial cell heterogeneity. *Cold Spring Harbor Perspectives in Medicine*, 2: a006429.

Albelda SM. 1991. Endothelial and epithelial cell adhesion molecules. *American Journal of Respiratory Cell and Molecular Biology*, 4: 195–203.

Appelmelk BJ et al. Why *Helicobacter pylori* has Lewis antigens. *Trends in Microbiology*, 8(12): 565–570.

Backert S et al. 2011. Molecular mechanisms of gastric epithelial cell adhesion and injection of CagA by *Helicobacter pylori. Cell Communication and Signaling*, 9: 28.

Bartlett AH, Park PW. 2015. Proteoglycans in host–pathogen interactions: Molecular mechanisms and therapeutic implications. *Expert Reviews in Molecular Medicine*, 12: e5.

Baum LG et al. 2014. Microbe-host interactions are positively and negatively regulated by galectin-glycan interactions. *Frontiers in Immunology*, 5: 284.

Berne C et al. 2018. Bacterial adhesion at the single-cell level. *Nature Reviews Microbiology*, 16(10): 616–627.

Bhella D. 2015. The role of cellular adhesion molecules in virus attachment and entry. *Philosophical Transactions of the Royal Society B*, 370: 20140035.

Bomsel M, Alfsen A. 2003. Entry of viruses through the epithelial barrier: Pathogenic trickery. *Nature Reviews Molecular Cell Biology*, 4: 57–68.

Boren T et al. 1993. Attachment of *Helicobacter pylori* to human gastric epithelium mediated by blood group antigens. *Science*, 262(1892): 1982–1985.

Boulanger MJ et al. 2010. Apicomplexan parasite adhesins: Novel strategies for targeting host cell carbohydrates. *Current Opinion in Structural Biology*, 20: 551–559.

Busch A, Phan G, Waksman G. 2015. Molecular mechanism of bacterial type 1 and P pili assembly. *Philosophical Transactions of the Royal Society A*. doi:10.1098/rsta.2013.0153.

Busscher HJ, van der Mei HC. 2012. How do bacteria know they are on a surface and regulate their response to an adhering state? *PLoS Pathogens*, 8(1): e1002440.

Chagnot C et al. 2012. Bacterial adhesion to animal tissues: Protein determinants for recognition of extracellular matrix components. *Cellular Microbiology*, 14(11): 1687–1696.

Chaudhry R et al. 2007. Adhesion proteins of *Mycoplasma pneumoniae*. *Frontiers in Bioscience*, 12: 690–699.

Chhatwal GS. 2002. Anchorless adhesins and invasins of Gram-positive bacteria: A new class of virulence factors. *Trends in Microbiology*, 10(5): 205–208.

Coburn J et al. 2013. Illuminating the roles of the *Borrelia burgdorferi* adhesins. *Trends in Microbiology*, 21(8): 327–379.

Coureuil M et al. 2017. A journey into the brain: Insight into how bacterial pathogens cross blood-brain barriers. *Nature Reviews Microbiology*, 15(3): 149–159.

Cummings R, Turco S. 2009. Parasitic infections. In: *Essentials of Glycobiology*, 2nd ed., Ch. 40, Cold Spring Harbor, NY: Cold Spring Harbor Laboratory Press.

Daneman R, Prat A. 2015. The blood–brain barrier. *Cold Spring Harbor Perspectives in Biology*, 7: a020412.

de Groot PWJ et al. Adhesins in human fungal pathogens: Glue with plenty of stick. *Eukaryotic Cell*, 12(4): 470–481.

Doran KS et al. 2013. Concepts and mechanisms: Crossing host barriers. *Cold Spring Harbor Perspectives in Medicine*, 3: a010090.

Fletcher M (Ed.). 1996. *Bacterial Adhesion: Molecular and Ecological Diversity* (Wiley Series in Ecological and Applied Microbiology). 1st ed. Wiley-Liss, New York.

Foster TJ et al. 2014. Adhesion, invasion and evasion: The many functions of the surface proteins of *Staphylococcus aureus*. *Nature Reviews Microbiology*, 12: 49–62.

Foster TJ, Höök M. 1998. Surface protein adhesins of *Staphylococcus aureus*. *Trends in Microbiology*, 6(12): 484–488.

Gallegosa B et al. 2014. Lectins in human pathogenic fungi. *Revista Iberoamericana de Micología*, 31(1): 72–75.

Garcia B et al. 2016. Surface proteoglycans as mediators in bacterial pathogens infections. *Frontiers in Microbiology*. doi:10.3389/fmicb.2016.00220.

García MA et al. 2015. *Entamoeba histolytica*: Adhesins and lectins in the trophozoite surface. *Molecules*, 20: 2802–2815.

Gómez-Arreaza A et al. 2014. Extracellular functions of glycolytic enzymes of parasites: Unpredicted use of ancient proteins. *Molecular and Biochemical Parasitology*, 193(2): 75–81.

Gow NAR et al. 2012. *Candida albicans* morphogenesis and host defence: Discriminating invasion from colonization. *Nature Reviews Microbiology*, 10: 112–122.

Henderson B et al. 2011. Fibronectin: A multidomain host adhesin targeted by bacterial fibronectin-binding proteins. *FEMS Microbiological Reviews*, 35(1): 147–200.

Henderson B, Martin A. 2011. Bacterial virulence in the moonlight: Multitasking bacterial moonlighting proteins are virulence determinants in infectious disease. *Infection and Immunity*, 79(9): 3476–3491.

Hospenthal MK et al. 2017. A comprehensive guide to pilus biogenesis in Gram-negative bacteria. *Nature Reviews Microbiology*, 15: 365–379.

Jacques M. 1996. Role of lipo-oligosaccharides and lipopolysaccharides in bacterial adherence. *Trends in Microbiology*, 4(10): 408–410.

Jerse AE, Rest RF. 1997. Adhesion and invasion by the pathogenic neisseria. *Trends in Microbiology*, 5(6): 217–221.

Juge N. 2012. Microbial adhesins to gastrointestinal mucus. *Trends in Microbiology*, 20(1): 30–39.

Katze MG, Korth MJ, Law GL, Nathanson N. 2016. *Viral Pathogenesis: From Basics to Systems Biology*. 3rd ed. Academic Press, Elsevier, London, UK.

Kendall K, Kendall M, Rehfeldt F. 2011. *Adhesion of Cells, Viruses and Nanoparticles*. 1st ed. Springer, London, UK.

Kerr JR et al. 1999. Cell adhesion molecules in the pathogenesis of and host defence against microbial infection. *Journal of Clinical Pathology: Molecular Pathology*, 52: 220–230.

Kline KA et al. 2009. Bacterial adhesins in host-microbe interactions. *Cell Host & Microbe*, 5: 580–592.

Kline KA et al. 2010. A tale of two pili: Assembly and function of pili in bacteria. *Trends in Microbiology*, 18(5): 224–232.

Kozik A et al. 2015. Fibronectin-, vitronectin- and laminin-binding proteins at the cell walls of *Candida parapsilosis* and *Candida tropicalis* pathogenic yeasts. *BMC Microbiology*, 15: 197.

Krause DC. 1998. *Mycoplasma pneumoniae* cytadherence: Organization and assembly of the attachment organelle. *Trends in Microbiology*, 6(1): 15–18.

Linden SK et al. 2008. Mucins in the mucosal barrier to infection. *Mucosal Immunology*, 1(3): 183–197.

Lipke PN. 2018. What we do not know about fungal cell adhesion molecules. *Journal of Fungi (Basel)*, 4(2): 59.

Loker ES, Hofkin BV. 2015. *Parasitology: A Conceptual Approach*. Garland Science, New York.

Mandlik A et al. 2007. Pili in Gram-positive bacteria: Assembly, involvement in colonization and biofilm development. *Trends in Microbiology*, 16(1): 33–40.

Marsh PD (Ed.). 2016. Chapter 5, Dental plaque. In: *Marsh and Martin's Oral Microbiology*, 6th ed., Elsevier, Edinburgh, UK.

McGuckin MA et al. 2011. Mucin dynamics and enteric pathogens. *Nature Reviews Microbiology*, 9(4): 265–278.

Nash AA, Dalziel RG, Fitzgerald JR. 2015. *Mim's Pathogenesis of Infectious Disease*. 6th ed. Elsevier, London, UK.

Nathanson N. 2007. *Viral Pathogenesis and Immunity*. 2nd ed. Academic Press, Elsevier, London, UK.

Nobbs AH et al. 2009. *Streptococcus* adherence and colonization. *Microbiology and Molecular Biology Reviews*, 73(3): 407–450.

Ofek I, Doyle RJ. 1994. *Bacterial Adhesion to Cells and Tissues*. 1st ed. Chapman & Hall, New York.

Pizarro-Cerdá J, Cossart P. 2006. Bacterial adhesion and entry into host cells. *Cell* 124: 715–727.

Poole J et al. 2018. Glycointeractions in bacterial pathogenesis. *Nature Reviews Microbiology*, 16(7).

Proft T, Baker EN. 2009. Pili in gram-negative and gram-positive bacteria – Structure, assembly and their role in disease. *Cellular and Molecular Life Sciences*, 66: 613–635.

Ribet D, Cossart P. 2015. How bacterial pathogens colonize their hosts and invade deeper tissues. *Microbes and Infection*, 17: 173–183.

Robbins JR, Bakardjiev AI. 2012. Pathogens and the placental fortress. *Current Opinion in Microbiology*, 15(1): 36–43.

Schneider-Schaulies J. 2000. Cellular receptors for viruses: Links to tropism and pathogenesis. *Journal of General Virology*, 81: 1413–1429.

Schwarz-Linek U et al. Fibronectin-binding proteins of Gram-positive cocci. *Microbes and Infection*, 8: 2291–2298.

Sharon N, Ofek I. 2000. Safe as mother's milk: Carbohydrates as future anti-adhesion drugs for bacterial diseases. *Glycoconjugate Journal*, 17: 659–664.

Silva S et al. 2011. Adherence and biofilm formation of non-*Candida albicans Candida* species. *Trends in Microbiology*, 19(5): 241–247.

Singh B et al. 2010. Vitronectin in bacterial pathogenesis: A host protein used in complement escape and cellular invasion. *Molecular Microbiology*, 78(3): 545–560.

Singh B et al. 2012. Human pathogens utilize host extracellular matrix proteins laminin and collagen for adhesion and invasion of the host. *FEMS Microbiology Reviews*, 36: 1122–1180.

Sperandio B, Fischer N, Sansonetti PJ. 2015. Mucosal physical and chemical innate barriers: Lessons from microbial evasion strategies. *Seminars in Immunology*, 27(2): 111–118.

Stewart PL, Nemerow GR. 2007. Cell integrins: Commonly used receptors for diverse viral pathogens. *Trends in Microbiology*, 15(11): 500–507.

Telford JL et al. 2006. Pili in gram-positive pathogens. *Nature Reviews Microbiology*, 4(7): 509–519.

Trivedi K et al. 2011. Mechanisms of meningococcal colonisation. *Trends in Microbiology*, 19(9): 456–463.

Tronchin G et al. 2008. Adherence mechanisms in human pathogenic fungi. *Medical Mycology*, 46: 749–772.

Viguier M et al. 2014. Galectins in epithelial functions. *Tissue Barriers*, 2: e29103.

Weidenmaier C et al. 2012. *Staphylococcus aureus* determinants for nasal colonization. *Trends in Microbiology*, 20(5): 243–250.

Weidenmaier C, Peschel A. 2008. Teichoic acids and related cell-wall glycopolymers in Gram-positive physiology and host interactions. *Nature Reviews Microbiology*, 6(4): 276–287.

Wilson BA, Salyers AA, Whitt DD, Winkler ME. 2011. *Bacterial Pathogenesis: A Molecular Approach*. 3rd ed. ASM Press, Washington, DC.

Wilson M, McNab R, Henderson B. 2002. *Bacterial Disease Mechanisms: An Introduction to Cellular Microbiology*. Cambridge University Press, Cambridge, UK.

Chapter 5: Facilitated Cell Entry

INTRODUCTION

In this chapter, we examine the methods employed by pathogens to cross intact skin and mucosae. As described in Chapter 4, the skin and mucosae are often called barrier epithelia because they protect the body from physical and chemical damage, infection, dehydration, and heat loss. The process used by pathogens to cross barrier epithelia is frequently called *invasion*, but the term invasion implies force on the part of the pathogen, whereas in reality almost all pathogens trick barrier epithelial cells into taking them up. Once pathogens have crossed the barrier epithelia, they can migrate across the endothelium and invade the circulatory and lymphatic systems, causing systemic infections. We will see that the different classes of pathogen, bacteria, fungi, viruses and parasites employ remarkably similar strategies to enter epithelial cells that are the barrier to their invasion of deeper tissue and organs.

The infectious cycle of almost all human pathogens begins with attachment and colonisation of the barrier epithelia, as discussed in Chapter 4. In the majority of instances colonization is followed by traversal of the barrier epithelia (skin and mucosae) as the initial phase in local and systemic spread of the agent. The simplest way for pathogens to cross the barrier epithelia is to take advantage of breaches in the integrity of the epithelium caused by trauma. Epithelial damage may be as trivial as a microscopic wound or as severe as a traumatic wound. Surgical wounds or, indeed, the placement of intravascular catheters provides entry directly into the dermis and the circulatory system, by-passing the epithelial barrier. The invasive action of a catheter is replicated in nature by the biting mouthparts of arthropod vectors such as mosquitoes and ticks that penetrate and traverse the skin. Animal or human bites may also breach the skin. Intact skin is a formidable barrier to infectious agents, so the vast majority of pathogens choose to enter the human body *via* the natural portals of entry: the eyes, nose, mouth, genitourinary tract and anus. The most common routes of entry are the respiratory and the gastrointestinal tracts. Here pathogens encounter the mucosal epithelia. The human respiratory tract is estimated to encompass an area of 50–100 m^2 and the digestive tract some 30–40 m^2. In contrast, the skin of a human is estimated to cover an area of about 1.5–2.0 m^2.

The mucosal epithelia function to permit the uptake of nutrients, excretion of waste products and exchange of oxygen and carbon dioxide and in some areas are only a single cell thick. All mucosal surfaces are bathed by glandular secretions, produced by exocrine glands located adjacent to the mucosae,

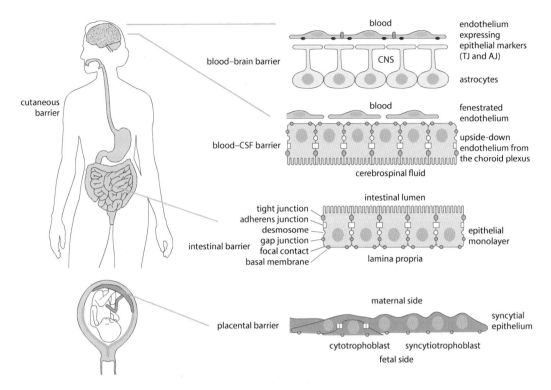

Figure 5.1 Example of host barriers. From top to bottom: The blood-brain barrier is made by brain endothelial cells that express tight junctions. The blood–cerebrospinal fluid (CSF) barrier is constituted by a monolayer of epithelial cells from the choroid plexus, which separates the blood located in a fenestrated endothelium from the CSF. The intestinal barrier is a mucosal barrier made by a monolayer of epithelial cells. The placental barrier is constituted by both syncytial trophoblastic cells and mononuclear cytotrophoblasts. (From Doran, K.S. et al. *Cold Spring Hard Perspect Med.*, 3, a010090, 2013.)

which flush the epithelial surfaces and contain many anti-microbial factors. Given the large surface area and delicacy of the mucosal epithelium, it is not surprising that the human immune system commits the majority of its innate and adaptive immune resources to their protection in the form of an integrated system known as the mucosal immune system. **Figure 5.1** shows some of the barriers traversed by pathogenic microorganisms. These barriers including the skin were described in Chapter 4.

CROSSING INTACT SKIN

Intact skin is a formidable barrier for pathogens to breach. The surface layers of the skin, the epidermis, are composed of strata of squamous epithelial cells, the outermost of which are dead, keratinised squames. Most of the surface of the skin contains hair follicles, but in humans, the ventral surface of the fingers, palms, soles of the feet, lips, labia minora, and glans penis lack hair. These areas are termed glabrous skin. Sebaceous glands are found all over the skin except for the palms of the hands and soles of the feet. Sebaceous glands are either associated with hair follicles as pilosebaceous units or found alone. Sweat glands are found in all areas of human skin. The skin surface is protected by a number of mechanisms that include its resident microbiota, an acidic pH (4.0–5.0) and continuous shedding of the surface cells, termed desquamation, that carries away microorganisms that have adhered to surface squames. It has been estimated that the total surface of the skin is replaced every month by desquamation. In addition, sebum, the secretion of sebaceous glands, contains an array of antimicrobial factors including the muramidase lysozyme that hydrolyses peptidoglycan; various antimicrobial peptides, including dermacidin, psoriasin, RNase 7, human β-defensins 1, 2, and 3 and cathelicidin; fatty acids and the antibody immunoglobulin A.

There is no evidence that pathogenic bacteria or viruses are capable of penetrating intact, keratinised, squamous epithelium. Rather, they breach the skin *via* hair follicles, sweat glands, or sites of microscopic or macroscopic damage. Certain eukaryotic pathogens can penetrate intact, keratinised, squamous epithelium and use hydrolytic enzymes to do so. However, given the opportunity, these organisms take advantage of natural (or otherwise) breaches in the skin. Because the mucosal epithelia is not keratinised (except for the gingivae, hard palate and dorsum of the tongue) and is but a single cell thick in some areas, organisms capable of digesting intact skin are likely able to digest intact mucosal epithelium too.

Enzymatic degradation

Many microorganisms both saprophytes and parasites secrete degradative enzymes such as proteases, lipases and glycosidases. These hydrolytic enzymes function to acquire nutrients from the environment and, in the case of human parasites, to break down tissues. Three major groups of fungi, *Candida albicans*, the dermatophytes (*Epidermophyton* species, *Trichophyton* species, and *Microsporum* species) and *Malassezia* species are capable of infecting and destroying skin, nails and hair shafts. These fungi produce a variety of hydrolytic enzymes including lipases, phospholipases, acid proteases, elastase, and keratinases. Although penetration of the skin by dermatophytes is usually limited to the non-living, cornified layer of the epidermis, there is evidence of invasion of the stratum granulosum (see Chapter 4, Figure 4.1). In the case of protozoan, nematode, trematode, and cestode parasites, the larvae of the nematodes *Anclostoma duodenale*, *Necator americanus* and *Stronglyoides stercoralsis* can penetrate intact skin as well as enter through hair follicles or cracks in the skin. The ability to invade intact skin by these larvae is thought to be mediated by hydrolytic enzymes. In the nematodes, hydrolases are among the extracellular secreted products (ESP) released from specialised larval glands. These larvae produce all of the known classes of proteases and can degrade collagen, fibronectin, laminin, elastin and hyaluronic acid. The cercaria larvae of shistosomes *(S. mansoni, S. japonicum, S. haematobium)* possess seven serine proteases (elastases) and five metalloproteases secreted from post- and pre-acetabular glands. These glands in shistosomes comprise five pairs of large secretory cells located close to the ventral sucker (acetabulum). Three pairs are termed post-acetabular and two pre-acetabular. Trophozoites of the pathogenic and opportunistic free-living amoebae, *Acanthamoeba* species and *Balamuthia mandrillaris*, appear to have the capacity to invade intact skin as well as intact mucus membranes. Again, they secrete hydrolytic enzymes that include phospholipases, serine and cysteine proteases, and metalloproteases.

Several different lipases are produced by members of the bacterial genus *Staphylococcus*. *S. aureus,* the foremost pathogen in this genus, produces two lipases that were initially thought to contribute to breakdown of the skin by degrading triglycerides to release free fatty acids. However, it is now believed that *S. aureus* lipases function in the persistence of the bacterium on the barrier epithelia by degrading antimicrobial lipids such as sapienic and linoleic acids.

CROSSING INTACT MUCOSAL EPITHELIUM

Entry *via* microfold (M) cells

The mucosal epithelia overlying the gut-associated and pharynx-associated lymphoid tissues contain cells known as microfold (M) cells, which sample the contents of the lumen of the gut and the oropharynx. M cells take up and transcytose microorganisms as well as food and macromolecules. Transcytosis is a type of endocytosis in which macromolecules and particles are internalised in vesicles on one side of a cell and discharged on the other side. Following transcytosis through M cells, the cargo is discharged into an intraepithelial pocket known as the sub-epithelial dome, which contains macrophages and dendritic cells (DCs) (**Figure 5.2**).

Several bacteria such as *Yersinia enterocolitica*, *Shigella flexneri* and *Salmonella enterica* serovar typhi and *S. enterica* serovar typhimurium and viruses such as enteroviruses and HIV use this mechanism as a means of entry. In addition to M cells, there are DCs in the lamina propria of mucosal epithelia that extend fingers of cytoplasm (dendrites) between enterocytes (intestinal epithelial cells) and into the lumen of the gut where they can sample food and microorganisms. Pathogens entering the sub-epithelial tissues are engulfed and destroyed by resident phagocytic cells such as macrophages. These cells, called sentinel cells, together with the epithelium itself, alert the adaptive immune system to the presence and nature of pathogens. Phagocytosis is the major mechanism by which the host destroys and clears invading pathogens, but some pathogens are able to utilise phagocytosis to facilitate their spread. They are able to do this because they have devised mechanisms to avoid destruction in the phagolysosome of these cells and use the phagocyte as a *Trojan horse*. (see Chapter 8 Evasion of human innate immune system, *Circumvention of phagocytosis*).

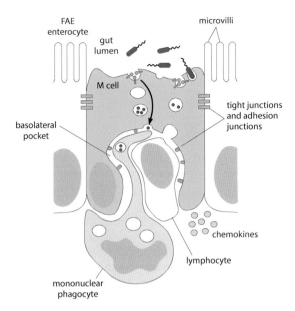

Figure 5.2 The morphological features of M cells. FAE, Follicle-associated epithelium. (From Mabbott, N.A. et al., *Mucosal Immunol.*, 6, 666–677, 2013.)

Enzymatic degradation

A number of parasites utilise hydrolytic enzymes to invade mucosal epithelia. *Trichomonas vaginalis* employs cysteine proteases in the enzymatic destruction of vaginal, cervical and prostatic epithelium. *Entamoeba histolytica* uses cysteine proteases and amoebapores, which are small proteins including saposin-like protein (SAPLIP), stored in lysosomes (vesicles), that are able to form pores to destroy the barrier epithelium. Saposins are small proteins that activate various lysosomal lipases. The intermediate flagellate form of *Naegleria fowleri* enters its human host *via* the olfactory neuroepithelium. The amoebae produce sucker-like appendages that nibble away at cells.

Trichinella spiralis and *Dracunculus medinensis* can invade the small intestine presumably by enzymatic degradation as both have been shown to possess metallo- and serine proteases. The larvae of the flukes *Fasciola hepatica*, *Opisthorchis sinensis* and *Paragonimus westermani* invade the duodenum utilising proteolytic enzymes including cathepsin L, cathepsin B, asparaginyl endopeptidase, cysteine proteases, trypsin-like serine proteases and carboxypeptidases. It is likely that the larval forms of the nematode *Ascaris lumbricoides*, various *Toxocara* species and *Baylisacaris procyonis* penetrate the mucosal epithelium of the small intestines by employing secreted degradative enzymes.

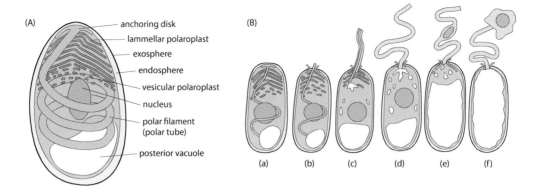

Figure 5.3 (A) Diagram of a microsporidian spore showing the major structures discussed in the text. (B) Polar tube eversion during spore germination: (a) Dormant spore, showing polar filament (black), nucleus (gray), polaroplast and posterior vacuole, (b) polaroplast and posterior vacuole swelling, anchoring disk ruptures, and polar filament begins to emerge, everting as it does so, (c) polar filament continues to evert, (d) once the polar tube is fully everted, the sporoplasm is forced into and (e) through the polar tube, (f) sporoplasm emerges from the polar tube bound by new membrane. (From Keeling, P.J. and Fast, N.M., *Ann. Rev. Microbio.*, 56, 93–116, 2002.)

Polar tube formation

Microsporidia are obligate intracellular fungal parasites that use a unique organelle called the polar tube to inject their spore contents (sporoplasm) inside the target mucosal epithelial cell. In the environment, the organism exists as a spore with a double-layered wall. Following adhesion of the spore to the host cell membrane, the polar tube is discharged from the anterior part of the spore – probably as the result of increased osmotic pressure – and acts like a needle to pierce the host cell membrane, delivering the sporoplasm to the host cell cytoplasm (**Figure 5.3a** and **b**).

Moving junction

Apicomplexan parasites are a large group of protozoans that are characterised by having a special organelle called an apical complex. Most of the apicomplexa are single-celled and spore-forming and include the human pathogens *Babesia* species, *Plasmodium* species, *Cryptosporidium parvum*, *Cyclospora cayetanensis*, *Isospora belli*, and *Toxoplasma gondii*. The apicomplexa use a structure known as a moving junction to form a tight union between the parasite and the host cell membrane through which the parasite passes to enter the host cell (**Figure 5.4**). Although these parasites enter different types of host cells, they employ the same method of host cell entry. The host cell membrane invaginates in response to the forward movement of the parasite, which becomes progressively internalised in a vacuole termed the parasitophorous vacuole. A common feature of apicomplexan protozoa is the presence at their apex of secretory organelles termed micronemes and rhoptries. Proteins secreted from these organelles are involved in parasite adhesion, motility and internalization. Two groups of proteins have been associated with internalization, the rhoptry neck proteins (RONs) and the

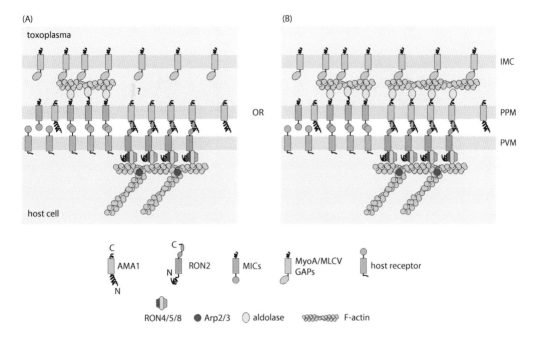

Figure 5.4 Working models of the molecular interactions at the moving junction (MJ): (A) The MJ and flanking regions depicting one of two working models for the molecular interactions at the parasite–host interface. The motive force shown in panel (a) comes from the actin/myosin motor of the parasite driving the parasite forward via interactions between the MICs and the host surface. (B) The MJ provides the driving force instead of the MICs. Here AMA1 is bridged to polymerized parasite actin through an interaction with aldolase and, as in the model in (a), interacts with the host cytoskeleton through the RONs. Abbreviations: IMC, inner membrane complex; PPM, parasite plasma membrane; PVM, parasitophorous vacuole membrane; HPM, host plasma membrane; MICs, micronemes; RON, rhoptry neck protein; ARP, actin-related protein. (From Tyler, J.S. et al. *Trends Parasitol.*, 27, 410–420, 2011.)

apical membrane antigen (AMA) 1. Some RON proteins form a ring-like structure in the host cell membrane through which the protozoan enters, whereas others together with the AMA1 protein and microneme proteins are located along the surface of the parasite adjacent to the membrane of the host cell and comprise the moving junction. These proteins anchor the parasite membrane to the host cell membrane so that the actin-myosin motor of the parasite can propel it forward into the host cell (**Figure 5.5**).

Paracytosis

Paracytosis is the term given to the mechanism by which microorganisms cross the barrier epithelia *via* the tight intercellular junctions that seal adjacent epithelial cells together. Cells of the mucosal epithelia are sealed together by a number of intercellular components (**Figure 5.6**). The most apical is the tight junction below which are the adherens junction, desmosomes and gap junctions. The first three comprise the apical junction complex and consist of transmembrane proteins coupled to cytoplasmic adaptors and the actin cytoskeleton. Transmembrane proteins of tight junctions are the junction components most vulnerable to pathogenic microorganisms when they contact the mucosal epithelium because they are the most apically located. Examples

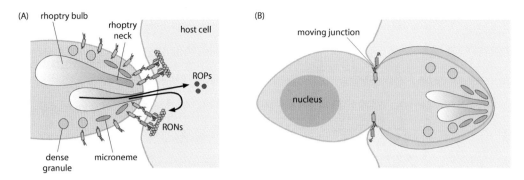

Figure 5.5 Schematic representation of early and mid-invasion steps by apicomplexans. (A) A schematic representation of the apical tip of an apicomplexan parasite, in this case a *Toxoplasma* tachyzoite, at the initial stage of invasion and indicating the relative location of the secretory organelles: the micronemes, rhoptries (showing the relative location of the bulb and neck compartments), and dense granules. (B) Schematic representation of a partially invaded parasite. As the parasite penetrates the host cell, the MJ is coincident with a visible constriction on the parasite. It remains unclear whether this constriction is the result of forces generated by the host cytoskeleton, which the parasite must displace to enter the cell, or by internal forces generated by the operation of the actin/myosin motors that power invasion at the site of entry. (From Tyler, J.S. et al. *Trends Parasitol.*, 27, 410–420, 2011.)

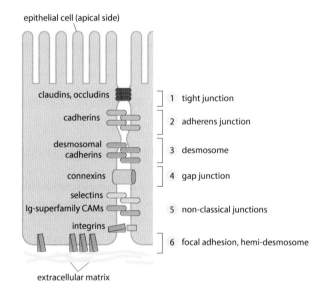

Figure 5.6 Components of intercellular junctions. Cell adhesion molecules (CAMs) and junctional complexes are abundant in epithelial tissues. (1) Tight junctions build a seal between adjacent cells and are connected to actin filaments. (2) Adherens junctions are plaques of cadherins linked to the actin cytoskeleton. (3) Desmosomes are formed by desmosomal cadherins, linked to intermediate filaments. (4) Gap junctions connect the cytoplasm of two adjacent cells and are linked to microfilaments. (5) Selectins, Ig-superfamily CAMs, and other CAMs promote homophilic adhesion outside of junctions. Integrins bind in a heterophilic manner. (6) Focal adhesions (linked to actin) and hemi-desmosomes (linked to intermediate filaments) are cell–matrix junctions that are formed by integrins. (From Schnell, U. et al., *Biochim. Biophys. Acta*, 1828, 1989–2001, 2013.)

of pathogens using paracytosis to cross the nasal epithelium are the human respiratory bacterial pathogens *S. pneumoniae* and *H. influenzae*. These bacteria induce a Toll-like receptor (TLR)-dependent down-regulation of claudins 7 and 10. TLR are pattern recognition receptors (PRR) found largely on cells of the immune system, particularly professional phagocytes but also on epithelial cells. TLRs recognise microbe-associated molecular patterns (MAMPS), which are conserved structural molecules found in bacteria, fungi, viruses and parasites. Claudins are tight junction components key to the maintenance of epithelial barrier integrity and their down-regulation compromises the integrity of the tight junction, allowing ingress of these bacteria.

Enteropathogenic *Escherichia coli*, enterohaemorrhagic *E. coli* and other enteropathogens such as *Salmonella enterica* serovar Typhimurium and *Shigella flexneri* employ a microscopic syringe termed a secretion system, of which there are several types, to inject bacterial proteins, called effectors, into the cytosol of the host cell. These injected proteins facilitate the entry of these bacteria into host cells. The principal role of these effectors is to manipulate the actin cytoskeleton to promote uptake by endocytosis. The secretion systems of gram-negative and gram-positive bacteria are described later in this chapter.

One effect of the bacterial proteins injected by secretion systems can be the disruption of intercellular tight junctions. The mechanism(s) by which this occurs is(are) not fully understood.

Helicobacter pylori (a bacterium that can cause peptic ulcers) adheres to the intercellular contact areas of the gastric and duodenal epithelium and targets the tight junctions using a type 4 secretion system (T4SS). The T4SS requires attachment of the bacterium to the host cell either by direct cell-to-cell contact or *via* a bridge-like structure similar to that used by bacteria to transfer DNA by conjugation. The T4SS bridge can transport and receive both DNA and proteins. In *H. pylori*, two proteins secreted by the T4SS, CagA and VacA, have been implicated in destruction of tight junctions. *Clostridium perfringens* is an anaerobic gram-positive spore-forming rod that causes necrotising infections of muscle but also an acute, generally mild form of food poisoning. Type A strains of *C. perfringens* produce an enterotoxin (a toxin that acts on the intestinal wall). The enterotoxin is released from the bacterium during its transition from vegetative cell to spore in the small intestine and binds to claudins 3 and 4 in the tight junction between enterocytes impairing their integrity. Another *Clostridium* species, *C. difficile*, that causes pseudomembranous colitis uses two exotoxins (toxin A and toxin B) to target occludins. The Lancefield group B streptococcus (*S. agalactiae*) may also cross the intestinal mucosa *via* paracytosis, although no clear mechanism has been established.

Reoviruses and rotaviruses use protein components of the tight junction as their ligands for cell entry. Reoviruses bind to junctional adhesion molecule A (JAM-A), whereas rotaviruses utilise the ability of the V8 subunit of their protein spikes to alter the location of claudin-3 and occludin, which leads to the disruption of the tight junction. Rotaviruses also produce an enterotoxin, termed non-structural protein 4 (NSP4), which blocks the formation of tight junctions. Coxsackieviruses bind to the coxsackievirus and adenovirus receptor (CAR) and use it as a co-receptor in cell entry *via* intercellular tight junctions.

Several pathogenic fungi, including *Candida albicans,* can traverse epithelial barriers by proteolytic degradation of intercellular tight junctions. It has been suggested that *Entamoeba histolytica* and *Toxoplasma gondii* may cross the epithelium *via* paracytosis based on the observation that they collect at epithelial intercellular junctions.

Endocytosis

Endocytosis is a type of energy-requiring active transport by which cells take up molecules and particles into their cytoplasm by engulfing them. There are five different types of endocytosis: macropinocytosis, caveolae-mediated endocytosis, clathrin-mediated endocytosis, clathrin-independent endocytosis and phagocytosis. All involve reorganisation of the actin cytoskeleton and intersection with a common endosomal network. We will briefly consider the different forms of endocytosis with the exception of phagocytosis, which is a specialised type of endocytosis performed by professional phagocytic cells such as neutrophils, monocyte/macrophages and DCs. Phagocytosis is addressed in Chapter 8.

Macropinocytosis: Macropinocytosis occurs in response to growth factors or, in the case of antigen-presenting cells, it is a constitutive process. It is a mechanism by which cells sample soluble molecules and particulates in their surroundings. The membrane of the cell produces lamellipodia following actin reorganisation. Some of the lamellipodia retract but some fold over so that their tip fuses with the plasma membrane at the base of the lamellipodia, creating large, irregular shaped vesicles (see **Figure 5.7**). The macropinosomes pinch off from the plasma membrane and traffic deeper into the cell where they enter the endocytic pathway and fuse with lysosomes. In some cases, macropinosomes migrate back to the cell membrane. The process of macropinocytosis is shown in Figure 5.7.

Caveola-mediated endocytosis: Caveolae are a specialised type of lipid raft that forms sub-microscopic, plasma membrane pits consisting of a family of cell membrane proteins called caveolins of which there are three members (caveolin 1, 2 and 3). Caveolins 1 and 3 are essential for the formation of caveolae. Cargo binds to the caveolar membrane by receptor–ligand interaction. Caveolin oligomerisation is necessary for formation of caveolar endocytic vesicles. Each caveola contains roughly 140–150 caveolin 1 (CAV1) molecules. Cytoplasmic proteins, called cavins, work together with caveolins to regulate the formation of caveolae

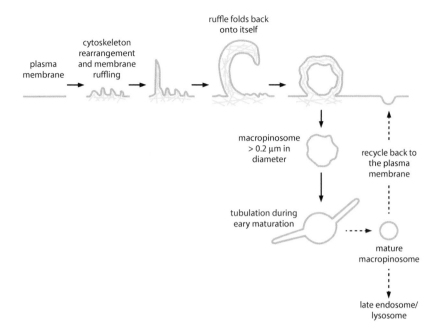

Figure 5.7 Early maturation of macropinosomes involve extensive tubulation resulting in mature macropinosomes that are more spherical. Content of the macropinosomes are then either degraded at the late endosome/lysosome or recycled back to the plasma membrane. (From Lim, J.P. and Gleeson, P.A., *Immunol. Cell Biol.*, 89, 836–843, 2011.)

Calveolins are linked to the actin cytoskeleton by a cross-linking protein called filamin, and dynamin is recruited to the neck of the caveloae. Dynamin is a GTPase that is involved in sealing and cutting off caveolae and other types of vesicle from the cell membranes. Caveolae have various functions that include protection of cells from mechanical stress given that they can act as mechanosensors and in cell signalling and lipid regulation. Viral pathogens such as SV40, polyoma, echovirus1, HIV, and respiratory syncytial virus and certain gram-negative bacteria expressing type 1 pili enter host cells *via* caveolae. Caveolin-mediated endocytosis of a virus particle is shown in **Figure 5.8**.

Clathrin-mediated endocytosis: Clathrin-mediated endocytosis (CME) is thought to begin with the formation of a plasma membrane pit in which motifs (ligands) on soluble molecules or particulates such as microorganisms bind to cell membrane receptors, recruiting a highly conserved protein called AP2. However, more recently it has been suggested that CME begins when a presumptive nucleation module binds to the cell membrane lipid phosphatidylinositol-4,5-biphosphate (PtdIns(4,5)P$_2$). The nucleation module contains proteins that have curvature-inducing ability that is necessary for the continued development of the pit into a vesicle. The module proteins recruit AP2, which binds to the cytoplasmic domain of cell membrane receptors, sometimes *via* adaptor proteins, and to PtdIns(4,5)P$_2$. AP2 is central to CME because it binds clathrin and most of the accessory proteins involved in the process.

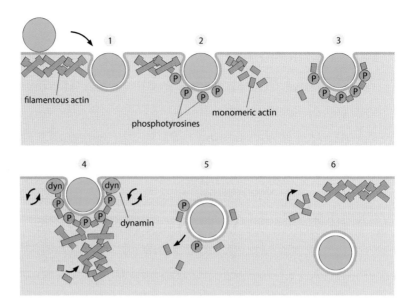

Figure 5.8 Initial stages of SV40 internalization via caveolae. After binding to the membrane, virus particles are mobile until trapped in caveolae, which are linked to the actin cytoskeleton (1). In the caveolae, SV40 particles trigger a signal transduction cascade that leads to local protein tyrosine phosphorylation and depolymerization of the cortical actin cytoskeleton (2). Actin monomers are recruited to the virus-loaded caveolae and an actin patch is formed (3). Concomitantly, dynamin is recruited to the virus-loaded caveolae and a burst of actin polymerization occurs on the actin patch (4). Virus-loaded caveolae vesicles are now released from the membrane and move into the cytosol (5). After internalization, the cortical actin cytoskeleton returns to its normal pattern (6). (From Pelkmans, L. and Helenius, A., *Traffic*, 3, 311–320, 2002.)

The protein clathrin is then recruited from the cytosol to the developing vesicle. Clathrin is a homotrimer with three bent arms that polymerises to form a cage around the vesicle, resulting in its stabilisation. The vesicle is sealed and released from the plasma membrane by dynamin. Once the vesicle is pinched off from the plasma membrane, the clathrin cage disassembles and the vesicle enters the endocytic pathway. CME is illustrated in **Figure 5.9**.

Clathrin-independent endocytosis: Clathrin-independent endocytosis (CIE) encompasses several pathways that do not involve clathrin and may or may not utilise dynamin. Examples of CIE that use dynamin is RhoA-dependent IL-2 receptor endocytosis. This pathway is also by several other cytokines, also. Another is endocytosis of the epidermal growth factor receptor (EGFR). The clatherin-independent carrier (CLIC)/glycosylphosphatidylinositol-anchored protein (GPI-AP) enriched compartment (GEEC) Pathway, ADP-ribosylatin factor 6 (ARF6)-Associated Pathway, and flotillin-mediated endocytosis are examples of dynamin- and clathrin-independent endocytic processes. CIE mechanisms in non-polarized and polarized epithelial cells is shown in **Figure 5.10**.

Figure 5.9 Clathrin-mediated endocytosis. (From https://biochem.wisc.edu/labs/bednarek/endocytosis-and-vaem-research-bednarek-lab.)

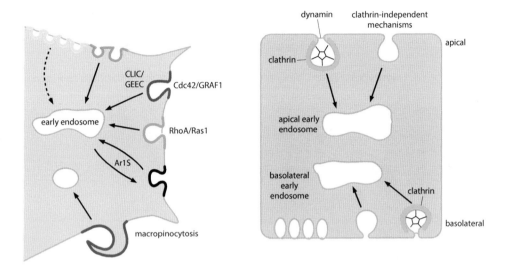

Figure 5.10 Clathrin-independent endocytosis. Endocytic mechanisms in non-polarized and polarized epithelial cells. In non-polarized cells (left) several clathrin-independent endocytic mechanisms have been described. Flotillins have been described to be involved in dynamin-dependent and dynamin-independent endocytic pathways. The small GTPase Cdc42, together with GRAF1, is involved in a dynamin-independent mechanism, also described as the CLIC/GEEC pathway. Two other proteins of the group of small Rho GTPases, RhoA and Rac1, are involved in a dynamin-dependent pathway. Arf6 has been described to be involved in a clathrin-independent mechanism; however, its direct function may be related to recycling processes. Cargo that is internalized by these endocytic mechanisms enters early endosomes. In polarized epithelial cells (right) regulation of endocytic pathways is more complex as apical clathrin-independent endocytosis has been found to be regulated by a number of factors/signals that do not affect basolateral uptake. (From Sandvig, K. et al., *Curr. Opin. Cell Biol.*, 23, 413–420, 2011.)

Reorganisation of the actin cytoskeleton and endosomal trafficking

Actin filaments (**Figure 5.11**) are necessary for almost all movements made by a eukaryotic cell, especially movement involving the plasma membrane (**Figure 5.12**). The protein actin is an ATPase and exists as a free moment (G actin) and an actin filament consists of two intertwined actin molecules (F actin). Actin filaments have structural polarity with a plus and minus end and can extend from either end to lengthen the filament. In a resting cell,

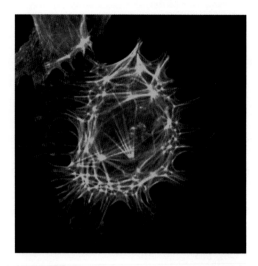

Figure 5.11 Actin filaments in a cell. (From https://mistertrack.wordpress.com/tag/jega/.)

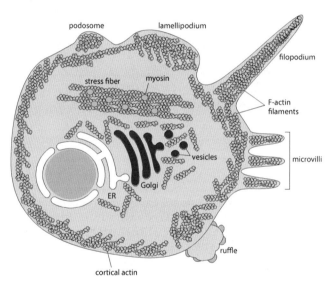

Figure 5.12 Actin filaments can be organized to produce a range of cellular extensions, including podosomes, lamellipodia, filopodia, microvilli and large membrane ruffles. Podosomes contain several actin-binding proteins, signalling molecules and metalloproteinases (black dots). ER, endoplasmic reticulum. (From Taylor, M.P. et al., *Nat. Rev. Microbiol.*, 9, 427–439, 2011.)

actin is concentrated in a mesh at the perimeter of the cell just under the cell membrane where it is known as cortical actin. There are a large number of actin-binding proteins that control the polymerisation of actin, some preventing it, whereas others promote it. The actin-binding proteins can be activated by the interaction of extracellular ligands with their cell membrane receptors. Such is the case in endocytosis. The signals generated by cell membrane receptors binding ligands are transduced to a group of monomeric GTP-binding proteins known as the Rho family that act as on/off switches. These Rho-GTPases are shown in **Figure 5.13**. Activation of different Rho GTPases drive polymerisation of actin to form different structures as shown in Figure 5.12.

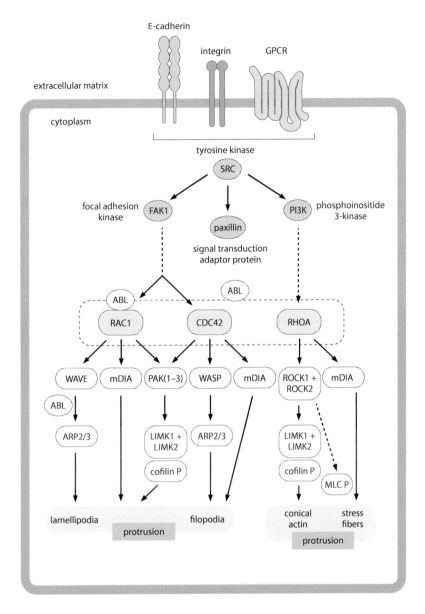

Figure 5.13 RHO-family GTPase-mediated modelling of the actin cytoskeleton. E-cadherins, integrins or guanylyl-nucleotide-binding protein (G protein)-coupled receptors (GPCRs) activate SRC (a tyrosine-protein kinase), which phosphorylates focal adhesion kinase 1 (FAK1). FAK1 promotes the formation of protrusive actin structures by activating RHO-family GTPases such as RAC1 and cell division cycle 42 (CDC42). The downstream effectors of RHO-family GTPases are: Wiscott–Aldrich syndrome protein (WASP)–WASP-family verprolin-homologous protein (WAVE) proteins, Diaphanous-related formins (mDIA proteins; also known as DRF or DIAPH proteins) and kinases such as PAKs and RHO-associated protein kinases (ROCKs). WASP–WAVE proteins stimulate the activation of the ARP2/3 complex. PAKs and ROCKs contribute to the formation of actin filaments by inactivating cofilin via phosphorylation of LIM domain kinases (LIMKs). mDIA stimulates the nucleation and extension of parallel actin filaments. RHOA is initially inactivated by integrin signalling via paxillin. Phosphoinositide 3-kinase (PI3K) is phosphorylated by SRC and then activates RHOA, leading to the formation of stress fibres. ABL tyrosine kinases negatively regulate the RHO–ROCK signalling pathway while activating RAC1 and WASP–WAVE. ROCK proteins also stimulate the phosphorylation of myosin regulatory light chain (MLC), thus contributing to the contractility of actin–myosin. Red 'T bars' show where viruses target the actin cytoskeleton. (From Taylor, M.P. et al., *Nat. Rev. Microbiol.*, 9, 427–439, 2011.)

After internalisation of a microbe by any of the types of endocytosis describe above the vesicle formed by the plasma membrane containing the microbe and its receptor enter a common endocytic network. After scission from the plasma membrane, the vesicle fuses with early endosomes that are located at the perimeter of the cell. These endosomes are called early or sorting endosomes. The maturation of early endosomes to late endosomes is associated with progressive acidification effected by the proton pump V-ATPase in the endosomal membrane. The terminal pH of the late endosome may reach as low as pH 4.5. In addition, the maturation of the endosome is marked by changes in the composition of its membrane. The role of the endosomal network is to direct the cargo within the endosome to various sites within the cell. For example, the cell membrane receptors and other membrane molecules that became pinched off to form the vesicle are returned quickly to the plasma membrane as recycling endosomes. Other endosomes are targeted to the trans-Golgi network and some are directed to fuse with lysosomes so that their contents can be degraded. Lysosomes are vesicles that contain acid hydrolases such as proteases, nucleases, glycosidases, lipases, phosphatases, sulfatases, and phospholipases. The endosome-lysosome system is shown in **Figure 5.14**.

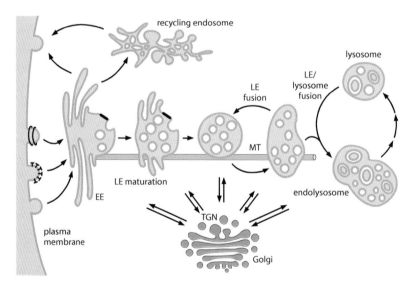

Figure 5.14 The endosome/lysosome system. The primary endocytic vesicles deliver their contents and their membrane to EEs in the peripheral cytoplasm. After a period of about 8–15 min during which the EEs accumulate cargo and support recycling to the plasma membrane (directly or via recycling endosomes in the perinuclear region), conversion of the Early endosome (EEs) to late endosome (LE) takes place. Thus, as the endosomes are moving towards the perinuclear space along microtubules (MT), the nascent LE are formed inheriting the vacuolar domains of the EE network. They carry a selected subset of endocytosed cargo from the EE, which they combine en route with newly synthesized lysosomal hydrolases and membrane components from the secretory pathway. They undergo homotypic fusion reactions, grow in size, and acquire more intra luminal vesicles (ILVs). Their role as feeder system is to deliver this mixture of endocytic and secretory components to lysosomes. To be able to do it, they continue to undergo a maturation process that prepares them for the encounter with lysosomes. The fusion of an endosome with a lysosome generates a transient hybrid organelle, the endolysosome, in which active degradation takes place. What follows is another maturation process; the endolysosome is converted to a classical dense lysosome, which constitutes a storage organelle for lysosomal hydrolases and membrane components. (From Huotari, J. and Helenius, A., *EMBO J.*, 30, 3481–3500, 2011.)

EXPLOITATION OF ENDOCYTOSIS PATHWAYS BY PATHOGENS

Bacteria

Pathogenic bacteria use two methods of actin cytoskeleton reorganisation to induce endocytosis by mucosal epithelial cells. These are called the zipper mechanism and the trigger mechanism. **Figure 5.15** Gram-negative bacteria that use either of these mechanisms employ a type 3 secretion system (T3SS), described below. Both mechanisms interfere with phosphoinositide (PI) metabolism and/or guanosine triphosphate hydrolase (GTPase) signalling. The zipper mechanism is also employed for cell entry by certain fungal pathogens (see below).

ZIPPER MECHANISM

Two bacterial pathogens that employ the zipper mechanism are species of the gram-negative genus *Yersinia* and the gram-positive rod, *Listeria monocytogenes*. The zipper mechanism uses bacterial adhesins that bind receptors in the host cell surface, typically integrins and cadherins that indirectly interact with the actin cytoskeleton (see Figure 13). Cadherins

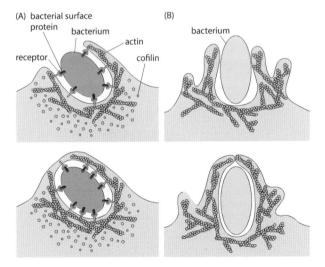

Figure 5.15 Zipper and trigger mechanisms for entry into host cells: (A) Zipper Mechanism: Bacterial surface proteins stimulate cell–cell and cell–matrix adhesion proteins to cause the membrane to engulf bacteria and then close in a zipper-like manner to internalise the pathogen. (B) Trigger Mechanism: Bacterial effector proteins secreted *via* the T3SS activate signalling pathways that modify the cytoskeleton underlying the membrane, creating membrane extrusions that engulf the pathogens. Cofilin is an actin-binding protein that disassembles actin filaments. (From How do bacteria modulate the host cytoskeleton? MBINFO, https://www.mechanobio.info/topics/pathogenesis/pathogen-subversion-of-host-cytoskeletal-machinery/.)

derive their name from the contractional 'calcium-dependent adhesion' and they play an important role in cell adhesion by forming adherens junctions that bind cells together. Cadherins are linked to the actin cytoskeleton by proteins called catenins. Integrins mediate cell-extracellular matrix (ECM) interactions with collagen, fibrinogen, fibronectin, and vitronectin. The α5β1 integrin is a common ligand targeted by pathogens. Some integrins are found only on certain surfaces of epithelium. For example, α1 integrins are restricted to the basolateral surface of enterocytes; however, they are expressed on the apical surface of M cells. Therefore, enteric pathogens can enter *via* the apical (luminal) surface of M cells, reach the basement membrane, move laterally and enter enterocytes *via* their basolateral surfaces.

The bacterial pathogens *Y. enterocolitica* and *Y. pseudotuberculosis* possess an adhesin called invasin that binds integrins with higher affinity than the natural ligands of integrin. The binding of integrins by invasin activates several small GTPases, including Rac1 and Arf 6. Rac1 is a part of the Rho family of master actin regulators that link membrane receptors to the cytoskeleton by regulating signal transduction pathways (see Figure 5.13). Arf6 (adenosine diphosphate [ADP]-ribosylation factor-6) is a member of the ADP ribosylation factor family of GTP-binding proteins that are involved in trafficking of biological membranes. The Rac1-Arf6 molecular complex recruits a kinase that modulates phosphatidylinositol metabolism such that actin is polymerised and remodelled to allow the epithelial cell membrane to flow over and around the bacterial cell, enclosing it in a vacuole. The interaction of invasin with integrin causes the accumulation of integrin receptors along the bacteria–host interface, thus appearing to zip *Yersinia* into the host cell membrane (**Figure 5.16**).

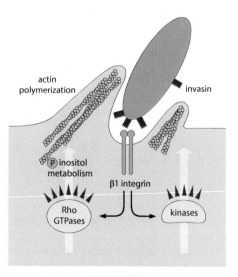

Figure 5.16 The *Yersinia* outer membrane invasin interacts with β1 integrin receptors leading to activation of the small Rho GTPase Rac1, which modulate the phosphatidylinositol metabolism to induce actin rearrangements at the site of bacterial entry, promoting internalization. (From Pizarro-Cerdá, J. and Cossart, P., *Cell*, 124, 715–727, 2006.)

The adhesins of *Listeria monocytogenes* involved in host cell entry are termed internalins and there are two, InIA and InIB (**Figure 5.17**). The receptors on the host cell surface for InIA and InIB are E-cadherin and the hepatocyte growth factor receptor, Met, respectively. The ligation of InIA to E-cadherin recruits α- and β-catenin. The binding of InIB to Met initiates the recruitment of phosphoinositol-3 kinase (PI3K), which activates Rac1 and the polymerisation of actin. Actin reorganisation involves two small GTPases of the Rho family, Rac1 and Cdc42, and cortactin, a cortical actin-binding protein.

Pathogenic gram-positive cocci enter epithelium and endothelium using a zipper-like mechanism similar to that described above. In this instance, fibronectin is bound by fibronectin-binding proteins on the bacterial surface and by α5β1 integrin on the host cell plasma membrane. Integrin-coupled intracellular signalling results in actin reorganisation and bacterial uptake. In the case of *Staphylococcus aureus*, the extracellular adherence protein (Eap) acts cooperatively with the staphylococcal fibronectin-binding proteins. In *Streptococcus pyogenes*, the fibronectin binding proteins implicated in endocytic bacterial cell uptake are M protein, FbaB and SfbI. Fibronectin-binding proteins are also involved in the uptake of *S. agalactiae* the Lancefield group B streptococcus. Another ECM molecule, vitronectin, serves as the bridge between vitronectin-binding proteins, such as the choline-binding protein PspC on the surface of *Streptococcus pneumoniae*,

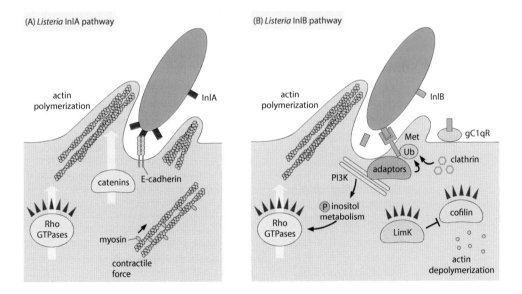

Figure 5.17 *Listeria* is taken up by target cells combining two molecular pathways. (A) In the InIA-dependent pathway, the bacterial protein InIA interacts with the cell adhesion molecule E-cadherin and promotes the subversion of cell adherens junction machinery (including β- and α-catenins) to induce entry. Actin polymerization is induced largely by the RhoGTPase Rac1. (B) In the InIB-dependent pathway of *Listeria*, the bacterial protein InIB interacts with the molecule gC1qR, and with the signaling receptor Met, which recruits several molecular adaptors, which recruits PI3K (involved in the activation of the RhoGTPase Rac1 and the polymerization of actin), and also the ubiquitination of Met and the endocytosis of the receptor via a clathrin-dependent mechanisms. (From Pizarro-Cerdá, J. and Cossart, P., *Cell*, 124, 715–727, 2006.)

and avβ3 integrin on the host cell surface. Uptake of pneumococci by host cells *via* vitronectin requires reorganisation of the actin cytoskeleton mediated by integrin-linked kinase (ILK), phosphatidylinositol 3-kinase (PI3K), and protein kinase B (Akt). *Neisseria gonorrhoeae*, a venereal pathogen, employs an outer membrane protein specific for heparan sulphate to bridge vitronectin bound to avβ5 or avβ3 integrins in its uptake by zippering.

Trigger mechanism

The trigger mechanism employs the T3SS. Secretion systems are nano-machines that function like micro syringes to allow certain pathogenic bacteria to inject proteins called effectors into the host cell cytoplasm or the cytoplasmic surface of the cell membrane. There are seven recognised types of secretion system (see **Table 5.1**).

The T3SSs, also known as injectosomes, and T4SSs are used by several gram-negative bacterial pathogens. The effectors have multiple functions in pathogenesis, one of which is to induce the uptake of the pathogen by host cells, notably mucosal epithelial cells, by endocytosis, but they can also be involved in paracytosis and in membrane pore formation. The T3SS injectisome shares an evolutionary origin with the flagellum. There are seven families of T3SS injectisomes (and they are listed in **Table 5.2**).

Two T3SS – T3SS-1 and T3SS-2 – are found in *Salmonella enterica* serovar Typhimurium where the genes encoding proteins involved in cell entry are found on a pathogenicity island termed SPI-1 (*Salmonella* pathogenicity island 1). Pathogenicity islands are a type of mobile genetic element that contain clusters of genes involved in pathogenesis. They may be found integrated into the chromosome or found on extrachromosomal elements such as plasmids and are acquired by horizontal gene transfer. T3SS-1 is involved in host cell entry. There are several effectors that form the T3SS-1: SipB and SipC create a pore in the host cell membrane through which effectors enter the host cell cytoplasm and drive actin polymerisation and bundling. Another Sip protein, SipA, prevents the polymerised actin from disassembling. SopE and SopE2 activate the GTPases Cdc42 and Rac1 and initiate formation of the host cell membrane pseudopods (ruffles) required for engulfment of bacteria. As is the case for all pathogens that enter cells, after engulfment, the host cell membrane rapidly resumes its normal contour indicating that the actin remodelling process is short lived. For *Salmonella,* it appears that continuity of the plasma membrane is restored by means of the activity of a bacterial effector, SptP, which hydrolyses GTP (**Figure 5.18**).

Shigella species are unable to invade the apical surface of enterocytes because the ligands for their adhesins are not expressed on this surface. However, they can enter through the basolateral surfaces of enterocytes

type	name	structure	sec dependence	example molecules secreted	bacteria	other information
type I	T1SS	• three component system consisting of ATPase binding cassette (ABC) transporter, a membrane fusion protein and a pore-forming outer membrane protein.	yes	• proteases • heme-binding proteins • proteins with repeats-in-toxins (RTX) motif	seen in most pathogenic gram negative bacteria	• simplest chaperone dependent system • substrate typically contains a C-terminal signal sequence that is recognised by the T1SS and remains uncleaved.
type II	T2SS	• large multi-protein machinery, made up of a number of distinct protein subunits known as the general secretory proteins (GSPs) set into four main components- outer membrane complex, the inner membrane complex, the secretion ATPase and the pseudopilus.	no	transport of multiple virulence factors	seen in most pathogenic gram negative bacteria	• two step system • proteins are first translocated across the inner membrane through the general secretion system Sec or Tat and then the proteins are folded in the periplasm and subsequently transported across the outer membrane through the T2SS
type III	T3SS	• structurally and genetically they are related to bacterial flagella. • Multi-protein structure organised into three components – Basal body, Inner rod and needle complex. • T3SS has a core of nine proteins that are highly conserved among all known systems, eight of these proteins is shared with the flagellar apparatus found in many bacteria and are evolutionarily related to flagellin	yes	specific cellular toxins	• Salmonella species • Shigella species • Yersinia pestis • Pathogenic Escherichia coli • Pseudomonas aeruginosa	• the hallmark of T3SS is the needle complex also called injectisomes when the ATPase is excluded. • bacterial proteins that need to be secreted pass from the bacterial cytoplasm through the needle directly into the host cytoplasm
type IV	T4SS	• the structure is homologous to conjugation machinery of bacteria. • structure is not well defined except for a subtype-IVa • structure consist of 11–13 core proteins that form a channel.	yes	• transporting both DNA and protein • CagA toxin by H.pylori	• Helicobacter pylori • Bordetella pertussis • Legionella pneumophila • Agrobacterium tumefaciens	• found in both Gram-negative and Gram-positive bacteria. • they can transport molecules in one step or two step.

(Continued)

type	name	structure	sec dependence	example molecules secreted	bacteria	other information
type V	T5SS	structure consists of an inner membrane and a β-barrel pore in the outer membrane.	no	• universal transport of several types of protein molecules	• seen in many different environmental and pathogenic bacteria.	• grouped into three main categories based on structural organisation 1. type Va-Autotranporters) 2. type Vb-Two partner system 3. type Vc-Oligomeric coiled-coil system • substrates to be transported contains a N terminal sec signal sequence.
type VI	T6SS	• large multi-protein machinery, made up of a 13 core proteins. • the basic structure consists of a cytosolic cylinder similar to tail of a bacteriophage and a membrane associated structure that anchors it to cytoplasm.	no	transporting both DNA and protein	• *Yersinia pestis* • *Salmonella species* • *E.coli* • *Rhizobium species*	• found in both Gram-negative and Gram-positive bacteria. • they have a possible evolutionary origin from bacteriophage.
type VII	T7SS	• nothing much is known. They are characterised by their ESX structure. • the predicted structure consists of a cytoplasmic membrane spanning channel made up of five subunits.	–	• not clear but required for *Staphylococcus aureus* virulence	• *S. aureus* • *Mycobacterium species*	• most recently identified structure. • seen predominantly in gram positive organisms

Table 5.1 Types 1–7 secretion systems (http://varuncnmicro.blogspot.com/2016/11/btb-11-bacterial-secretion-systems.html.)

injectisome families	species	description
chlamydiales	*Chlamydia trachomatis*	obligate intracellular human pathogen (trachoma, genital infections)
	Chlamydia pneumoniae	obligate intracellular human pathogen (acute respiratory disease)
Hrp1	*Enterobacter agglomerans*	pathogenic
	Vibrio parahaemolyticus	human pathogen (seafood-borne gastroenteritis)
Hrp2	*Burkholderia pseudomallei*	human pathogen (meloidosis)
SPI-1	*Salmonella enterica*	human pathogen (gastroenteritis)
	Shigella flexneri	human pathogen (dysenteria)
	Burkholderia pseudomallei	human pathogen (meloidosis)
	Chromobacterium violaceum	emerging human pathogen (evoking meloidosis)
	Yersinia enterocolltica	human pathogen (gastroenteritis, mesenteric adenitis)
SPI-2	*Escherichia coli* EPEC	human pathogen (gastroenteritis)
	Escherichia coli EHEC	human pathogen (ureamia, haemolysis)
	Salmonella enterica	human pathogen (gastroenteritis)
	Chromobacterium violaceum	emerging human pathogen (evoking meloidosis)
	Yersinia pestis	rodent and human pathogen (plague)
	Yersinia pseudotuberculosis	rodent and human pathogen
	Edwardsiella tarda	human pathogen (gastroenteritis)
Ysc	*Yersinia pestis*	rodent and human pathogen (plague)
	Yersinia pseudotuberculosis	rodent and human pathogen
	Yersiinia enterocolitica	human pathogen (gastroenteritis, mesenteric adenitis)
	Pseudomonas aeruginosa	animal, insect and human (cystic fibrosis, burned, immunocompromized patients) pathogen
	Vibrio parahemolyticus	human pathogen (seafood-borne gastroenteritis)
	Bordetella pertussis	human pathogen (whooping cough)

Table 5.2 Type 3 injectisome families. (From Song, I. and Cornelis, G.R., The type III secretion systems, in *Bacterial Secreted Proteins: Secretory Mechanisms and Role in Pathogenesis*, Wooldrige, K. (Ed.), Caister Academic Press, Norfolk, UK, 2009.)

or they can be taken up by microfold cells (M cells). M cells are part of the gut-associated lymphoid tissues that sample the luminal content of the bowel (see Figure 5.2). Within minutes of contact with the basolateral surface of an enterocyte, the *Shigella* T3SS delivers multiple effectors both into the host cell and around the bacterium itself. These effectors, encoded on a 220 kb plasmid, function similarly to those described for the *Salmonella* T3SS above. The *Shigella* effectors involved in cell entry include the invasion plasmid antigens IpaA, IpaB1, and IpC; the invasion plasmid gene IpgD; and the cysteine protease VirA. IpaC moves from the host cell cytoplasm and integrates into the host cell membrane where it induces the polymerisation of actin. VirA indirectly stimulates the Rho GTPase, Rac1, favouring actin polymerisation, and IpgD affects phosphoinositol metabolism and promotes membrane ruffling. IpaA activates vinculin, inducing actin depolymerisation and recovery of the plasma membrane after bacterial entry. The mechanism of *Shigella* uptake is shown in **Figure 5.19**.

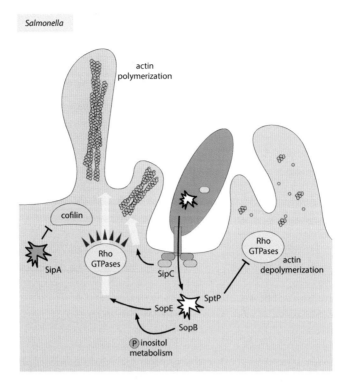

Figure 5.18 The *Salmonella* T3SS-1: SipC drives actin polymerisation and actin-filament bundling; SopE activates Rho GTPases, initiating actin polymerisation and membrane ruffle formation; SopB modulates inositol-polyphosphate metabolism, activating indirectly the same Rho GTPases as SopE; and SipA blocks the actin depolymerisation factor cofilin also, favouring membrane ruffle formation. SptP plays a role once internalisation has taken place, inactivating the Rho GTPases, inhibiting actin polymerisation, and helping the closure of the plasma membrane over the internalised bacteria. (From Pizarro-Cerdá, J. and Cossart, P., *Cell*, 124, 715–727, 2006.)

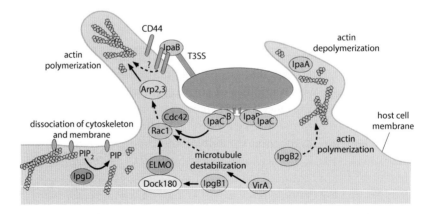

Figure 5.19 *S. flexneri* type III-secreted effectors. Injection of the Mxi-Spa-secreted effectors IpaC, IpgB1, and VirA by *S. flexneri* induces Rac1/Cdc42-dependent actin polymerization and the formation of large membrane ruffles. The binding of IpaB to the CD44 receptor and the activity of IpgB2 might also trigger cytoskeleton remodeling or membrane ruffling, respectively. The phosphoinositide 4-phosphatase IpgD promotes the disconnection of the actin cytoskeleton from the cytoplasmic membrane, thus facilitating the structural reorganization of the entry site. IpaA mediates the localized depolymerization of actin, which is required to close the phagocytic cup. PIP2, phosphatidylinositol-4,5-biphosphate; PIP, phosphatidylinositol-5-phosphate. (From Schroeder, G.N. and Hilbi, H., *Clin. Microbiol. Rev.*, 21, 134–156, 2008.)

pathotype	diseases	symptoms	virulence factors
Enteric E. coli			
EnteroPathogenic *E. coli* (EPEC)	diarrhoea in children	watery diarrhoea and vomiting	Bfp, Intimin, LEE
EnteroHaemorrhagic *E. coli* (EHEC)	haemorrhagic colitis, HUS	bloody diarrhoea	shiga toxins, intimin, Bfp
EnteroToxigenic *E. coli* (ETEC)	traveler's diarrhoea	watery diarrhoea and vomiting	heat-labile and sheat-stable toxins, CFAs
EnteroAggregative *E. coli* (EAEC)	diarrhoea in children	diarrhoea with mucus and vomiting	AAFs, cytotoxins
Diffusely Adherent *E. coli* (DAEC)	acute diarrhoea in children	watery diarrhoea, recurring UTI	Daa, AIDA
EnteroInvasive *E. coli* (EIEC)	shigellosis-like	watery diarrhoea; dysentery	shiga toxin, hemolysin, cellular invasion, Ipa
Adherent Invasive *E. coli* (AIEC)	associated with Crohn disease	persistent intestinal inflammation	type 1 fimbriae, cellular invasion
Extraintestinal E. coli (*ExPEC*)			
UroPathogenic *E. coli* (UPEC)	lower UTI and systemic infections	cystitis, pyelonephritis	type 1 and P fimbriae; AAFs, hemolysin
Neonatal Meningitis *E. coli* (NMEC)	neonatal meningitis	acute meningitis, sepsi	S fimbrie; K1 capsule

Table 5.3 *Escherichia coli* pathogenic types. Bfp, Bundle-forming pili; LEE, Locus for enterocyte effacement; HUS, haemolytic-uraemic syndrome; CPA, colonization factor antigen; AAF, aggregative adherence fimbria; Daa, diffuse adhesin; A1DA, adhesin involved in diffuse adherence; Ipa, Invasion plasmid antigen. (From Allocati, N. et al., *Int. J. Environ. Res. Public Health*, 10, 6235–6254, 2013.)

Pathogenic *E. coli* can be divided into eight pathovars as shown in the **Table 5.3**. However, all of the *E. coli* pathovars use a T3SS to enter mucosal epithelium, although there are some differences between the pathovars in the initial events involved in their entry into host cells. Host cell entry by both enteropathogenic *Escherichia coli* (EPEC) and enterohaemorrhagic *Escherichia coli* (EHEC) involves the formation of attaching and effacing (A/E) lesions. An A/E lesion is marked by localised destruction of the microvilli of enterocytes and the attachment of bacteria to the denuded apical enterocyte membrane, often in a cup-like pedestal structure. Supporting the pedestal is a dense pillar of actin. For EPEC, initial adhesion to the enterocyte is mediated likely by bundle-forming and type-I pili. Adhesion of EHEC is likely mediated by type-I and type-IV pili and perhaps by flagella. The mechanism of pedestal formation induced by EPEC and EHEC is very similar. The T3SS effector EspB inhibits the interaction of myosins with actin, resulting in the destruction of the microvilli (effacement). The effector Tir (translocated intimin receptor) enters the host cell cytoplasm and becomes displayed on the cell membrane surface where it serves as the ligand for the *E. coli* outer membrane adhesin intimin to facilitate tight adhesion. The binding of Tir by intimin results in its clustering (**Figure 5.20a**). The clustered Tir initiates actin reorganisation to create a finger-like projection of the host cell cytoplasmic membrane called a pedestal to which *E. coli* is tethered (**Figure 5.20b**). The EPEC and EHEC T3SS is encoded by genes on a 35 kb pathogenicity island named the locus of enterocyte effacement (LEE).

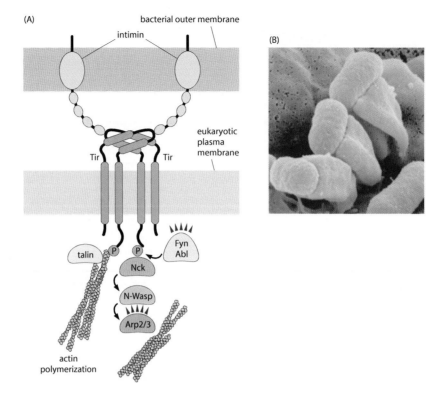

Figure 5.20 Enterpathogenic *E. coli* (EPEC) entry into cells. Tir/intimin interaction. (A) EPEC, *via* its TTSS, injects the protein Tir into the cytosol of target cells, where it integrates into the host-cell plasma membrane, dimerises, and functions as a receptor for the bacterial outer membrane intimin. Tir/intimin interaction promotes Tir phosphorylation by Fyn and Abl, inducing the recruitment of the protein adaptor Nck, which in turn recruits N-WASP and the Arp2/3 complex, leading to actin polymerisation and the formation of structures known as pedestals. Actin binding proteins such as talin are recruited to the pedestal, stabilising the structure. (B) Scanning electron micrograph of EPEC perched on top of pedestals on HeLa cells. (Reprinted with permission from Pizarro-Cerdá, J. and Cossart, P., *Cell*, 124(4), 715–727, 2006.)

Uropathogenic *E. coli* (UPEC) can invade a number of host cell types in addition to the terminally differentiated superficial facet cells of the bladder epithelium. The entry of UPEC into bladder epithelial cells is mediated by type 1 pili, which are terminated with the FimH adhesin that bind mannose saccharides onto uroplakin and $\alpha3\beta1$ integrin complexes. Uroplakins (UPIa, UPIb, UPII and UPIII) are glycoproteins presented on the surface of uroepithelium. Receptor-ligand binding results in actin clustering at the sites of these interactions. Bacterial uptake as a result of actin reorganisation is mediated by the action of kinases and Rho-family GTPases.

Type 4 secretion systems: There are three types of T4SS. One of the three types is used to transfer toxic effector proteins or protein complexes into the cytoplasm of the host cell. Among pathogenic bacteria employing a

T4SS are *Helicobacter pylori, Bordetella pertussis, Legionella pneumophila, Brucella* species and *Bartonella* species. In the case of *H. pylori*, one of the functions of its T4SS is to inject effectors that target the tight junctions between gastric epithelial cells to facilitate bacterial entry by paracytosis. The T4SS is encoded by the cag pathogenicity island and forms a needle-like device to deliver the oncoprotein CagA into gastric epithelial cells. Inside the cell, CagA is phosphorylated by the cellular Src and Abl kinases. Phosphorylated CagA is targeted to the apical junctional complex of gastric epithelial cells. Different domains of CagA are able to inhibit cell adhesion, cell polarity, and cell migration. The T4SS of *H. pylori* is shown in **Figure 5.21**.

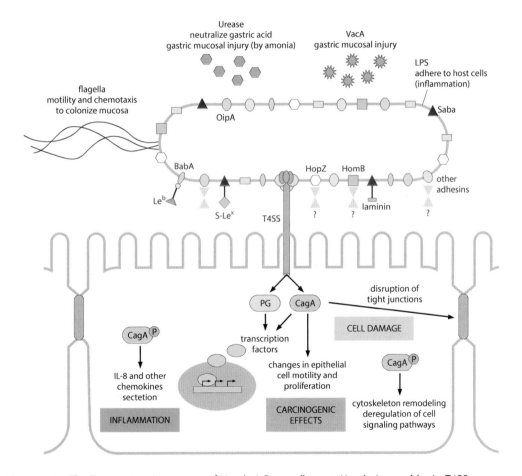

Figure 5.21 The Type 4 secretion system of *H. pylori*. Once adherent, *H. pylori* assembles its T4SS, a needle-like device, at its surface enabling the translocation of bacterial effector proteins into the cytoplasm of host gastric epithelial cells, such as the CagA oncoprotein and peptidoglycan. The intracellular delivery of CagA activates numerous host signalling pathways, inducing epithelial responses with carcinogenic potential. Le^b, fucosylated Lewis b blood group antigen; LPS, lipopolysaccharide; P, phosphorylated; PG, peptidoglycan; S-Le^x, sialyl-dimeric-Lewis X glycosphingolipid. (From Oleastro, M. and Ménard, A., *Biology*, 2, 1110–1134, 2013, Figure 3.)

Viruses

Endocytosis is the mechanism by which many enveloped and non-enveloped viruses enter cells. In some cases, they do so by binding to common cell membrane receptors such as integrins, heparan sulphate and other proteoglycans or glycolipids. However, in most cases, the interaction of the virus with its cell membrane receptor is specific and this interaction determines which cells are permissive for virus uptake. Some viruses display lectins on their envelope that bind sialic acid or the sugar moieties of glycolipids. A list of viral receptors on epithelial cells is shown in **Table 5.4**. All of the endocytic pathways are utilised by enveloped viruses for cell entry and all involve actin reorganisation. Clathrin-mediated endocytosis is the most common mechanism of virus uptake, and this involves both existing clathrin-coated domains in the host cell plasma membrane and clathrin assembly at the point of contact between the virus particle and the cell membrane. Several viruses are taken up by macropinocytosis, a process used by the cell to take up fluid into vacuoles. Other viruses use caveolar-lipid raft endocytosis as a method of entry by binding integrins, or saccharides on gangliosides in the host cell membrane. Lymphocytic choriomeningitis virus (LCMV), human papilloma virus type 16 (HPV16) and influenza virus all enter cells by crosslinking various molecules in the host cell plasma membrane that act as receptors to induce formation of a vesicle. This method of entry does not require clathrin or caveolins. The various types of endocytosis used by viruses are shown in **Figure 5.22**.

Fungi

Unfortunately, relatively little is known about the pathogenesis of the fungi that infect humans. They are, as a whole, opportunists that take advantage of reduced host resistance. The genus *Candida* and the species *C. albicans* is the principal human pathogen, so most is known about this organism. *Candida albicans* is dimorphic and can switch between a yeast form and a hyphal form. Hyphae are the invasive form of the fungus. The fungus *Candida albicans* can enter host epithelial and endothelial cells by two mechanisms. The first of these is by force, using destructive proteases, and this mechanism has been discussed earlier in the chapter. In addition, *C. albicans* is taken up by clathrin-mediated endocytosis. This process is mediated by the *Candida* invasins Als3 (agglutinin-like sequence 3) and Ssa1 (a heat shock protein) that bind to plasma membrane cadherins. Internalisation by epithelium involves binding of the fungus to E-cadherin and epidermal growth factor receptor (EGFR) and the related protein human epidermal growth factor receptor (HER2). Several fungal pathogens can infect the lungs following inhalation and are thought to be taken up by respiratory epithelium by endocytosis. However, the fungal surface molecules and epithelial cell membrane receptors have not been identified. For example, *Cryptococcus neoformans* is known to be uptaken by pulmonary epithelial cells in culture, and *Aspergillus fumigatus* is uptaken by type II pneumocytes and

virus	family	characteristics	epithelial tropism	attachment carbohydrate*	protein*
human					
herpes simplex virus (HSV-1, -2)	Herpesviridae-α	enveloped dsDNA	retinal pigment epithelial cell, cornea	HSPG	nectin 1, HVEM
varicella-zoster virus	Herpesviridae-α	enveloped dsDNA	gastrointestinal tract, retinal pigment epithelial cell	HSPG	Man6-P/IGFII-R, nectin 1
human cytomegalovirus (HCMV)	Herpesviridae-β	enveloped dsDNA	retinal pigment epithelial cell	HSPG	–
Epstein-Barr virus (EBV)	Herpesviridae-γ	enveloped dsDNA	–	–	CR2 (CD21), poly Ig-receptor
vaccinia virus	Poxviridae	enveloped dsDNA	rhinopharynx, skin	–	–
human immunodeficiency virus 1 (HIV-1)	Retroviridae	enveloped ssDNA	gastrointestinal and genital tracts	galactosylceramide	CCR5
respiratory syncytial virus (RSV)	Paramixoviridae	enveloped ssDNA	pulmonary and respiratory tracts	HSPG	ICAM1, VLDLR
sendai virus	Paramixoviridae	enveloped ssRNA	bronchial tract, upper airway	sialyloligosaccharide	–
measles virus	Paramixoviridae	enveloped ssRNA	respiratory tract	–	CD46, CD46/moesin
Black Creek canalvirus	Bunyaviridae	enveloped ssRNA	pulmonary tract	–	integrin-β₃
influenza virus	Orthomyxoviridae	enveloped ssRNA	bronchial epithelium	sialyloligosaccharide	–
vesicular stomatitis virus (VSV)	Rhabdoviridae	enveloped ssRNA	bronchial epithelium	GlcNAc	–
rotavirus	Reoviridae	naked dsRNA	intestinal tract	–	integrins-α₂β₁, -α₄β₁, -α₄β₃
reovirus-1 and -3	Reoviridae	naked dsRNA	intestinal tract including M cells	sialyloligosaccharide	tight-junction-associated protein
human papilloma virus (HPV)	Papillomaviridae	naked dsDNA	mucosa, oesophagus, skin	HSPG	–
adeno-associated virus (AAV)	Parvoviridae	naked ssDNA	airway	HSPG	–
Jamestown Canyon (JC) virus	Polyomaviridae	naked dsDNA	colorectal tract, neuroepithelial cells	sialyloligosaccharide	–
adenovirus	Adenoviridae	naked dsDNA	airway, ocular and gastrointestinal tracts	HSPG	integrin-α,β₅, CAR
coxsackievirus	Picornaviridae	naked ssRNA	airway	–	CAR
poliovirus	Picornaviridae	naked ssRNA	gastrointestinal tract	–	Ab D171, PVR, PRR1,2
rhinovirus major group	Picornaviridae	naked ssRNA	respiratory tract	–	ICAM1
rhinovirus minor group	Picornaviridae	naked ssRNA	respiratory tract	–	LDLR family
echovirus and human parechovirus (HPeV)	Picornaviridae	naked ssRNA	intestinal tract	DAF	integrin-α₂β₁

Table 5.4 Viral receptors on epithelial cells. *Together the attachment carbohydrate and the protein form the epithelial receptor. CAR, coxsackie adenovirus receptor; CFA, carcino-embryonic antigen; DAF, decay acceleration factor; ds, double stranded; HLA-1, human leukocyte antigen 1; HSPG, heparan sulphate proteoglycan; ICAM1, intracellular adhesion molecule 1; Ig, immunoglobulin; LDLR, low density lipoprotein receptor; Man6-P/IGFII-R, mannose 6-phosphate/insulin-like growth factor receptor: PRR, polio virus related receptor; PVR, polio virus protein receptor; ss, single stranded; VLDLR, very low density lipoprotein receptor. (From Bomsel, M. and Alfsen, A., Nat. Rev. Mol. Cell Biol., 4, 57–68, 2003.)

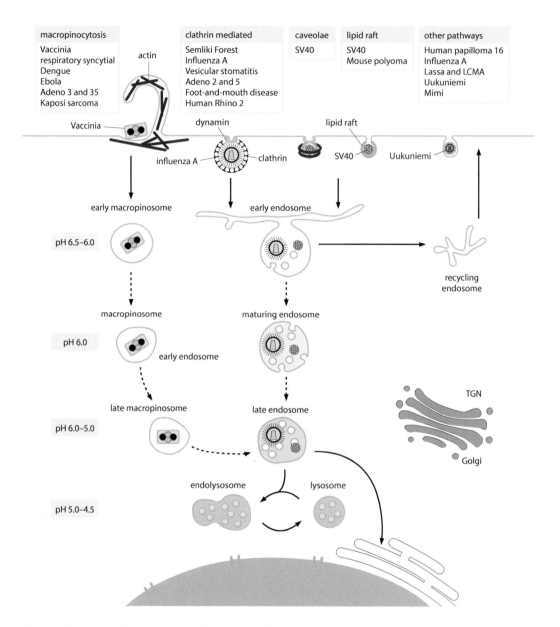

Figure 5.22 Endocytic pathways used by viruses. (From Pascale Cossart, P. and Helenius, A., *Cold Spring Harb. Perspect. Biol.*, 6, a016972, 2014.)

tracheal epithelial cells by endocytosis, thought to involve a zipper-like mechanism. Human fungal pathogens that are believed to be up-taken by epithelial cells are shown in **Table 5.5**.

MICROTUBULE REORGANISATION

Microtubules are heterodimers of two proteins, α and β tubulin. They are an important component of the cytoskeleton and function to modulate actin (see **Figure 5.23**). Microtubules are also involved in organelle

organism	target host cell	fungal structure that induces invasion	host cell receptor	mechanism	microfilaments required	microtubules required	result of invasion
C. albicans	epithelial cells	unknown	unknown	zipper	yes	unknown	host cell damage, fungal egress.
	endothelial cells	Als3p expressed by hyphae	N-cadherin and other proteins	zipper	yes	yes	host cell damage, fungal egress.
Cr. neoformans	epithelial cells	capsule (GXM)	unknown	unknown			host cell damage, fungal egress.
	brain endothelial cells	capsule (GXM)	unknown	membrane ruffling	yes	unknown	fungal egress (transcytosis), minimal host cell damage.
A. fumigatus	epithelial cells	conidia	unknown	zipper	yes	yes	delayed fungal germination, host cell damage, inhibition of apoptosis.
	endothelial cells	hyphae	unknown	zipper	yes	unknown	host cell damage, fungal egress.
R. oryzae	endothelial cells	hyphae	unknown	unknown	yes	unknown	host cell damage, fungal egress.
S. schenckii	endothelial cells	yeast	90-kDa and 135-kDa proteins	unknown	unknown	unknown	intracellular persistence of the organisms, no host cell damage
H. capsulatum	epithelial cells	yeast	unknown	unknown	unknown	unknown	some organisms persist, others are killed
P. brasiliensis	epithelial cells	30-kDa and 43-kDa proteins expressed by yeast	unknown	unknown	yes	yes	intracellular persistence of the organisms, induction of host cell apoptosis

Table 5.5 Fungal uptake by epithelial and endothelial cells. (From Filler, S.G. and Sheppard, D.C., *PloS Pathog.*, 2006, doi:10.1371/journal.ppat.0020129.)

microtubule network   nucleus

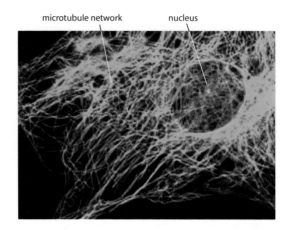

Figure 5.23 The microtubule network. (From https://micro.magnet.fsu.edu/cells/microtubules/microtubules.html.)

movement and segregation of chromosomes during mitosis. It is likely that microtubules are involved in almost all mechanisms of pathogen uptake where the reorganisation of the actin cytoskeleton plays the major role. In this section, we will limit discussion to those microorganisms in which microtubule reorganisation plays a part in uptake by epithelial cells. The role of microtubules in the pathogenesis of intracellular pathogens will be considered in Chapter 8. As we have seen earlier in this chapter, injectisome (T3SS and T4SS) modulation of the actin cytoskeleton by gram-negative bacteria is achieved by secretion of effectors that act *via* the small GTPases in the cytoplasm of the host cell. In contrast, manipulation of microtubules is achieved by interaction of effectors with microtubule subunits directly or by recruiting accessory proteins involved in the microtubule system. Frequently, microtubule reorganisation occurs simultaneously with reorganisation of the actin cytoskeleton and is mediated by certain of the effectors injected by the injectisome. In order to reorganize the actin cytoskeleton to allow the formation of lamellipodia and pseudopodia for their uptake pathogens destabilize microtubules.

Shigella flexneri: The VirA effector protein of *Shigella flexneri* has two recognised functions. One function is the activation of the small GTPase Rac1, which induces membrane ruffles and lamellipodia central to the entry of the bacterium into the host cell. In addition, VirA destabilises the host cell microtubule network by creating a cytosolic tunnel through which the bacterium travels following uptake. VirA inhibits the polarisation of microtubules and induces their depolymerisation. The mechanism by which it accomplishes this remains unclear, but VirA is a cysteine protease that can degrade the α-subunit of the tubulin heterodimer.

Enteropathogenic *Escherichia coli*: As described earlier in the chapter EPEC adhere to enterocytes and induce the formation of attaching and effacing lesions. Among the effectors injected by its T3SS are EspG and EspG2 which destabilise microtubules below the site of adhesion. Similar to VirA, EspG and EspG2 inhibit microtubule polymerisation and destabilise microtubule polymerisation. It has been proposed that the mechanism of EspG/EspG2-mediated microtubule destruction is proteolysis akin to VirA. VirA is closely related to EspG/EspG2, but no proteolytic activity of EspG/EspG2 has been detected. The function of EspG/EspG2 may be to increase permeability of the tight junctions between enterocytes. This may be the result of the effect of microtubule destruction on actin reorganisation.

TRANSCYTOSIS

There are three pathways that endosomes can follow once they are pinched off from the cell membrane. In the case of epithelial cells, endocytosis occurs on the apical surface of the plasma membrane. One pathway that endosomes may follow is to return to the apical cell membrane. This pathway is called the recycling pathway. The second pathway involves endosomes moving deeper into the cytosol to fuse with lysosomes where their cargo is degraded by hydrolytic enzymes. In the third pathway, endosomes can traffic from the apical surface to the basolateral surface of the cell, a process termed transcytosis. Transcytosis from the basolateral surface to the apical surface also occurs. The M cell that was discussed earlier in this chapter is an example of a cell that is dedicated to the process of transcytosis. Transcytosis by M cells is the mechanism by which *Salmonella* Typhimurium and *Shigella flexneri* access the lamina propria. *Neisseria meningitidis*, *N. gonorrhoeae* and *Listeria moncytogenes* have been shown to cross epithelial cells to gain access to the lamina propria by transcytosis. *N. meningitidis* crosses non-ciliated respiratory epithelium, and *N. gonorrhoeae* the urethral epithelium, by a transcellular route without disrupting the barrier function of the cell. This appears to be mediated by the actin cytoskeleton and the microtubule network. The IgA1 protease secreted by these bacteria appears to prevent the vacuole containing the bacteria from fusing with lysosomes by cleaving the lysosome-associated membrane protein 1 (LAMP-1), which is essential for endosome-lysosome fusion. In addition to its ability to escape from the phagosome using the pore-forming enzyme listeriolysin (LLO), *Listeria monocytogenes* appears to be able to traverse villus epithelial cells by transcytosis using the microtubule system. Transcytosis is dependent on internalin A (InlA) but not on LLO and the actin assembly-inducing protein (ActA). Several viruses such as HIV-1 and EBV employ transcytosis to cross the epithelial barrier.

KEY CONCEPTS

- With few exceptions, human pathogens must cross the barrier epithelia to invade deeper tissues and organs.

- The vast majority of pathogens cross mucosae rather than skin because mucosae are less formidable barriers.

- Respiratory epithelia are the most common target of human pathogens, followed by the intestinal mucosa.

- Subversion of the various forms of endocytosis is the most common means of cell entry.

- Endocytosis is initiated by the binding of molecules on the surface of the microbe by receptors (complementary molecules) on the cell membrane of host cells.

- Entry into host epithelial cells almost always involves remodelling of the actin cytoskeleton and probably the microtubule system.

- Certain gram-negative bacteria employ nanosyringes called type 3 and type 4 secretion systems to inject protein effectors into the host cell cytoplasm where they target small Rho GTPases that are central to the control of the actin cytoskeleton.

- Parasites have highly sophisticated mechanisms of cell entry, including polar tube formation and moving junction.

BIBLIOGRAPHY

Aktories K. 2011. Bacterial protein toxins that modify host regulatory GTPases. *Nature Reviews Microbiology*, 9(7): 487–498.

Alexander EH, Hudson MC. 2001. Factors influencing the internalization of *Staphylococcus aureus* and impacts on the course of infections in humans. *Applied Microbiology and Biotechnology*, 56: 361–366.

Allocati N et al. 2013. *Escherichia coli* in Europe: An overview. *International Journal of Environmental Research and Public Health*, 10: 6235–6254.

Backert S et al. 2011. Molecular mechanisms of gastric epithelial cell adhesion and injection of CagA by *Helicobacter pylori*. *Cell Communication and Signaling*, 9: 28.

Barocchi MA et al. 2012. Cell entry machines: A common theme in nature? *Nature Reviews Microbiology*, 3(4): 349–358.

Baum J et al. 2008. Host-cell invasion by malaria parasites: Insights from *Plasmodium* and *Toxoplasma*. *Trends in Parasitology*, 24(12): 557–563.

Bomsel M, Alfsen A. 2003. Entry of viruses through the epithelial barrier: Pathogenic trickery. *Nature Reviews Molecular Cell Biology*, 4(1): 57–68.

Cianfanelli FR et al. 2016. Aim, load, fire: The type VI secretion system, a bacterial nanoweapon. *Trends in Microbiology*, 24(1): 51–62.

Clementi CF, Murphy TF. 2011. Non-typeable *Haemophilus influenzae* invasion and persistence in the human respiratory tract. *Frontiers in Cellular and Infection Microbiology*, 1: Article 1.

Collins A et al. 2011. Structural organization of the actin cytoskeleton at sites of clathrin-mediated endocytosis. *Current Biology*, 21(14): 1167–1175.

Cossart P, Helenius A. 2014. Endocytosis of viruses and bacteria. *Cold Spring Harbor Perspectives in Biology*, 6: a016972.

Cossart P, Sansonetti PJ. 2004. Bacterial invasion: The paradigms of enteroinvasive pathogens. *Science*, 304: 242–248.

Cowman AF, Crab BS. 2006. Invasion of red blood cells by malaria parasites. *Cell*, 124: 755–766.

Dando SJ et al. 2014. Pathogens penetrating the central nervous system: infection pathways and the cellular and molecular mechanisms of invasion. *Clinical Microbiology Reviews*, 27(4): 691–726.

Doran KS et al. 2013. Concepts and mechanisms: Crossing epithelial barriers. *Cold Spring Harbor Perspectives in Medicine*, 3(7): a010090.

Elkin SR et al. 2016. Endocytic pathways and endosomal trafficking: A primer. *Wiener Medizinische Wochenschrift*, 166(7–8): 196–204.

Feldmesser M et al. 2001. Intracellular parasitism of macrophages by *Cryptococcus neoformans*. *Trends in Microbiology*, 9(6): 273–278.

Filler SG, Sheppard DC. 2006. Fungal invasion of normally non-phagocytic host cells. *PLOS Pathogens*, 2(12): e129.

Flannagan RS et al. 2009. Antimicrobial mechanisms of phagocytes and bacterial evasion strategies. *Nature Reviews Microbiology*, 7(5): 355–366.

Fredlund J, Enninga J. 2014. Cytoplasmic access by intracellular bacterial pathogens. *Trends in Microbiology*, 22(3): 128–137.

Garzoni C, Kelley WL. 2009. *Staphylococcus aureus*: New evidence for intracellular persistence. *Trends in Microbiology*, 17(2): 59–65.

Goebel W, Gross R. 2001. Intracellular survival strategies of mutualistic and parasitic prokaryotes. *Trends in Microbiology*, 9(6): 267–273.

Gonzalez V et al. 2009. Host cell entry by apicomplexa parasites requires actin Ppolymerization in the host cell. *Cell Host & Microbe*, 5: 259–272.

Gouin E et al. 2005. Actin-based motility of intracellular pathogens. *Current Opinion in Microbiology*, 8: 35–45.

Gow NAR et al. 2012. *Candida albicans* morphogenesis and host defence: Discriminating invasion from colonization. *Nature Reviews Microbiology*, 10: 112–122.

Green ER, Mecsas J. 2016. Bacterial secretion systems—An overview. *Microbiology Spectrum*, 4(1). doi:10.1128/microbiolspec.VMBF-0012-2015.

Guttman JA, Finlay BB. 2009. Tight junctions as targets of infectious agents. *Biochimica et Biophysica Acta*, 1788: 832–841.

Ham H et al. 2011. Manipulation of host membranes by bacterial effectors. *Nature Reviews Microbiology*, 9(9): 635–646.

Helenius A. 2011. Endosome maturation. *The EMBO Journal*, 30: 3481–3500.

Houben ENG et al. 2014. Take five—Type VII secretion systems of mycobacteria. *Biochimica et Biophysica Acta*, 1843: 1707–1716.

Huotari J, Helenius A. 2011. Endosome maturation. *The EMBO Journal*, 30: 3481–3500.

Hybiske K, Stephens RS. 2008. Exit strategies of intracellular pathogens. *Nature Reviews Microbiology*, 6(2): 99–110.

Kaksonen M, Roux A. 2018. Mechanisms of clathrin-mediated endocytosis. *Nature Reviews Molecular Cell Biology*, 19(5): 313–326.

Kats LM et al. 2008. Protein trafficking to apical organelles of malaria parasites—Building an invasion machine. *Traffic*, 9: 176–186.

Katze MG, Korth MJ, Law GL, Nathanson N. 2016. *Viral Pathogenesis: From Basics to Systems Biology*. 3rd ed. Academic Press, Elsevier, London, UK.

Keeling PJ, Fast NM. 2002. Microsporidia: Biology and evolution of highly reduced intracellular parasites. *Annual Reviews of Microbiology*, 56: 93–116.

Kopecko DJ et al. 2001. *Campylobacter jejuni*—Microtubule-dependent invasion. *Trends in Microbiology*, 9(8): 389–396.

Kreikemeyer B et al. 2004. The intracellular status of *Streptococcus pyogenes*: Role of extracellular matrix-binding proteins and their regulation. *International Journal of Medical Microbiology*, 294: 177–188.

Laskay T et al. 2003. Neutrophil granulocytes—Trojan horses for *Leishmania major* and other intracellular microbes? *Trends in Microbiology*, 11(5): 210–214.

Lim JP, Gleeson PA. 2011. Macropinocytosis: An endocytic pathway for internalising large gulps. *Immunology and Cell Biology*, 89: 836–843.

Liss V, Hensel M. 2015. Take the tube: Remodelling of the endosomal system by intracellular *Salmonella enterica*. *Cellular Microbiology*, 17(5): 639–647.

Loker ES, Hofkin BV. 2015. *Parasitology: A Conceptual Approach*. Garland Science, New York.

Mabbott NA et al. 2013. Microfold (M) cells: Important immunosurveillance posts in the intestinal epithelium. *Mucosal Immunology*, 6: 666–677.

Maza PK et al. 2017. *Candida albicans*: The ability to invade epithelial cells and survive under oxidative stress is unlinked to hyphal length. *Frontiers in Microbiology*, 8: Article 1235.

McMahon HT, Boucrot E. 2011. Molecular mechanism and physiological functions of clathrin-mediated endocytosis. *Nature Reviews Molecular Cell Biology*, 12(8): 517–533.

Menard R et al. 1996. Bacterial entry into epithelial cells: The paradigm of *Shigella*. *Trends in Microbiology*, 4(6): 220–226.

Mercer J, Helenius A. 2009. Virus entry by macropinocytosis. *Nature Cell Biology*, 11(5): 510–520.

Mostowy S, Cossart P. 2012. Septins: The fourth component of the cytoskeleton. *Nature Reviews Molecular Cell Biology*, 13(3):183–194.

Nash AA, Dalziel RG, Fitzgerald JR. 2015. *Mim's Pathogenesis of Infectious Disease*. 6th ed. Elsevier, London, UK.

Nathanson N. 2007. *Viral Pathogenesis and Immunity*. 2nd ed. Academic Press, Elsevier, London, UK.

Nowak SA, Chou T. 2009. Mechanisms of receptor/coreceptor-mediated entry of enveloped viruses. *Biophysical Journal*, 96: 2624–2636.

Ogawa M et al. 2008. The versatility of *Shigella* effectors. *Nature Reviews Microbiology*, 6: 11–16.

Oleastro M, Ménard A. 2013. The role of *Helicobacter pylori* outer membrane proteins in adherence and pathogenesis. *Biology*, 2: 1110–1134.

Pascale Cossart P, Helenius A. 2014. Endocytosis of Viruses and Bacteria. *Cold Spring Harbor Perspectives in Biology*, 6: a016972.

Pelkmans L, Helenius A. 2002. Endocytosis via caveolae. *Traffic*, 3: 311–320.

Pizarro-Cerdá J, Cossart P. 2006. Bacterial adhesion and entry into host cells. *Cell*, 124: 715–727.

Pöhlmann S, Simmons G (Eds.). 2013. *Viral Entry into Host Cells* (Advances in Experimental medicine and Biology 790). Landes Bioscience, Austin, TX.

Popova NV et al. 2013. Clathrin-mediated endocytosis and adaptor proteins. *Acta Naturae*, 5(3 (18)): 62–73.

Radhakrishnan GK, Splitter GA. 2012. Modulation of host microtubule dynamics by pathogenic bacteria. *Biomolecular Concepts*, 3(6): 571–580.

Ray K et al. 2009. Life on the inside: The intracellular lifestyle of cytosolic bacteria. *Nature Reviews Microbiology*, 7(5): 333–340.

Reis RS, Horn F. 2010. Enteropathogenic *Escherichia coli*, *Samonella*, *Shigella* and *Yersinia*: Cellular aspects of host-bacteria interactions in enteric diseases. *Gut Pathogens*, 2: 8.

Sandvig K et al. 2011. Clathrin-independent endocytosis: Mechanisms and function. *Current Opinion in Cell Biology*, 23: 413–420.

Schnell U et al. 2013. EpCAM: Structure and function in health and disease. *Biochimica et Biophysica Acta*, 1828(8): 1989–2001.

Schroeder GN, Hilbi H. 2008. Molecular pathogenesis of *Shigella* spp.: Controlling host cell signaling, invasion, and death by Type III secretion. *Clinical Microbiology Reviews*, 21(1): 134–156.

Sendi P, Procter RA. 2008. *Staphylococcus aureus* as an intracellular pathogen: The role of small colony variants. *Trends in Microbiology*, 17(2): 54–58.

Sibley LD. 2011. Invasion and intracellular survival by protozoan parasites. *Immunological Reviews*, 240: 72–91.

Sinha B, Fraunholz M. 2010. *Staphylococcus aureus* host cell invasion and post-invasion events. *International Journal of Medical Microbiology*, 300: 170–175.

Song K et al. 2016. Septins as modulators of endo-lysosomal membrane traffic. *Frontiers in Cell and Developmental Biology*, 4: Article 124.

Song I, Cornelis GR. 2009. The type III secretion systems. In: Wooldrige K (Ed.), *Bacterial Secreted Proteins: Secretory Mechanisms and Role in Pathogenesis.* Caister Academic Press, Norfolk, UK.

Stevens JM et al. 2006. Actin-dependent movement of bacterial pathogens. *Nature Reviews Microbiology*, 4: 91–101.

Swidergall M, Filler SG. 2017. Oropharyngeal candidiasis: Fungal invasion and epithelial cell responses. *PLOS Pathogens*, 13(1): e1006056.

Taylor MP et al. 2011. Subversion of the actin cytoskeleton during viral infection. *Nature Reviews Microbiology*, 9(6): 427–429.

Tobias A, Oelschlaeger TA, Hacker JH (Eds.). 2000. *Bacterial Invasion into Eukaryotic Cells* (Subcellular Biochemistry Volume 33). Kluwer Academic/Plenum Publishers, New York.

Torraca V, Mostowy S. 2016. Septins and bacterial infection. *Frontiers in Cell and Developmental Biology*, 4: Article 127.

Tyler JS et al. 2011. Focus on the ringleader: The role of AMA1 in apicomplexan invasion and replication. *Trends in Parasitology*, 27(9): 410–420.

Tyler KM et al. 2005. Responsive microtubule dynamics promote cell invasion by *Trypanosoma cruzi*. *Cellular Microbiology*, 7(11): 1579–1591.

van Schaik EJ et al. 2013. Molecular pathogenesis of the obligate intracellular bacterium *Coxiella burnetii*. *Nature Reviews Microbiology*, 11: 561–573.

Wachtler B et al. 2012. *Candida albicans*—epithelial interactions: Dissecting the roles of active penetration, induced endocytosis and host factors on the infection process. *PLoS One*, 7(5): e36952.

Wessler S et al. 2011. Regulation of the actin cytoskeleton in *Helicobacter pylori*-induced migration and invasive growth of gastric epithelial cells. *Cell Communication and Signaling*, 9: 27.

Wilson BA, Salyers AA, Whitt DD, Winkler ME. 2011. *Bacterial Pathogenesis: A Molecular Approach*. 3rd ed. ASM Press, Washington, DC.

Wilson M, McNab R, Henderson B. 2002. *Bacterial Disease Mechanisms: An Introduction to Cellular Microbiology*. Cambridge University Press, Cambridge, UK.

Wooldrige K (Ed.). 2009. *Bacterial Secreted Proteins: Secretory Mechanisms and Role in Pathogenesis*. Caister Academic Press, Norfolk, UK.

Wroblewski LE, Peek Jr RM. 2011. Targeted disruption of the epithelial-barrier by *Helicobacter pylori*. *Cell Communication and Signaling*, 9: 29.

Yang W et al. 2014. Fungal invasion of epithelial cells. *Microbiological Research*, 169: 803–810.

Yoshida N et al. 2011. Invasion mechanisms among emerging food-borne protozoan parasites. *Trends in Parasitology*, 27(10): 499–456.

Yoshida S, Sasakawa C. 2003. Exploiting host microtubule dynamics: A new aspect of bacterial invasion. *Trends in Microbiology*, 11(3): 139–143.

Chapter 6: Exotoxins and Endotoxins

BACTERIAL EXOTOXINS

Introduction

In bacteria, exotoxins are proteins that are secreted during the exponential phase of growth or are released when the bacterium undergoes lysis or during the transition from a vegetative organism to a spore or the reverse. In bacteria, exotoxins may be *true virulence factors*, as in the case of tetanospasmin (the neurotoxin produced by *Clostridium tetani*), botulinum toxin (produced by *C. botulinum*), and the diphtheria toxin (produced by *Corynebacterium diphtheriae*). In the case of these exotoxins, inactivating the gene encoding the toxin renders the bacterium avirulent. Furthermore, neutralising antibodies directed against these toxins are protective and found in the blood of patients who have recovered from these intoxications. Toxoids are toxins treated with formalin or heat to eliminate their toxigenicity while retaining their antigenicity and are used as vaccines. Exotoxins not shown to be essential for virulence by gene inactivation, but which contribute to pathogenesis, are more appropriately termed *determinants of pathogenesis*. Some of these exotoxins may prove to contribute significantly to virulence once the effects of gene inactivation are explored. Exotoxins usually are encoded by extra-chromosomal genetic elements such as plasmids, prophages, and pathogenicity islands. Exotoxins bind to host cell membrane receptors whether or not their site of action is the cell membrane itself or cytosolic cell machinery. In this chapter, we classify bacterial exotoxins by their site of action, *viz.*, membrane-acting toxins in which binding of the toxin to its cell membrane receptor inhibits signal transduction pathways, membrane-damaging toxins that create pores or otherwise perturb the cytoplasmic membrane, and intracellular toxins that affect signal transduction pathways. Exotoxins are cytotoxins when they kill the cells to which they bind and leukotoxins when they kill leukocytes. **Table 6.1** summarises bacterial exotoxins by their site of action, and **Table 6.2** lists pathogenic bacteria that produce exotoxins and their modes of action.

site of action	structure	family	receptor	delivery
membrane acting	superantigens		MHC Class II TCR	
	heat stable		membrane-bound guanylate cyclase	
pore forming	α Helical β Barrel	cytolysin A haemolysins aerolysins CDCs RTX MARTX	cholesterol membrane sugars membrane proteins cholesterol β$_2$ integrin family non-specific	
cytosolic/signalling	A/B		GM1 ganglioside	T3SS Clathrin-coated pit endocytosis

Table 6.1 Classification of bacterial toxins by site of action. Abbreviations: CDC, cholesterol-dependent cytolysins; MARTX, multifunctional auto-processing repeats-in-toxin toxins; RTX, repeats-in-toxin toxins; TCR, T-cell receptor; T3SS, type 3 secretion system.

Membrane-acting toxins

Membrane-acting toxins fall into two groups: superantigens (SAs) and the stable toxin (ST) family.

Superantigens (SAs)

SAs bridge major histocompatibility (MHC) class II molecules on professional antigen-presenting cells (APCs) and the T-cell receptor (TCR) on CD4$^+$ T cells outside the peptide-binding groove of both receptors and cause the release of cytokines from the T cell that act on macrophages to release pro-inflammatory cytokines (**Figure 6.1**). Different SAs bind to different families of TCR, recognising a shared motif on the variable domain of the beta chain. Bacteria that produce SAs are *Staphylococcus aureus*; various Lancefield group streptococci but, particularly, the group A streptococcus *S. pyogenes*, *Yersinia pseudotuberculosis* and *Mycoplasma arthritidis*. SAs are implicated in the pathogenesis of syndromes caused by these bacteria such as rheumatic fever, streptococcal toxic shock syndrome, staphylococcal toxic shock syndrome, Kawasaki disease, and autoimmune disease. *Staphylococcus aureus* produces a large number of SAs, some of which act on the intestinal mucosa. Exotoxins that target the gastrointestinal tract are called enterotoxins. Staphylococcal enterotoxins (SEs) are secreted proteins that are soluble in water, are remarkably resistant to heat and to most proteolytic enzymes, and retain their activity in the digestive tract after ingestion. The mechanism(s) of action of SEs is/are not completely clear. It is possible that SEs stimulate the vagus nerve in the intestines, transmitting a signal to the vomiting centre in the brain. SEs are able

bacteria	type of toxin	structure	family	name
Staphylococcus aureus	enterotoxin	superantigen		staphylococcal exotoxins
		pore-forming β barrel	haemolysin	haemolysins leukocidins
Clostridium tetani				tetanospasmin
C. botulinum				botulinum toxin
Clostridium perfringens		pore-forming β barrel	aerolysin	β-toxin δ-toxin
	enterotoxin	pore-forming β barrel	aerolysin	enterotoxin
Vibrio cholera		pore-forming β barrel	haemolysin	V. cholerae cytotoxin (VCC)
	cytotoxin	pore-forming β barrel	RTX	VcRtxA
			MARTX	VcRtxA
		cytosolic/signalling	A/B	cholera toxin (CT)
Vibrio vulnificus		pore-forming β barrel		V. vulnificus haemolysin (VVH)
Corynebacterium diphtheriae				diphtheria toxin
Enterotoxigenic E. coli (ETEC)	enterotoxin	heat stable		STa STb
		cytosolic/signalling	A/B	heat-labile enterotoxin (LT)
		cytosolic/signalling	A/B	shiga-like toxins (Stx1 and Stx2)
ETEC enteroaggregative E. coli (EAEC), enterohaemorrhagic E. coli (EHEC), enterotoxigenic E. coli (ETEC), enteropathogenic E. coli	enterotoxin	heat stable		enteroaggregative Escherichia coli heat-stable toxin1
Escherichia coli, Salmonella enterica, Shigella flexneri		pore-forming α helical	Cly A	
Shigella dysenteriae	enterotoxin	cytosolic/signalling	A/B	shiga toxin (Stx)
B. pertussis and related Bordetella species		pore-forming β barrel	RTX	adenylate cyclase toxin (ACT)
Bacillus anthracis		cytosolic/signalling	A/B	anthrax toxin (AT)
Bordetella pertussis		cytosolic/signalling	A/B	pertussis toxin (PT)

Table 6.2 Toxins produced by common pathogenic bacteria. MARTX, multifunctional auto-processing repeats-in-toxin toxins; RTX, repeats-in-toxin toxins; VcRtxA, *Vibrio cholera* RTX toxin.

to cross the barrier epithelium by transcytosis and induce local and systemic immune responses. SEs bind to submucosal mast cells, resulting in the release of inflammatory mediators such as histamine, leukotrienes, and neuroenteric peptide (substance P). These mediators induce vomiting and may induce inflammation in the gastrointestinal tract. Diarrhoea associated with SEs probably results from inhibition of water and electrolyte reabsorption in the small intestine, as is the case with non-SA enterotoxins.

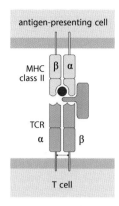

Figure 6.1 Superantigens (SAs; red) bind to a motif on the Vβ domain of the T cell receptor (TCR; blue) and to the outer face of MHC class II molecules (yellow). The motif on the Vβ domain of the TCR is shared by as much as 20% of CD4+ T cells, so an SA can activate a significant fraction of T cells regardless of the specificity of their TCR. Binding of a SA causes the release of large amounts of cytokines, the most important of which is interferon-gamma (IFN-γ). IFNγ activates macrophages to release the pro-inflammatory cytokines interleukin-1β (IL-1β), tumour necrosis factor-alpha (TNFα) and IL-6. When released in the blood at high levels, these cytokines can cause shock and multi-organ failure. SE, staphylococcal enterotoxin; TSST-1, toxic shock syndrome toxin 1. (Adapted from Figure 6.25 Murphy, K. and Weaver, C., *Janeway's Immunobiology*, 9th ed., Garland Science, New York, 2016.)

Heat-stable exotoxins (STs)

STs are heat-stable exotoxins that are resistant to heat denaturation (100°C/30 min). They play a central role in enteric disease caused by various *Escherichia coli* pathovars. Enterotoxigenic *E. coli* (ETEC) produce the heat-stable enterotoxins STa and STb. ETEC also produces a related enterotoxin (EAST1) that is also produced by enteroaggregative *E. coli* (EAEC), enterohaemorrhagic *E. coli* (EHEC), ETEC, and enterpathogenic *E. coli* pathovars. STs bind to, and activate, cell membrane guanylate cyclase, which leads to an increase in the intracellular level of cyclic GMP, activating protein kinase-G and resulting in the secretion of water and electrolytes from intestinal cells. A schematic of the mode of action of *E. coli* heat-stable toxins is shown in **Figure 6.2**.

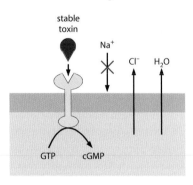

Figure 6.2 Schematic showing the mode of action of *E. coli* heat-stable toxin. (Taken and adapted from http://slideplayer.com/slide/7893538/, Slide #16.)

MEMBRANE-DAMAGING EXOTOXINS

Exotoxins that damage cell membranes do so by forming pores through the cell membrane. Pore-forming toxins (PFTs) constitute the largest group of bacterial exotoxins and they are listed in **Table 6.3**. By creating channels of various sizes through the cytoplasmic membrane, they compromise the ability of the host cell to control the influx and efflux of small molecules and

pore-forming toxin families (PFT)	family	class	organisms	stoichiometry	receptor
cytolysin A (ClyA also known as HlyE)	ClyA	α	E. coli, S. enterica, S. flexneri	12	cholesterol
non-haemolytic tripartite enterotoxin (Nhe)	ClyA	α	B. cereus	–	cholesterol
haemolysin BL (Hbl)	ClyA	α	B. cereus	7-8	cholesterol
α-haemolysin (Hla)	Haemolysins	β	S. aureus	7	PC/ADAM10/disintegrin
γ-haemolysin (Hlg)	Haemolysins	β	S. aureus	8	PC
leukocidins (for example, HlgACB, LukED)	Haemolysins	β	S. aureus	8	CCR5, CXCR1, CXCR2, CCR2, C5aR C5L2
necrotic enteritis toxin B (NetB)	Haemolysins	β	C. perfringens	7	cholesterol
δ-toxin	Haemolysins	β	C. perfringens	7	monosialic ganglioside 2 (GM2)
V. cholerae cytolysin (VCC)	Haemolysins	β	V. cholerae	7	glycoconjugates
V. vulnificus haemolysin (VVH)	Haemolysins	β	V vulnificus	7	glycerol, N-acetyl-D-galactosamine
α-toxin	Aerolysin	β	Clostridium spp.	–	GPI-anchored proteins
ε-toxin	Aerolysin	β	C. perfringens	7	HAVCR1
enterotoxin (CPE)	Aerolysin	β	C. perfringens	6	claudin
perfringolysinO (PFO)	CDCs	β	C. perfringens	30–50	cholesterol
suilysin (SLY)	CDCs	β	S. suis	30–50	cholesterol
intermedilysin (ILY)	CDCs	β	S. intermedius	30–50	cholesterol, CD59
listeriolysin O (LLO)	CDCs	β	L. monocytogenes	30–50	cholesterol
lectinolysin (LLY)	CDCs	β	S. mitis	30–50	cholesterol, CD59
anthrolysin O (ALO)	CDCs	β	B. anthracis	30–50	cholesterol
streptolysin O (SLO)	CDCs	β	S. pyogenes	30–50	cholesterol
Bth-MACPF (BT 3439)	MACPF	β	B. thetaiotaomicron	>30	–
HlyA	RTX	α?	E. coli	–	–
bifunctional haemolysin-adenylyl cyclase toxin (CyaA)	RTX	α?	B. pertussis	–	–

Table 6.3 Pore-forming toxin families. Cly, cytotoxin; ADAM10, disintegrin and metalloproteinase domain-containing protein 10; C5aR, C5a receptor; CCR5, CC-chemokine receptor type 5; CDC, cholesterol-dependent cytolysin; CXCR1, CXC-chemokine receptor type 1; HAVCR1, hepatitis A virus cellular receptor 1; IM, bacterial inner membrane; LPS, lipopolysaccharide; MACPF, membrane attack complexcomponent/perforin; MARTX, multifunctional autoprocessing repeats-in-toxin; OM, bacterial outer membrane; PC, phosphatidylcholine; PFT, pore-forming toxin; RTX, repeats in toxin. (From Dal Peraro, M. and van der Goot, F.G., *Nat. Rev. Microbiol.*, 14, 77–92, 2016.)

electrolytes. The creation of pores in the cell membrane affects cell function and may result in cell death by osmotic lysis or apoptosis. PFTs comprise two major groups depending on whether their secondary structure consists of α-helices or β-barrels. The receptors for all PFTs are cell membrane proteins, lipids or sugars to which the exotoxins bind and oligomerise, exposing hydrophobic domains that insert through the membrane. The general mechanism of pore formation by α-helical exotoxins is that soluble PFT components (protomers) insert into the cell membrane concurrent with oligomerisation, leading to the formation of an incomplete or complete pore, both of which are functional. In the case of β-barrel PFTs, oligomerisation occurs on the surface of the cell membrane, producing a structure termed a pre-pore, which undergoes rearrangement and full membrane insertion. The molecular mechanisms of pore formation are illustrated in **Figure 6.3**.

α-helical pore-forming exotoxins

Within the α-helical group of PFTs are the cytolysin A (ClyA) family. These exotoxins are produced by certain clones of *Escherichia coli*, *Salmonella enterica*, *Shigella flexneri* and *Bacillus cereus*. Their receptor is cholesterol

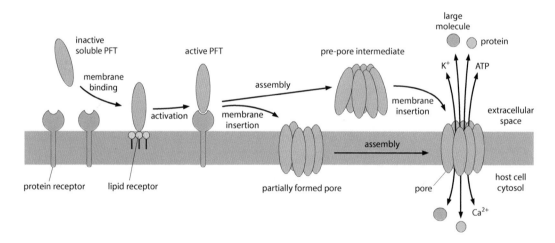

Figure 6.3 Schematic representation of the pore formation pathway of pore-forming toxins (PFTs). Soluble PFTs are recruited to the host membrane by protein receptors and/or specific interactions with lipids. Upon membrane binding, the toxins concentrate and start the oligomerisation process, which usually follows one of two pathways. In the pathway followed by most β-PFTs, oligomerisation occurs at the membrane surface, producing an intermediate structure known as a pre-pore (mechanism 1), which eventually undergoes conformational rearrangements that lead to concerted membrane insertion. In the pathway followed by most α-PFTs, PFT insertion into the membrane occurs concomitantly with a sequential oligomerisation mechanism, which can lead to the formation of either a partially formed, but active, pore (mechanism 2), or the formation of complete pores. In both α-PFT and β-PFT pathways, the final result is the formation of a transmembrane pore with different architecture, stoichiometry, size and conduction features, which promotes the influx or efflux of ions, small molecules and proteins through the host membrane. (From Dal Peraro, M. and van der Goot, F.G., *Nat. Rev. Microbiol.*, 14, 77–92, 2016.)

in the cell membrane. ClyA exotoxins are secreted as an α-helical protomer (a monomeric subunit of an oligomeric protein) that contains a short hydrophobic hook termed the β-tongue. Upon binding to cholesterol, the β-tongue is exposed and inserted into the lipid bilayer whereupon the amphipathic *N*-terminus of each protomer, up to this point facing away from the cell membrane, is inverted and inserted into it. There follows the sequential insertion of other protomers that oligomerise to form a pore with the amphipathic helices facing inwards to form the bore of the pore.

β-barrel pore-forming exotoxins

Bacterial β-barrel pore-forming toxins are secreted as water-soluble monomeric proteins and assemble into β-barrel–shaped pores/channels through the membranes of target cells, causing cell death and lysis. The β-barrel exotoxins fall into the following three groups: haemolysins, aerolysins and the cholesterol-dependent cytotoxins (CDCs).

Haemolysins: The haemolysin family is the best characterised of the β-barrel pore-forming toxin families. Haemolysins include exotoxins from *Staphylococcus aureus* (α-haemolysin, γ-haemolysin and the leukocidins, including the Panton-Valentine leukocidin), *Clostridium perfringens* (β-toxin and δ-toxin), *Vibrio cholerae* (*V. cholerae* cytotoxin [VCC]) and *Vibrio vulnificus* (*V. vulnificus* haemolysin [VVH]).

The *S. aureus* haemolysins are models for all β-barrel exotoxins, and their mechanism of action is shown in **Figure 6.4**. Staphylococcal PFTs are composed of a single component (α-haemolysin shown in Figure 6.4) and creates a seven-subunit pore, or two components (γ-haemolysin) and forms an eight-subunit pore. Pore formation begins by exposure of a pre-stem loop that inserts into the membrane where it combines with pre-stem loops of other protomers to form a β-barrel. In the case of *S. aureus* α-haemolysin, membrane binding initiates withdrawal of the pre-stem loop to create a partial

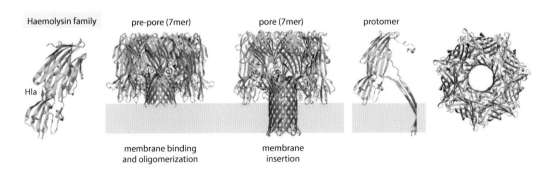

Figure 6.4 *Staphylococcus aureus* α-haemolysin (α-toxin; Hla) as a model of β-barrel pore-forming toxins. Protomer, the structural unit of an oligomeric protein. (Adapted from Dal Peraro, M. and van der Goot, F.G., *Nat. Rev. Microbiol.*, 14, 77–92, 2016.)

β-barrel that gives rise to a pre-pore. The mature pore forms by insertion of the complete β-barrel into the membrane.

Aerolysins: The exotoxin aerolysin was first identified in *Aeromonas hydrophila*, an aquatic, motile, gram-negative, facultatively anaerobic rod that can grow at temperatures as low as 4°C. This bacterium is an opportunistic pathogen associated with diarrhoeal diseases and deep wound infections in humans. The exotoxins of various *Clostridium* species belong to the aerolysin family. These include the α-toxin of *C. perfringens* and *C. septicum* and the ε-toxin and enterotoxin of *C. perfringens*. The distinctive feature of members of the aerolysin family is a pair of β-strands separated by a pre-stem domain/loop.

Cholesterol-dependent cytolysins: Cholesterol-dependent cyto-toxins (CDCs) are produced predominantly by gram-positive bacteria belonging to the genera *Bacillus*, *Clostridium*, *Listeria* and *Streptococcus* but have also been found in the gram-negative rod *Enterobacter lignolyticus*. CDC monomers consist of four domains rich in β sheets. A highly conserved tryptophan-rich 11-amino acid sequence is present at the end of the fourth domain. This tryptophan-rich, 11-amino acid sequence appears to mediate the binding of some CDCs to cholesterol and may, in fact, bind directly to cholesterol and penetrate a short distance into the cytoplasmic membrane.

RTX exotoxins

Repeats in toxin (RTX) exotoxins are produced by many genera of gram-negative bacteria (**Table 6.4**). As a family, they share two features. The first is that all are exported by means of a type 1 secretion system (T1SS) in which toxin secretion occurs through a pore created by three outer membrane proteins that span the gram-negative envelope. The second is the presence of glycine-rich and aspartic acid-rich nonapeptide repeats at the *C*-terminus of the toxin that gives this superfamily of

RTX toxin	bacterium
HlyA	Uropathogenic *Escherichia coli*
EhxA	Enterohemorrhagic *Escherichia coli*
CyaA	*Bordetella pertussis*
PvxA	*Proteus vulgaris*
MmxA	*Morganella morganii*
LtxA	*Aggregatibacter actinomycetemcomitans*
VcRtxA	*Vibrio cholera*
VvRtxA	*Vibrio vulnificus*
RTX cytotoxin	*Kingella kingae*

Table 6.4 RTX toxins of human bacterial pathogens. (From Linhartova, I. et al., *FEMS Microbiol. Rev.*, 34, 1076–1112, 2010.)

toxins its name, *repeats in toxin*. Their biological activity depends on the binding of Ca^{2+} in the nonapeptide repeats. RTX cytotoxins can be divided into the pore-forming leukotoxins and the multifunctional auto-processing repeats-in-toxin toxins (MARTX), such as the VcRtxA toxin from *V. cholerae*. Some RTX exotoxins bind members of the β_2 integrin family on the host cell membrane, whereas others do not appear to utilise a specific receptor. In this instance, the exotoxin binds to the cytoplasmic membrane by electrostatic interactions and then inserts into the membrane. However, a lectin-like interaction between the exotoxin and cell membrane sugars has been suggested. The mechanism of membrane insertion and pore formation of RTX exotoxins are currently unclear, although oligomerisation is suspected. The mechanism of host cell damage mediated by RTX exotoxins is likely their ability to dysregulate cytosol Ca^{2+} concentration, leading to the release of inflammatory mediators and, ultimately, to cell lysis.

MARTX exotoxins

MARTX is the acronym of a family of exotoxins that make up a separate family of RTX exotoxins. MARTX exotoxins are large toxins that differ in structure from the RTX exotoxin family described in the previous section. These exotoxins are produced by certain species of gram-negative bacteria in the genera *Vibrio*, *Aeromonas*, *Proteus* and *Yersinia*. The prototype of the MARTX exotoxins is $MARTX_{Vc}$, (VcRtxA) produced by *Vibro cholerae*. The $MARTX_{Vc}$ toxin is much larger than other RTX toxins, with a molecular mass of over 485 kDa, but the MARTX exotoxin of *V. vulnificus* is even larger, with a predicted molecular mass of 556 kDa. About one-quarter of the amino acids in the primary structure of the exotoxins are glycine-rich repeats. The repeats of MARTX toxins are of three types. The first two types of repeats are located at the *N*-terminal end of the molecules and consist of a 20-amino acid consensus sequence and a 19-amino acid consensus sequence, whereas the third type of repeat is at the *C*-terminus of the molecule. The third type of repeat is an 18-amino acid consensus sequence that shares a central core glycine-7X-glycine-XX-asparagine motif instead of the nonapeptide RTX repeat discussed previously. In contrast to the conserved sequences at the *N*- and *C*-termini, the primary sequence in the central region of the MARTX toxins are quite different and appear to have a mosaic structure related to the toxic activities of particular toxins. For example, in $MARTX_{Vc}$ the central region contains three domains a Rho-inactivation domain (RID), an actin cross-linking domain (ACD) and a cysteine protease domain (CPD). The CPD cleaves and releases the RID and ACD domains into the cytoplasm, disrupting the actin cytoskeleton (**Figure 6.5a** and **b**). Unlike RTX exotoxins, MARTX exotoxins do not form pores and, accordingly, they lack a conserved hydrophobic domain for generating a pore. Instead, $MARTX_{Vc}$ causes the rounding up of non-polarised cells and the loss of the integrity of the paracellular tight junctions of polarised cells. In both cases, the cells remain viable.

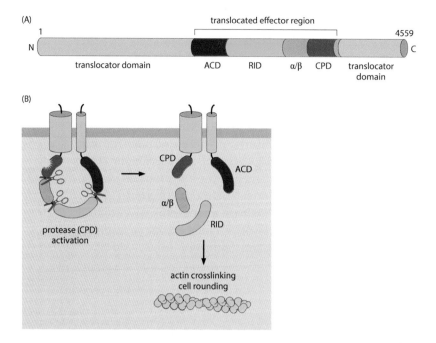

Figure 6.5 Please refer to text. (A) Domain organisation of *V. cholera* MARTX toxin. Conserved glycine-rich repeat regions in the *N*- and *C*-termini of MARTX toxins (MARTX conserved, red); actin crosslinking domain (ACD, orange); Rho-inactivating domain (RID, green); α/β hydrolase domain (α/β, purple), CPD (pink). Cleavage sites are shown. (B) Model of CPD-mediated activation of *V. cholera* MARTX toxin. The *N*- and *C*-terminal MARTX$_{Vc}$-conserved regions insert into the plasma membrane and form a pore that permits translocation of the toxin central region across the cell membrane. Following toxin entry, the CPD binds inositol hexakisphosphate, resulting in activation of its protease activity. Processing of MARTX$_{Vc}$ by the CPD releases the RID and α/β hydrolase domains into the cytosol, while the ACD and CPD remain tethered to the membrane. Cleavage of MARTX$_{Vc}$ activates the actin cross-linking activity of the ACD. (Adapted from Shen, A., *Toxins*, 2, 963–977, 2010.)

INTRACELLULAR EXOTOXINS

AB exotoxins

AB toxins are toxins that contain a binding (B) component that binds to a receptor on the cytoplasmic membrane of the target cell and an A component that is the toxic moiety. As such, they are termed binary toxins. There are two forms of AB toxins, those that have a single A and B unit and those with a single A component and five B components (AB$_5$). The B component(s) is(are) responsible for translocation of the toxic moiety (A) into the cytosol of the cell. In the case of AB$_5$ exotoxins, the A subunit enters the cytoplasm *via* a pore created by the B subunits (see the previous section on membrane damaging toxins). The A component is a catalytic component that has enzyme activity that targets the metabolic machinery of the target cell. The A and B components can be part of a single polypeptide that is cleaved to liberate the A component, or the A and B components can be separate polypeptides or polypeptides that are attached by disulphide or by non-covalent bonds. Other than entry through a pore, uptake of the toxin A component occurs by receptor-mediated endocytosis utilising clatherin-coated pits or, in some cases, the toxin is delivered to the target cell cytoplasm by means of a type 3 secretion system (T3SS) [discussed in Chapter 5 facilitated cell entry]. The A component of several AB toxins is an ADP-ribosyl transferase, which ribosylates G actin, resulting in disorganisation of the cytoskeleton and cell death. Some A components are proteases that hydrolyse mitogen-activated protein kinase kinases (MAPKK) that inhibit cell signalling, whereas others increase the level of intracellular cyclic AMP (cAMP), which results in oedema and immunosuppression.

Pseudomonas aeruginosa exotoxin A (ExoA): ExoA is secreted as single protein that has three domains, catalytic, receptor binding and translocation. Once the toxin is secreted by the bacterium the terminal lysine is cleaved from it to a produce an REDL (arginine-glutamic acid-aspartic acid-leucine) amino acid sequence at the C-terminus. This allows the toxin to bind to KDEL receptors in the Golgi apparatus once the toxin has been internalised. KDEL receptors initiate the mechanism by which proteins are transported from the Golgi to the ER. KDEL receptors recognize the amino acid sequence lysine-aspartic acid-glutamic acid-leucine at the C-terminus of proteins. The cell membrane receptor of ExoA is CD91 (α2-macroglobulin receptor/low-density lipoprotein receptor-related protein). ExoA reaches the endoplasmic reticulum by one of two routes, either the KDEL-receptor mediated pathway or the lipid-dependent sorting pathway. KDEL (lysine-aspartic acid-glutamic acid-leucine) is a target peptide sequence at the C-terminus of proteins to prevent them from being secreted from the endoplasmic reticulum (ER) or returning protein that have left the ER. ExoA molecules bound to the CD91 receptor are taken up by clathrin-coated pits. ExoA releases from the CD91 receptor in the early endosome and undergoes a conformational change that allows the protease furin to cleave it into two pieces, one of which contains the ADP-ribosylation activity. From the late endosome the ADP-ribosylase reaches the trans-Golgi network (TGN) by means of a Rab9 (a GTPase)-regulated pathway. In the TGN the ADP-ribosylase binds to the KDEL receptor and by this means the toxin is transported to the ER. Also, CD91-bound ExoA can be taken into the target cell with detergent-resistant microdomains by caveolin-mediated endocytosis mediated by Rab5. After cleavage in the early endosome, the ADP-ribosylase reaches the trans-Golgi network. From the TGN the ADP ribosylase traffics to the ER using a lipid-dependent sorting pathway. ExoA employs the ER-associated protein degradation pathway (ERAD) to enter the cytosol from the ER. In the cytosol the ExoA ADP-ribosylase ADP-ribosylates eukaryotic elongation factor-2 (eEF-2) on the ribosomes arresting protein synthesis which leads to apoptosis of the cell.

AB$_5$ exotoxins

Cholera toxin: The A component of cholera toxin (CT) consists of two subunits – CTA1 and CTA2 – joined by a disulphide bond. The CTA1 subunit is an ADP-ribosyltransferase, and the CTA2 subunit inserts the CTA1 subunit into the centre of a ring formed by five CTB subunits. The B subunits assemble by hydrogen bonding and charge–charge interactions. The cellular receptor for CT is GM1 ganglioside, particularly on enterocytes in the small intestine. The toxin is internalised by endocytosis using clatherin- and calveolin-coated vesicles which enter the endosomal pathway. Acidification of the endosome results in the release of the CTA1 subunit from the toxin and its entry into the cytoplasm. The CTA1 subunit also has a motif that targets it to the endoplasmic reticulum (ER). By ADP ribosylating adenylate cyclase, the level of cAMP is increased in the cell, resulting in the loss of water and electrolytes. The net affect is a profuse and protracted watery diarrhoea termed 'rice-water'.

Heat-labile enterotoxin: Heat-labile enterotoxin (LT) is produced by ETEC and is closely related to CT. Both toxins cause elevation of the level of cAMP in enterocytes. The heat-labile enterotoxin from different strains of ETEC fall into two groups: LTI and LTII, with LTII being subdivided into LTIIa and LTIIb. Like CT, LTs bind to GM1 gangliosides but also bind other related ligands. In the cytoplasm, the disulphide bonds between LTA1 and LTA2 are reduced, releasing the toxic subunit A1. The toxic component is an ADP ribosylase.

Shiga and shiga-like toxins: The Shiga toxin family is composed of the Shiga toxin (Stx) produced by *Shigella dysenteriae* and the Shiga-like toxins Stx1 and Stx2 produced by strains of EHEC and ETEC, respectively. The Stx toxin has a single A subunit which consists of two parts, A1 and A2, bound by a single disulphide bond and five B subunits. The A and B subunits associate by non-covalent interactions in an asymmetric fashion, binding to only three of the five B subunits. This quaternary structure is found in toxins that bind to glycoprotein or glycolipid ligands on the target cell membrane. The ligand for Stx is globotrioylceramide (Gb3), and binding triggers uptake of the toxin by endocytosis. The intracellular trafficking of Shiga and Shiga-like toxins is shown in **Figure 6.6**. The toxin undergoes retrograde sorting in early endosomes, during which Shiga toxin-containing vesicles bud from the early endosomes and migrate to the trans-Golgi network (TGN), where the A subunit is cleaved from the B subunits by the protease furin. However, the A subunit remains associated with the B subunits. The toxin then migrates to the ER. In the ER, the disulphide bond between A1 and A2 is reduced, and A1 retro-translocates into the host cell cytoplasm. The A1 fragment is an RNA *N*-glycosidase that modifies ribosome translation thereby shutting down protein synthesis. This mode of action is termed ribotoxic stress. The binding of Shiga toxin to globotrioylceramide

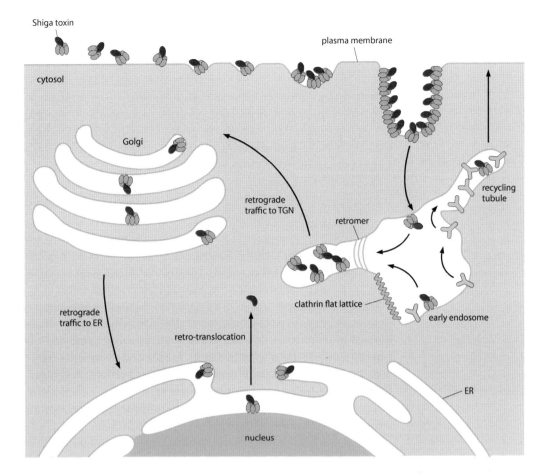

Figure 6.6 Shiga toxin binding to the plasma membrane induces local spontaneous curvature, membrane-mediated clustering and the toxin-driven formation of endocytic invaginations. The toxin then undergoes retrograde sorting in early endosomes, in which retrograde tubules are formed in a clathrin-dependent manner, and Shiga toxins preferentially localise to these tubules. Shiga toxins bypass the late endocytic pathway and are transferred directly from the early endosome to the trans-Golgi network (TGN) and, from there, on to the endoplasmic reticulum (ER). Finally, Shiga toxins use the ER-associated degradation (ERAD) machinery to enter the host cell cytosol. (From Johannes, L. and Römer, W., *Nat. Rev. Microbiol.*, 8, 105–116, 2010.)

(Gb3) on the target cell membrane not only initiates endocytosis but recruits and activates Gb3-containing, glycolipid-enriched domains of several tyrosine kinases, which facilitate toxin entry *via* actin and microtubular cytoskeleton remodelling. There is a gradation of response to internalisation of Shiga toxin in different types of cells that bear the Gb3 ligand. Some cells such as endothelium, epithelium and cells in the central nervous system are killed, whereas other cell types, such as monocyte/macrophages, are resistant. These cells respond to cytosolic toxin by synthesising and releasing pro-inflammatory cytokines. Unfortunately, these cytokines upregulate Gb3 expression on endothelial cells and sensitise target cells to the cytotoxic effect of the toxins. Shiga toxins can induce apoptosis in certain types of cells by various mechanisms; however, the caspase pathway is centrally involved in programmed cell death. Apoptosis is the cause of vascular lesions and tissue damage when Shiga toxins enter the bloodstream. It has been reported that the StxB subunit can also initiate apoptosis.

Pertussis toxin: Pertussis toxin (PT) is an exotoxin produced by the respiratory pathogen *Bordetella pertussis*, the aetiologic agent of whooping cough. The primary target of PT is respiratory epithelial cells. PT is composed of five subunits (S1–S5). The A subunit is termed S1 and the B oligomer subunits S2–S5; the B oligomer has two copies of S4. Interestingly, S2 and S3 have significant homology to the mannose-binding lectin, an innate pattern recognition receptor (PRR) that recognises mannose and fucose arrays on certain classes of pathogen. Following binding of PT to its ligand on the target cell membrane, the intracellular trafficking is similar to that described above for Shiga toxins (Figure 6.8). During the trafficking process, the A subunit is activated by reducing a disulphide bond, possibly by the action of glutathione and ATP. PT catalyses the ADP-ribosylation of heterotrimeric G proteins, thus perturbing intracellular communication. PT also increases the release of insulin, causing hypoglycaemia. Interestingly, PT has no part in generating the non-productive paroxysmal cough that is characteristic of the disease pertussis, suggesting that a second toxin, a 'cough toxin', must exist. Indirect evidence for a second toxin comes from the fact that, while immunised individuals or individuals who have recovered from *B. pertussis* infection do not display effects of PT upon reinfection, they do exhibit the paroxysmal cough.

Anthrax toxin: Anthrax toxin (AT) is a toxin produced by the spore-forming, aerobic, gram-positive bacterium, *Bacillus anthracis*. AT is unusual in that it consists of three separate polypeptide subunits termed protective antigen (PA), oedema factor (EF), and lethal factor (LF). EF and LF have enzyme activity. PA initiates entry of LF and EF into the target cell by binding to either tumour endothelial marker 8 (TEM8 or ANTXR1) or capillary morphogenesis 2 (CMG2 or ANTXR2). Following binding, PA is cleaved by the enzyme furin into an active form that, together with the receptor, assembles into a heptameric pre-pore structure. The pre-pore can bind up to three LF and/or EF subunits, and this complex undergoes a conformational change which permits its uptake by endocytosis. Acidification of the endosome permits pore formation in the endosome membrane and allows unfolding of LF and EF and transport into the cytoplasm. LF is a zinc-dependent metalloprotease that affects signal transduction pathways leading to the inhibition of the release of pro-inflammatory cytokines and the induction of apoptosis. EF is a calcium-independent, calmodulin-dependent adenylate cyclase that increases cAMP, leading to fluid release from the cell.

AB exotoxins

***Clostridium botulinum* neurotoxins:** *C. botulinum* produces eight
antigenically distinct exotoxins (A, B, C_1, C_2, D, E, F and G). Of these,
types A, B and E are most commonly associated with human intoxication. The neurotoxins are produced as a single polypeptide consisting of two domains termed the heavy chain (100 kDa) and light chain
(50 kDa) that are linked by a disulphide bond. The heavy chain of the
toxin binds to both synaptotagmin II and the ganglioside GD1a at the
presynaptic surface of cholinergic neurones, and the toxin:receptor
complex is internalised by endocytosis usually *via* recycling synap-
tic vesicles. Acidification of the vesicles allows the disulphide bond
between the two chains to be cleaved and the LC to escape into the
cytoplasm. The LC acts as a zinc-dependent protease that interacts
with different proteins (synaptosomal-associated protein [SNAP] 25,
vesicle-associated membrane protein, and syntaxin) in the nerve ter-
minals to prevent fusion of acetylcholine vesicles with the cell membrane, resulting in blockage of neurotransmitter release.

***Clostridium tetani* neurotoxin:** Tetanospasmin is the neurotoxin pro-
duced by *C. tetani*. It has the same structural organisation as the
botulium toxins discussed in the previous section and its mode of
action is identical. Where the neurotoxins differ is that whereas botulinum toxins target excitatory neurons at the neuromuscular junction, tetanospasmin targets inhibitory neurons in the spinal cord.
Tetanospasmin binds to the presynaptic membrane of the neuromuscular junction, is endocytosed and migrates retro-axonally to
the spinal cord. Spastic paralysis induced by the toxin results from
the blockade of neurotransmitter release from spinal inhibitory
interneurons.

***Corynebacterium diphtheriae* diphtheria toxin:** Diphtheria toxin
(DT) is a single polypeptide chain consisting of an A and a B subunit linked by disulphide bonds. DT binds to the heparin-binding
epidermal growth factor precursor (HB-EGF precursor). Binding of
DT to the HB-EGF precursor triggers receptor-mediated endocytosis
of DT:receptor complexes *via* clathrin-coated pits and entry into the
endosomal pathway. Most of the DT that enters cells by this pathway
is degraded in endolysosomes, but some DT translocates from the
endosome to the cytosol before fusion with lysosomes. Translocation requires the acidic environment of the endosome to induce a
conformational change that allows the DT B subunit to insert into
the endosome membrane and form a channel through which the
A subunit migrates into the cytosol. The target of DT is elongation
factor 2 (EF-2). DT transfers the adenosine diphosphate ribose
(ADPR) moiety from NAD to EF-2, which inactivates EF-2 and inhibits chain elongation during protein synthesis. By inhibiting protein
synthesis, DT kills target cells.

FUNGAL TOXINS

Toxins produced by filamentous fungi growing in food are termed mycotoxins. Unlike bacterial exotoxins, mycotoxins are non-protein secondary metabolites. Secondary metabolites are small organic molecules produced by an organism and are not essential for their growth, development and reproduction. Mycotoxins produce diverse effects on human organ systems and, in addition, are mutagenic and teratogenic. The most important mycotoxins are aflatoxins, ochratoxin A, fumonisins, trichothecenes and zearalenone. Aflatoxins are produce by *Aspergillus flavus*, *A. parasiticus* and *A. nomius*. Peanuts, maize and cottonseed are the most common foodstuffs contaminated by these *Aspergillus* species. Of the roughly 14 types of aflatoxin reported, B1 is the most toxic. Aflatoxin causes acute hepatic necrosis, leading to cirrhosis and, possibly, carcinoma of the liver. In fact, aflatoxin is the most potent liver carcinogen known. Aflatoxin intoxication in children can lead to stunted growth and delayed development. Ochratoxin A is a nephrotoxin produced by *Aspergillus ochraceus*, *A. carbonaris* and *Penicillium verrucosum*. Ochratoxin also affects the central nervous system and suppresses the immune system. The fungus is found as a contaminant of grain, coffee and grapes. Fumonisins are one of several toxins produced by *Fusarium moniliforme* in addition to fusaric acid, fusarins, gibberellins, and moniliformin. Fumonisins are produced when *F. moniliforme* grows on maize. Fumonisin B1 is the most important toxin among about 15 types of fumonisins. The toxic effects of fumonisins appear to result from the ability of the toxin to inhibit ceramide synthase due to altered sphingolipid metabolism. Neural tube defects and oesophageal cancer have been attributed to these toxins. The trichothecene toxins are produced by *Fusarium graminearum* that grows on wheat and other cereals, but these toxins can be elaborated by several other genera of fungi. Trichothecenes can be absorbed through the skin and are capable of inhibiting protein, DNA and RNA synthesis, inhibiting mitochondrial function and inducing apoptosis. Zearalenone is an estrogenic toxin also produced by *Fusarium graminearum*. Zearalenone has been found in the blood of children who display precocious sexual development.

PARASITE EXOTOXINS

The parasitic protozoan, *Toxoplasma gondii,* secretes a pore-forming protein named *Toxoplasma gondii* porforin-like protein 1 (TgPLP1) that allows the parasite to escape from the parasitophorous vacuole. In addition, *Plasmodium falciparum* secretes a perforin-like protein called plasmodium perforin-like protein 2 (PPLP2) that facilitates exit of *Plasmodium falciparum* gametocytes through the red blood cell membrane.

ENDOTOXINS

Lipopolysaccharide and lipooligosaccharide: Lipopolysaccharide (LPS) is a glycolipid that is the major component of the outer leaflet of the outer membrane of most gram-negative bacteria (**Figure 6.7**). The LPS molecule is composed of a hydrophobic region termed lipid A that is linked to a non-repeating oligosaccharide termed the core. The core comprises an inner and an outer region. The inner core is largely conserved and contains 2-keto-3-deoxyoctonic acid (KDO) and heptose and contributes to the stability of the outer membrane. The outer core is somewhat more structurally diverse. The outer core is linked to a polysaccharide termed the O or somatic antigen (**Figure 6.8**). Some genera of gram-negative bacteria such as *Neisseria, Moraxella, Haemophilus, Bordetella* and some *Campylobacter*

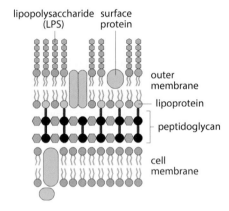

Figure 6.7 The gram-negative bacterial envelope showing lipopolysaccharide (LPS; endotoxin [turquoise circles]) in the outer leaflet of the outer membrane. (Adapted from Figure 2.9 in Murphy, K. and Weaver, C., *Janeway's Immunobiology*, 9th ed., Garland Science, New York.)

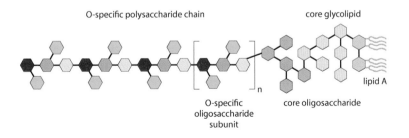

Figure 6.8 The LPS molecule is composed of a hydrophobic region termed lipid A that is linked to a non-repeating oligosaccharide termed the core. The core comprises an inner and an outer region. The inner core is largely conserved and contains 2-keto-3-deoxyoctonic acid (KDO) and heptose and contributes to the stability of the outer membrane. The outer core is somewhat more structurally diverse. The outer core is linked to a polysaccharide termed the O or somatic antigen. (Figure from https://www.pigprogress.net/Finishers/Articles/2016/1/The-hidden-dangers-of-lipopolysaccharides-2741096W/.)

species produce lipooligosaccharide (LOS) that lack the O antigenic poly-saccharide chain. Lipid A is the toxic moiety of LPS and LOS. The composition of lipid A is largely conserved among gram-negative bacteria and the lipid A of both LPS and LOS are microbe-associated molecular patterns (MAMPs) that are recognised by the pattern-recognition receptors (PRRs) Toll-like receptor 4 (TLR-4) and CD 14 *via* the LPS-binding protein (LPB). By virtue of its ability to bind these PRRs, lipid A of LPS and LOS is a potent immunostimulatory and pro-inflammatory molecule. Lipid A has profound effects on cells of the immune system that bear these PRRs such as dendritic cells, macrophages, mast cells and eosinophils. Gram-negative bacteria have a propensity to release outer membrane vesicles that are rich in lipid A, and when gram-negative bacteria enter the blood circulatory system, the systemic release of lipid A causes the release of the cytokine tumour necrosis factor-alpha (TNFα) from macrophages in the spleen and liver and other sites in the body. The result of the massive release of cytokines, sometimes referred to as a *cytokine storm*, is septic shock in which there is generalised increase in vascular permeability, leading to profound hypotension. Furthermore, TNFα activates the blood clotting cascade, leading to disseminated intravascular coagulation (DIC). Both hypotension and DIC lead to impaired perfusion of organs such as the lungs, heart, liver and kidneys. The O polysaccharides are antigenically diverse, being comprised of over 60 monosaccharides and over 30 non-carbohydrate components. Thus, the O polysaccharide can vary based on the sequence of monosaccharides ranging from homopolymers to heteropolymers, whether the polymers are linear or branched, the type of glycosidic linkage, the three-dimensional structure, and the presence or absence of non-polysaccharide components. This structural diversity is important in adhesion and evading host defences *via* antigenic variation, molecular mimicry and resistance of the bacterium to antibody and complement (see Chapter 9).

Although LOS glycolipids share the same toxic lipid A structures as LPS, they differ from LPS in the fact that they lack the O antigenic side chain, with LOS oligosaccharides limited to 10 monosaccharide units. In addition, LOS differs from LPS in that some LOS glycolipids are cross-reactive with human glycolipids, leading to what is termed molecular mimicry in which antibodies induced against LOS can bind structurally similar molecules on host cell membranes. A well-known example of this phenomenon is that certain persons infected with *Campylobacter jejuni* produce antibodies to an epitope on LOS that cross-react with the structurally identical GM1 gangliosides in nerve cell membranes. Antibodies, T cells and inflammatory cells attack the nerve cell, resulting in the peripheral neuropathy Guillain-Barré syndrome (Figure 6.3). Furthermore, LOS can be modified by the human host. One such modification is the sialylation of LOS that prevents complement deposition. Sialylation is accomplished by both bacterial and host sialyltransferases using the bacteria-derived or host-derived substrate cytidine 5'-monophosphate-*N*-acetylneuraminic acid (CMP-NANA).

KEY CONCEPTS

- Exotoxins of bacteria and parasites are secreted proteins.
- Toxins of fungi are non-protein secondary metabolites.
- Many toxins are enzymes that target cellular machinery.
- Many toxins form pores in the cell or endosomal membrane.
- LPS, a part of the outer leaflet of the outer membrane of gram-positive bacteria, is strongly pro-inflammatory.

BIBLIOGRAPHY

Aktories K. 2011. Bacterial protein toxins that modify host regulatory GTPases. *Nature Reviews Microbiology*, 9: 487–498.

Barth H et al. 2004. Binary bacterial toxins: Biochemistry, biology, and applications of common *Clostridium* and *Bacillus* proteins. *Microbiology and Molecular Biology Reviews*, 68(3): 373–402.

Dal Peraro M, van der Goot FG. 2016. Pore-forming toxins: Ancient, but never really out of fashion. *Nature Reviews Microbiology*, 14: 77–92.

Dowling RB, Wilson R. 1998. Bacterial toxins which perturb ciliary function and respiratory epithelium. *Journal of Applied Microbiology*, 85: 138S–148S.

Heuck AP et al. 2001. β-Barrel pore-forming toxins: Intriguing dimorphic proteins. *Biochemistry*, 40: 9065–9073.

Huber BT et al. 1996. Virus-encoded superantigens. *Microbiological Reviews*, 60(3): 473–482.

Johannes L, Römer W. 2010. Shiga toxins – From cell biology to biomedical applications. *Nature Reviews Microbiology*, 8: 105–116.

Linhartova I et al. 2010. RTX proteins: A highly diverse family secreted by a common mechanism. *FEMS Microbiology Reviews*, 34: 1076–1112.

Michalska M, Wolf P. 2015. *Pseudomonas* Exotoxin A: Optimized by evolution for effective killing. *Frontiers in Microbiology*, 6: 963.

Nash AA, Dalziel RG, Fitzgerald JR. 2015. *Mim's Pathogenesis of Infectious Disease*. 6th ed. Elsevier, London, UK.

Nathanson N. 2007. *Viral Pathogenesis and Immunity*. 2nd ed. Academic Press, Elsevier, London, UK.

Nieva JL et al. 2012. Viroporins: Structure and biological functions. *Nature Reviews Microbiology*, 10: 563–574.

Peraica M et al. 1999. Toxic effects of mycotoxins in humans. *Bulletin of the World Health Organization*, 77(9): 754–766.

Proft T, Fraser JD. 2003. Bacterial superantigens. *Clinical and Experimental Immunology*, 133: 299–306.

Raetz CRH, Whitfield C. 2002. Lipopolysaccharide endotoxins. *Annual Reviews of Biochemistry*, 71: 635–700.

Shen A. 2010. Autoproteolytic activation of bacterial toxins. *Toxins*, 2(5): 963–977.

Wilson BA, Salyers AA, Whitt DD, Winkler ME. 2011. *Bacterial Pathogenesis: A Molecular Approach*. 3rd ed. ASM Press, Washington, DC.

Wilson M, McNab R, Henderson B. 2002. *Bacterial Disease Mechanisms: An Introduction to Cellular Microbiology*. Cambridge University Press, Cambridge, UK.

Winer JB. 2008. Guillain-Barre syndrome. *British Medical Journal*, 337: a671.

Chapter 7: Extracellular Degradative Enzymes

INTRODUCTION

Degradative enzymes have been considered in earlier chapters wherever they are involved in mechanisms of pathogenesis. In this chapter, we examine degradative (hydrolytic) enzymes as classes, specifically proteases, glycosidases and phospholipases. All classes of microorganisms, covering the gamut from saprophyte to obligate pathogen, produce degradative enzymes. However, the significance of these enzymes in pathogenesis remains, with few exceptions, unclear. This is because there is a paucity of data that show that inactivating genes that specify degradative enzymes affects virulence in anything other than cells in culture or animal models. Neither of these models are necessarily extrapolatable to infections in human. The fact that saprophytic bacteria and fungi produce the widest range of degradative enzymes suggests that the primary function of these enzymes is to breakdown macromolecules to provide nutrients for the organism. For example, two saprophytic bacteria, *Clostridium perfringens* and *Pseudomonas aeruginosa*, whose natural habitat is soil and water but are also opportunistic bacterial human pathogens, produce a plethora of degradative enzymes. For example, 2.8% (155 genes) of the *P. aeruginosa* strain PAO1 genome (~5,568 genes) encode proteases with the following distribution: 84 (54%) serine proteases, 45 (29%) metalloproteases, 11 (7%) cysteine proteases, 5 (3%) threonine proteases, 3 (2%) aspartic proteases and 7 (5%) unassigned.

Glycosidases are important, primarily, in nutrient (carbon) acquisition as has been shown for commensal oral and gut bacteria. However, it is possible that these enzymes contribute to pathogenicity.

Lipases, particularly phospholipases, are considered a determinant of pathogenesis of many bacteria, fungi, viruses and parasites because the cell membrane is rich in phospholipids.

In addition to secreted degradative enzymes that degrade host molecules, both pathogenic prokaryotes and eukaryotes possess cytoplasmic and membrane proteases that play important roles in controlling expression and solubility of determinants of pathogenesis. However, the focus of this chapter is host-directed, secreted degradative enzymes. The role(s) of degradative enzymes in pathogenesis depend on the environments in which the bacteria that produce them operate. For example, pathogens attempting to cross mucosal surfaces encounter consortia of resident microorganisms both attached to the epithelial surface and in the planktonic phase. Each member of the consortium may produce one or more degradative enzymes that impact the activity of enzymes produced by the pathogen. Thus, the net effect of various degradative enzymes on host cells and molecules are difficult to predict, and are not at all comparable to a single enzyme–substrate interaction in a test tube.

PROTEASES

Many extracellular proteases are secreted in an inactive form termed a zymogen. Proteases may have a narrow or broad range of specificity. Although this chapter focuses on pathogen proteases that degrade host proteins, proteases produced by pathogens are involved in degrading pathogen surface proteins once they have performed their function(s). Furthermore, proteases may play a role in biofilm dispersal. Virtually all host proteins including those of the immune system can be targets of pathogen proteases. An overview of the proposed roles of bacterial and fungal proteases in pathogenesis is shown in **Figures 7.1** and **7.2**, respectively.

POTENTIAL ROLES OF MICROBIAL PROTEASES IN PATHOGENESIS

Tissue destruction and cell internalisation

Members of the aspartic protease family (Sap) produced by the dimorphic fungus *Candida albicans* are reported to play a role in adhesion to, and

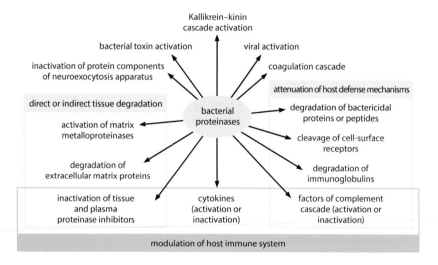

Figure 7.1 Modulation of host immune system by bacterial proteases. (From Maeda, H. and Yamamoto, T., *Biol. Chem. Hoppe-Seyler*, 377, 217–226, 1996.)

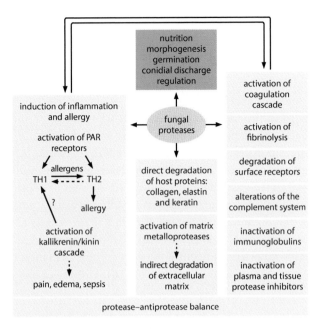

Figure 7.2 Schematic showing multiple functions of fungal proteases. (From Yike, I., *Mycopathologia*, 171, 299–323, 2011.)

invasion of, epithelia and evasion of the immune system. Several species of the genus *Plasmodium* produce cysteine and serine proteases that are involved in entrance into and exit from erythrocytes. Cysteine and particularly serine proteases are thought to play a role in the entry of *Trypanosoma cruzi* into the cells they infect. In addition, cysteine proteases are implicated in the internalisation of *Leishmania* species and tissue invasion and destruction by *Entamoeba histolytica* and *Naegleria fowleri*. Cysteine proteases are thought to play a role in the pathogenesis of *Trichomonas vaginalis* by allowing the parasite to degrade the mucus layer and reach the vaginal epithelial surface and in tissue invasion by the nematode *Strongyloides stercoralis* and the hookworm *Necator americanus*. Similarly, tissue invasion of the larval stages of cestodes and trematodes is likely mediated by proteases.

Inactivation of plasma protease inhibitors

Microbial proteases inactivate plasma serine protease inhibitors. Host plasma protease inhibitors control a variety of biological cascades that include coagulation and inflammation and three—alpha 1-antitrypsin, alpha 1-antichymotrypsin, and alpha 2-macroglobulin—are released in increased amounts from the liver during the systemic response to infection-induced inflammation, termed the acute phase response. These host-produced protease inhibitors, particularly α2 macroglobulin, are important in inhibiting microbial proteases. *Pseudomonas aeruginosa*, *Serratia marcescens*, and *Candida albicans* degrade α2 macroglobulin. *Streptococcus pyogenes* binds α2-macroglobulin and prevents the protease from degrading other surface proteins of the bacterium.

Activation of bradykinin-generating and blood-clotting cascades

Bacterial and fungal proteases can activate the bradykinin-generating and the blood clotting cascade. Bradykinin is a molecule that is important in vasodilation and the cascade can be interrupted by bacteria and fungi cleaving Hageman factor and/or prekallikrein. Bacteria and fungi can interrupt the blood clotting cascade by cleaving Hageman factor (factor XII), factor X and prothrombin and by inactivating clotting inhibitors such as antithrombin III. Vasodilatation allows recruitment of neutrophils and other immune cells from the blood but also allows easier access of pathogens to the blood circulatory system. Clotting blood vessels at the site of infection prevents microorganisms entering the blood circulatory system, so it is an advantage to pathogens if they can inhibit formation of this barrier. A diagram of the effects of proteases on the clotting cascade is shown in **Figure 7.3**.

Protease-activated receptor

Another target of microbial proteases is protease-activated receptors (PARs). PARs are expressed by epithelial cells, endothelial cells and leucocytes, and they regulate leucocyte function and pro-inflammatory signalling pathways. An effective inflammatory response is an essential component of host resistance to microbial infection. Thus, down-regulation of inflammation impairs the immune response to pathogenic microbes. Some bacterial proteases that affect PARs are shown in **Table 7.1**.

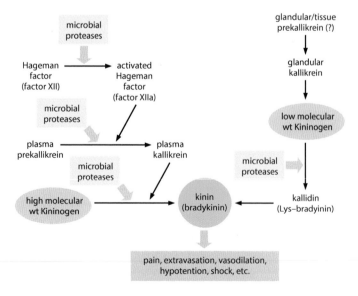

Figure 7.3 Effects of microbial proteases on the clotting cascade. (From Maeda, H. and Yamamoto, T., *Biol. Chem. Hoppe-Seyler*, 377, 217–226, 1996.)

bacteria and protease	targeted PAR and cells	effect
P. aeruginosa LasA (elastase)	PAR-2/respiratory epithelial cells	disarm PAR-2 in regard to any further activating proteolysis by activating protease
LepA	PAR-1, PAR-2 and PAR-3/PAR transfected COS-7 cells and human bronchiole epithelial cell line	activate NF-κB-driven promoter, induce IL-8 production through PAR-2 in EBC-1 cells
S. marcescens Serralysin	PAR-2/human lung squamous cell carcinoma (EBC-1 cells)	activate the critical transcription factors AP-1, C/EBPβ, and NF-κB, induced 1L-6 and IL-8 mRNA expression

Table 7.1   Bacterial proteases affecting PARS. (From Potempa, J. and Pike, R.N., *J. Innate Immun.*, 1, 70–87, 2009.)

Chemoattractant molecules

Chemoattractant molecules such as complement components C5a and C3a are targets of bacterial and fungal proteases as is formyl-methionyl -leucyl-phenylalanine (f-MLP). F-MLP is an *N*-formylated tripeptide that is a potent chemoattractant molecule for polymorphonuclear leucocytes (neutrophils) and macrophages. Some bacterial proteases that target components of the innate immune system are shown in **Table 7.2**.

Immunoglobulins

Immunoglobulins are a target of microbial proteases and, generally, they are cleaved at the hinge region, an area that is rich in the amino acid proline. Indeed, the structure of the immunoglobulin G (IgG) molecule was determined using proteases-pepsin from the mammalian stomach and from the papaya plant. These proteases cleave the heavy chain on either side of the disulphide bonds at the hinge region to generate, in the case of IgG, fragments that retain antigen-binding capacity either Fab or Fab'2 and a crystallisable fragment termed Fc. However, although cleavage of other immunoglobulin isotypes yields intact Fab fragments, the Fc fragment is generally degraded. Cleaving IgM and IgG eliminates their ability to activate the classical pathway of the complement cascade and their function as opsonins and reduces the strength of binding to antigens because the binding of monomeric Fab fragments to antigen is some 10^3- to 10^4-fold less than it is for the bivalent Fab'2 fragment.

The barrier epithelia are colonised with complex microbiotas, so proteolytic activity is high. Although IgG and IgM can both be actively transported across epithelia using the neonatal FcRn receptor and the polymeric immunoglobulin receptor (poly-Ig receptor), respectively, once released onto the mucosal surface, both are highly susceptible to proteolysis. Consequently, there is little intact IgG and IgM in external secretions. In contrast, polymeric secretory immunoglobulin A (SIgA), the principal antibody isotype in mucosal secretions, is much more resistant to proteolysis than are the other immunoglobulin isotypes.

pathogen	protease name(s)	targets
Staphylococcus aureus	staphopain A (ScpA)	contact activation (HMWK)
S. aureus	staphopain B (SspB)	1. contact activation (HMWK) 2. phagocytes (neutrophils and monocytes)
S. aureus	aureolysin	1. AMPs (LL-37) 2. cytokine receptors (IL-6R)
Streptococcus pyogenes	streptopain (SpeB)	1. contact activation (HMWK) 2. complement (C3, properdin) 3. AMPs (LL-37) 4. phagocytes (neutrophils) 5. cytokines [IL-1β precursor (pIL-1 β)] 6. phagocyte function (uPAR)
S. pyogenes	C5a peptidase (ScpA)	complement (C5a)
S. pyogenes	*Sly*CEP (ScpC)	chemokines (IL-8, KC, MIP-2, GCP-2, GROα)
Yersinia pestis	Pla (plasminogen activator)	fibrinolysis (plasminogen, α_2-antiplasmin)
Salmonella enterica serovar Typhimurium	PgtE (OmpE, protein E)	1. complement (C3b, C4b, C5) 2. AMPs (LL-37)
Escherichia coli	OmpT (opmtin)	AMPs (protamine)
E. coli O157:H7	StcE	complement (C1-inhibitor)
E. coli	ElaD	inflammatory signaling pathway (deubiquitination)
E. coli	Hbp (Tsh peptidase)	iron scavenging (Hb)
Enterococcus faecalis	gelatinase (GelE, coccolysin)	1. complement (C3) 2. AMPs (LL-37)
Yersinia pseudotuberculosis, Yersinia enterocolitica	YopJ/YopP	inflammatory signaling pathway
S. enterica serovar Typhimurium	SseL (ElaD)	inflammatory signaling pathway (deubiquitination)
Chlamydia trachomatis	*Chly*Dub1 and *Chly*Dub2	inflammatory signaling pathway (deubiquitination and de-NEDDylation)
Xanthomonas campestris	XopD	inflammatory signaling pathway (de-SUMOlylation)
Y. pestis	YopT	actin polymerization (small GTPases)
Pseudomonas syringae	AvrPphB	intracellular signaling (small GTPases)
Bacillus anthracis	anthrax lethal toxin	inflammatory signaling pathway (MKK)
Finegoldia magna	SufA	AMPs (LL-37)
Proteus mirabilis	ZapA (mirabilysin)	1. AMPs (LL-37, hBD1) 2. complement (C1q and C3)
Pseudomonas aeruginosa	elastase (LasB, pseudolysin)	1. AMPs (LL-37) 2. PAR-2 3. chemokines (RANTES, MCP-1, ENA-78) 4. cytokines (IL-6, IL-8, IFN-γ) 5. cytokine receptors (IL-6R) 6. phagocyte functions (uPAR, fMLP receptor)
P. aeruginosa	alkaline protease (aeruginolysin)	chemokines (RANTES, MCP-1, ENA-78)
P. aeruginosa	LasA protease (staphylolysin)	AMPs (IFN-γ, IL-2)
P. aeruginosa	LepA	PARs
Serratia marcescens	serralysin	1. PAR-2 2. cytokine receptors (IL-6R)
Treponema denticola	dentilisin (PrtP, trepolisin)	1. cytokines (IL-1β, IL-6, TNF-α) 2. complement (C3)
Legionella pneumophila	Msp peptidase	cytokines (IL-2)

Table 7.2 Bacterial proteases potential roles in pathogenesis. (From Potempa, J. and Pike, R.N., *J. Innate Immun.*, 1, 70–87, 2009.)

This is because a heavily glycosylated protein, the secretory component, that is derived from the poly-Ig receptor shields the hinge region. However, several mucosal pathogenic and commensal bacteria produce an IgA1 protease that is capable of cleaving subclass 1 of SIgA. Such bacteria include *Streptococcus pneumoniae, S. sanguinis, S. oralis, S. mitis* biovar 1 (some strains), *Neisseria meningitidis, N. gonorrhoeae, Haemophilus influenzae,* and *Ureaplasma urealyticum.* SIgA of subclass 2 is resistant to the action of IgA1 protease because it lacks a 13-amino acid sequence at the hinge region which contains the susceptible sites for cleavage. *Clostridium ramosum* an anaerobic gram-positive, spore-forming rod that is member of the commensal microbiota of the large intestine does cleave the SIgA2 heavy chains *C*-terminal to the 13-amino acid peptide deleted in the shortened hinge region. Because IgA1 protease is produced by mucosal bacteria, some of which are significant pathogens, IgA1 protease was considered to be an important determinant of pathogenesis capable of impairing the mucosal IgA antibody response. However, are the facts that several commensal bacteria also produce IgA1 protease and most mucosal secretions contain roughly equal concentrations of SIgA1 and SIgA2, and SIgA2 is resistant to IgA1 protease cleavage. Furthermore, when groups of adult male volunteers were infected *via* the urethra with either a wild-type strain of *Neisseria gonorrhoeae* that produces IgA1 protease or an isogenic mutant strain in which the gene encoding IgA1 protease was knocked out, no difference in any aspect of gonorrhoea infection between the groups was observed. The interaction of IgA1 protease with its substrate in mucosal secretions is likely far more complex than a simple binary interaction. Bacterial glycosidases, by virtue of their ability to remove sugars from hinge region carbohydrates, are capable of both increasing and decreasing the susceptibility of SIgA1 to IgA1 protease, depending on the extent of deglycosylation. Limited deglycosylation exposes the enzyme active site for more efficient cleavage, whereas extensive deglycosylation decreases cleavage of the hinge region of the heavy chains of the immunoglobulin. The above observations suggest that IgA1 protease probably exerts little effect on mucosal immunity. However, IgA1 protease can cleave substrates other than IgA1. A consensus sequence of the IgA1 heavy chain that is the site of protease cleavage is found in other proteins such as CD8 and the lysosome/late endosome (phagosome) membrane protein 1 (h-lamp-1). h-lamp-1 is thought to protect the membrane of lysosomes from the acid proteases it contains by forming a carbohydrate lining on the lumenal surface of the lysosome. Therefore, IgA1 protease may exert its effect by modulating the lysosomal environment of bacteria that secrete this enzyme and are internalised by epithelial cells and/or phagocytes. Many parasites also have cysteine proteases with the ability to degrade immunoglobulins. These include *Trichomonas vaginalis, Entamoeba histolytica, Giardia muris, Fasciola hepatica, F. gigantica, Taenia solium, T. rassiceps* and *Sprirometra* species.

MICROBE AND PARASITE GLYCOSIDASES

Many proteins that are part of the human immune system, such as mucins and immunoglobulins, are glycosylated and, as such, are targets of glycosidases of pathogenic microorganisms. Glycosylation of human proteins occurs in two forms, *O*-linked and *N*-linked. *O*-linked glycosylation involves the addition of *N*-acetyl-galactosamine (GalNAc) to serine or threonine residues. On the other hand, *N*-linked glycosylation involves the addition of *N*-acetylglucosamine (GlcNAc) to asparagine (Asn) that is present as a part of an Asn-X-Ser/Thr consensus sequence, where X is any amino acid except proline. Glycosidases are either exoglycosylases that cleave the terminal sugar of the glycan or endoglycosidases that cleave internal sugars of the glycan.

Deglycosylation of immunoglobulins

Several pathogenic bacteria target glycans at the hinge region and the CH2 domain of the heavy chains of, potentially, all immunoglobulin isotypes. This can result in the inability of the component C1q of the classical pathway of complement to bind to IgG and to impair binding of IgG to Fcγ receptors on professional phagocytes. *Streptococcus pyogenes* is an example of a bacterium that produces an endo-β-*N*-acetylglucosaminidase (EndoS) that cleaves the β1-4 bond between internal *N*-acetylglucosamine sugars of the IgG glycan to accomplish these effects.

Interestingly, glycosidases from the commensal α-haemolytic streptococcus *S. mitis* biovar 2 can progressively deglycosylate hinge region glycans of IgA1, resulting in increased susceptibility of the immunoglobulin to cleave by IgA1 protease. However, treatment with neuraminidase and endo-α-*N*-acetylgalactosaminidase decrease susceptibility to cleavage suggesting that glycosidases from endogenous bacteria may modulate the effect of IgA1 protease on exocrine IgA1. This is an example of the complexity of interactions between pathogenic and commensal microorganisms and the host.

Adhesion

As discussed in Chapter 4, *Adhesion to Host Surfaces*, adhesion to the barrier epithelium is a prerequisite for infection by almost all pathogens. Glycosidases can modify host cell molecules to serve as ligands for microbial adhesins. For example, *Streptococcus pneumoniae* possesses three cell wall–anchored exoglycosidases—neuraminidase, β-galactosidase, and β-*N*-acetylglucosaminidase—that cleave sialic acid, galactose, and *N*-acetylglucosamine, respectively. These are thought to act sequentially to cleave terminal sugars from host cell membrane glycoconjugates which may expose ligands for adhesion. Furthermore, they may play a role in the adhesion of the bacterium to brain microvascular cells so that

organism	glycosidase	specificity	known glycoprotein substrate(s)
S. pyogenes	EndoS	biantennary N-linked glycan on IgG	IgG
S. pyogenes	EndoS49	biantennary N-linked glycan on IgG	IgG
group C streptococci	EndoC	biantennary N-linked glycan on IgG	IgG
S. pneumoniae	NanA	sialic acid α2-3 or α2-6 linked to galactose	IgA1, hLF, hSC, α1-acid glycoprotein
S. pneumoniae	BgaA	galactose β1-4 linked to N-acetylglucosamine	IgA1, hLF, hSC, α1-acid glycoprotein
S. pneumoniae	StrH	N-acetylglucosamine β1 linked to mannose	IgA1, hLF, hSC, α1-acid glycoprotein
S. pneumoniae	EndoD	complex N-linked glycans	TF, fetuin, IgG
S. pneumoniae	EngSP	core-1 O-linked glycans	fetuin
E. faecalis	EndoE	biantennary and high mannose N-linked glycan (α-domain), biantennary N-linked glycan on IgG (β-domain)	RNaseB, hLF, IgG
E. faecalis	EngEF	core-1 and core-3 O-linked glycans	unknown
P. acnes	PPA1560	unknown	unknown
P. acnes	EngPA	core-1 and core-3 O-linked glycans	unknown
V. cholerae	NanH	sialic acid from higher-order gangliosides	GM$_1$ ganglioside

Table 7.3 TF, Transferrin; hSC, human secretory component; hLF, human lactoferrin. (From Garbe, J. and Collin, M., *J. Innate Immun.*, 4, 121–131, 2012.)

S. pneumoniae can cross the blood-brain barrier and infect the meninges. Examples of bacterial glycosidases that contribute to pathogenesis are listed in **Table 7.3**.

Viral neuraminidase cleaves terminal sialic acid residues from glycan structures on the membrane of infected cells and facilitates the release of virions and the spread of the virus. The best-known neuraminidase is that of influenza A virus. The influenza A neuraminidase is an exosialidase which cleaves the linkage between N-acetylneuraminic acid and an adjacent sugar. The influenza virus neuraminidase plays a role during budding of newly formed viral particles from the surface of the infected cell to prevent aggregation of viral particles, and it cleaves neuraminic acid residues from the respiratory tract mucins to facilitate virus movement to the target cell.

Protozoan parasites whose habitat is mucosal surfaces such as the intestines and the genitourinary tract produce many glycosidases, some of which are capable of degrading the mucin that hydrates and protects the barrier epithelia. The urogenital protozoan *Trichomonas vaginalis* produces a number of glycosidases that include α-N-acetylglucosaminidase, β-N-acetylgalactosaminidase, and β-galactosidase, some of which degrade mucin. The intestinal protozoans *Giardia lamblia* and *Entamoeba* produce β-N-acetylglucosaminidase, and *G. lamblia* produces β-N-acetylglucosaminidase and β-N-acetylgalactosaminidase (**Table 7.4**). These glycosidases help degrade mucin and allow access of the parasites to the mucosal epithelium.

enzyme	total activity (U mL⁻¹)ᵃ						
	T. foetus	T. vaginalis	G. lamblia	E. histolytica	L. donovani	T. brucei	T. cruzi
α-Galactosidase	4	4	0.2	–	–	–	–
β-Galactosidase	20	15	–	–	0.2	–	–
α-Glucosidase	2	2	–	–	0.2	0.2	0.5
β-Glucosidase	7	6	–	–	0.2	–	–
β-Glucuronidase	–	1	–	–	–	–	–
α-N-Acetyl Galactosaminidase	22	25	–	–	–	–	–
β-N-Acetyl Galactosaminidase	3	2	–	2	0.5	–	–
α-N-Acetyl Glucosaminidase	–	1	0.5	–	–	–	–
β-N-Acetyl Glucosaminidase	30	30	1.5	25	0.5	–	–
α-Mannosidase	1	2	0.5	0.5	–	–	–
β-Mannosidase	–	–	NT	NT	–	NT	1.5
β-Xylosidase	–	–	0.2	–	–	NT	NT

Table 7.4 Production of glucosidases by protozoans. ᵃEnzyme activity from 1 mL supernatant plus 1 mL cell lysate. '–' 0.1 U mL⁻¹ or below; NT not tested. (From Connaris, S. and Greenwell, P., *Glycoconj. J.*, 14, 879–882, 1997.)

MICROBE AND PARASITE PHOSPHOLIPASES

Phospholipases are considered a determinant of pathogenesis of many bacteria, fungi, viruses and parasites. The cell membrane, because it is rich in phospholipids, is the principal target of phospholipases and these enzymes can result in extensive tissue destruction. In addition to causing tissue destruction, phospholipases can induce the release of pro-inflammatory cytokines and interfere with cell signalling pathways by interacting with protein kinase B, also known as Akt, a serine/threonine-specific protein kinase. Protein kinase B is involved in several cellular activities including endocytosis and vesicular migration that are important in destroying internalised pathogens.

Bacterial phospholipases

Extracellular bacteria in which phospholipases are considered to contribute to pathogenesis include *Clostridium perfringens, Helicobacter pylori, Pseudomonas aeruginosa, Yersinia enterocolitica* and *Neisseria gonorrhoeae*. The α-toxin of *C. perfringens* is a phospholipase C (PLC) and is considered the principal toxin of this bacterium because it produces extensive cellular destruction and impairs blood supply at the site of infection, allowing proliferation of the bacterium. In the case of the gastric pathogen *H. pylori*, the phospholipase PldA, that is located in the outer membrane (OM), is able to reversibly alter the proportion of lysophospho-lipids in the OM. Lysophospholipid is the name given to any derivative of a phospholipid in which one or both acyl derivatives have been removed by hydrolysis. High levels of lysophospholipids are associated with higher haemolytic activity, increased release of urease and vacA, better adher-ence to epithelial cells and acid tolerance, all of which contribute to the pathogenesis of this bacterium.

Phospholipases also are implicated in the pathogenicity of several intra-cellular respiratory bacterial pathogens. Phospholipases of *Mycobac-terium tuberculosis* are thought to provide fatty acids as carbon sources for the bacteria in their intracellular location in alveolar macrophages, whereas phospholipases of *Legionella pneumophila* degrade lung sur-factant, providing nutrients and impairing lung function. Furthermore, a phospholipase is among the proteins injected into the host by the *L. pneumophila* Icm/Dot type 4 secretion system. The enteric pathogen *Salmonella enterica* serovar Typimurium when taken up by enterocytes and macrophages uses a phospholipase (SeeJ), an effector of its type III secretion system, to alter the phagosome/endosome membrane of cells by which it has been uptaken. SseJ localises at the cytoplasmic face of the phagosome/endosome, often termed the salmonella-containing vacuole (SCV), and esterifies cholesterol influencing the maturation of the SCV (**Figure 7.4**).

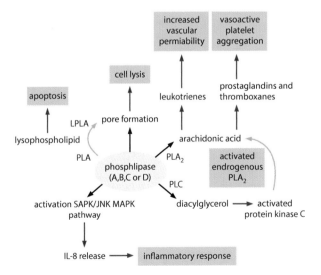

Figure 7.4 Eukaryotic signalling pathways induced by the action of bacterial phospholipases and corresponding effects on the host cell. Arrows do not nec-essarily indicate a direct induction. (From Bender, J. and Flieger, A., *Handbook of Hydrocarbons*, 68, 3241–3258, 2010.)

Fungal phospholipases

Pathogenic fungi that secrete phospholipase include *Candida albicans* and other *Candida* species, *Cryptococcus neoformans*, and *Aspergillus fumigatus*. *C. albicans* produces phospholipase B1 (PLB1), PLB2, and PLB5 and PLA, PLC, and PLD, and *C. neoformans* and *Cryptococcus gattii*, PLB1. PLBs are lysophospholipases. *Aspergillus fumigatus* secretes multiple extracellular phospholipases, including PLA, PLB, PLC, and PLD. PLB enzymes are thought to contribute to fungal invasion of host tissues by *C. albicans* and *C. neoformans*. For example, *C. neoformans* phospholipases are able to hydrolyse lung surfactant and the cell membrane of lung epithelium and are important in experimental the spread of *C. neoformans* from the lung to the brain *via* blood and lymph in experimental animals. PLBs are important for the survival of *C. neoformans* in macrophage phagosomes and escape from these phagocytes. PLBs are essential for the synthesis of bioactive anti-inflammatory eicosanoids by *C. neoformans* that have the potential to suppress the host immune system, promoting the survival, and dissemination of *C. neoformans* within the host. Although *Candida albicans* produces PLA, PLB, PLC and PLD, their role in human infections remains unclear. Examination of sera from patients with blood culture–positive invasive candidiasis showed antibodies reactive with PLB. However, although PLB is produced by the fungus during infection its role in pathogenesis remains unclear.

Parasite phospholipases

There is little direct evidence to support a role of phospholipases in human parasite infections. However, experiments performed *in vitro* implicate PLA in the penetration of host cells by *Toxoplasma gondii* and *Entamoeba histolytica*, which is, perhaps, not unexpected.

KEY CONCEPTS

- The principal role of degradative (hydrolytic) enzymes is to provide nutrients for the microorganism.
- Degradative enzymes may contribute, incidentally, to host damage and, thus, may be considered determinants of pathogenesis.
- With the exception of exotoxins that are enzymes, there is little direct evidence for the role of degradative enzymes in pathogenesis.

BIBLIOGRAPHY

Bandana K et al. 2018. Phospholipases in bacterial virulence and pathogenesis. *Advances in Biotechnology and Microbiology*, 10(5): 555798.

Bender J, Flieger A. 2010. Lipases as pathogenicity factors of bacterial pathogens of humans. In: Timmis KN (Ed.), *Handbook of Hydrocarbon and Lipid Microbiology*. Springer, Berlin, Germany.

Calderone RA. Cihlar RL. 2002. *Fungal Pathogenesis: Principles and Clinical Applications* (Mycology Book 14). CRC Press, Boca Raton, FL.

Connaris S, Greenwell P. 1997. Glycosidases in mucin-dwelling protozoans. *Glycoconjugate Journal*, 14(7): 879–882.

Djordjevic JT. 2010. Role of phospholipases in fungal fitness, pathogenicity, and drug development—Lessons from *Cryptococcus neoformans*. *Frontiers in Microbiology*, Article 125.

Flores-Díaz M et al. 2016. Bacterial sphingomyelinases and phospholipases as virulence factors. *Microbiology and Molecular Biology Reviews*, 80(3): 597–628.

Frees D et al. 2013. Chapter 7, Bacterial proteases and virulence. In: Dougan DA (Ed.), *Subcellular Biochemistry* (Book 66). Springer Science+Business Media, Dordrecht, the Netherlands.

Garbe J, Collin M. 2012. Bacterial hydrolysis of host glycoproteins—Powerful protein modification and efficient nutrient acquisition. *Journal of Innate Immunity*, 4: 121–131.

Ghannoum MA. 2000. Potential role of phospholipases in virulence and fungal pathogenesis. *Clinical Microbiology Reviews*, 13(1): 122–143.

Gómez-Arreaza A. 2014. Extracellular functions of glycolytic enzymes of parasites: Unpredicted use of ancient proteins. *Molecular & Biochemical Parasitology*, 193: 75–81.

Gupta SP (Ed.). 2017. *Viral Proteases and Their Inhibitors*. Academic Press/Elsevier, London, UK.

Hoge R et al. 2010. Weapons of a pathogen: Proteases and their role in virulence of *Pseudomonas aeruginosa*, In: Mendes-Vilas A (Ed.), *Current Research, Technology and Education Topics in Applied Microbiology and Microbial Biotechnology*, Formatex Research Center Series No. 2, Vol. 1, pp. 383–395.

Ingmer H, Brøndsted L. 2009. Proteases in bacterial pathogenesis. *Research in Microbiology*, 160(9): 704–710.

Istivan TS, Coloe PJ. 2006. Phospholipase A in gram-negative bacteria and its role in pathogenesis. *Microbiology*, 152: 1263–1274.

Kalaiselvi G. 2014. Role of phospholipase and proteinase as virulence factors of *Candida albicans* isolated from clinical samples of inpatients in a tertiary care hospital. *International Journal of Scientific Research and Reviews*, 3(2): 167–176.

Loker ES, Hofkin BV. 2015. *Parasitology: A Conceptual Approach*. Garland Science, New York.

Maeda H, Yamamoto T. 1996. Pathogenic mechanisms induced by microbial proteases in microbial infections. *Biological Chemistry Hoppe-Seyler*, 377: 217–226.

Miyoshi S-I. 2013. Extracellular proteolytic enzymes produced by human pathogenic *Vibrio* species. *Frontiers in Microbiology*, Article 339.

Nash AA, Dalziel RG, Fitzgerald JR. 2015. *Mim's Pathogenesis of Infectious Disease*. 6th ed. Elsevier, London, UK.

Parage MG et al. 2008. *Candida albicans*-secreted lipase induces injury and steatosis in immune and parechymal cells. *Canadian Journal of Microbiology*, 54: 647–659.

Park M et al. 2013. Lipolytic enzymes involved in the virulence of human pathogenic fungi. *Mycobiology*, 41(2): 67–72.

Pina-Vazquez C et al. 2012. Host-parasite interaction: Parasite-derived and induced proteases that degrade human extracellular matrix. *Journal of Parasitology Research*, 2012, Article ID 748206.

Potempa J, Pike RN. Corruption of innate immunity by bacterial proteases. *Journal of Innate Immunity*, 1: 70–87.

Schmiel DH, Miller VL. 1999. Bacterial phospholipases and pathogenesis. *Microbes and Infection*, 1(13): 1103–1112.

Sjögren J, Collin M. 2014. Bacterial glycosidases in pathogenesis and glycoengineering. *Future Microbiology*, 9(9): 1039–1051.

Stehr F et al. 2003. Microbial lipases as virulence factors. *Journal of Molecular Catalysis B: Enzymatic*, 22: 347–355.

Wiggins R et al. 2001. Mucinases and sialidases: Their role in the pathogenesis of sexually transmitted infections in the female genital tract. *Sexually Transmitted Infections*, 77: 402–408.

Wilson BA, Salyers AA, Whitt DD, Winkler ME. 2011. *Bacterial Pathogenesis: A Molecular Approach*. 3rd ed. ASM Press, Washington, DC.

Wilson M, McNab R, Henderson B. 2002. *Bacterial Disease Mechanisms: An Introduction to Cellular Microbiology*. Cambridge University Press, Cambridge, UK.

Yike I. 2011. Fungal proteases and their pathophysiological effects. *Mycopathologia*, 171(5): 299–323.

Chapter 8: Evasion of the Human Innate Immune System

INTRODUCTION

Traditionally the human immune system has been considered to comprise two parts, the innate immune system and the adaptive immune system. However, in reality innate and adaptive immunity interface seamlessly. The innate immune system is ancient, and elements of it are found in single-celled organisms. It responds immediately on contact with the infectious agent and is a low specificity, pattern recognition system. In comparison, the adaptive immune system evolved in jawed fish. The adaptive immune system is slow in onset but highly specific in its response and exhibits memory in that it responds more rapidly and with greater magnitude and specificity to a second and subsequent exposures/encounters with the same agent. In this chapter, we examine the ways in which various classes of pathogen attempt to evade the host immune system and focus on common themes employed by them. We begin with evasion of innate immunity and then proceed to consider evasion of adaptive immunity in the next Chapter.

ANTIMICROBIAL PEPTIDES

Overview

Antimicrobial peptides (AMPs) are sub–10-kDa, highly cationic hydrophobic peptides that are derived from pro-proteins by proteolytic cleavage. They are found in external secretions that protect the barrier epithelia by which they are produced. AMPs are also secreted by neutrophils, macrophages, mast cells, dendritic cells (DCs), endothelial cells, platelets and adipocytes. In humans, there are several classes of AMP: defensins (α-defensins, β-defensins), cathelicidin LL-37, lactoferricin, lactoferrampin, LF1-11, and histatins 1, 3 and 5. The skin antimicrobial peptide, dermcidin is anionic rather than cationic.

Some AMPs are widely distributed across mucosal surfaces, whereas others, such as the histatins, are found in saliva only. The AMPs produced by activated platelets are derived from chemokines by proteolysis and are termed thrombocidins and kinocidins. They are released directly into the bloodstream. These microbicidal molecules have activity against gram-negative and gram-positive bacteria, fungi, some enveloped viruses, and protozoan parasites. However, the microbicidal activity of dermcidin and the histatins appear to be limited to bacteria and fungi.

The precise mechanism by which AMPs insert into the cytoplasmic membrane to form pores or to disrupt the organisation of the cytoplasmic membrane is yet to be fully determined, and proposed models are shown in **Figure 8.1**. However, the mechanism(s) by which AMPs traverse the thick peptidoglycan layer of the gram-positive cell wall, the outer

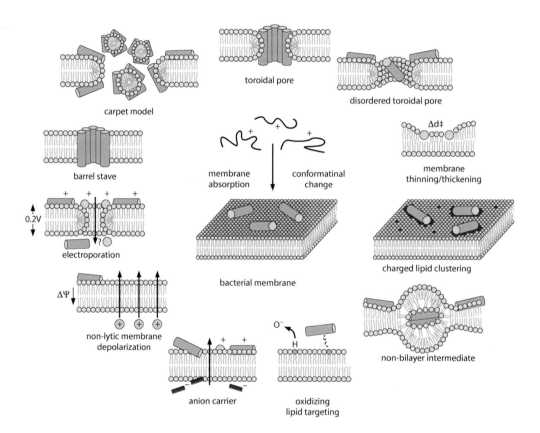

Figure 8.1 Potential mechanism of action of antimicrobial peptide (AMP). Events occurring at the bacterial cytoplasmic membrane following initial AMP adsorption. These events are not necessarily exclusive of each other. In the classical models of membrane disruption, the peptides lying on the membrane reach a threshold concentration and insert themselves across the membrane to form peptide-lined pores in the barrel-stave model, solubilise the membrane into micellar structures in the carpet model, or form peptide-and-lipid lined pores in the toroidal pore model. In the revised disordered toroidal pore model, pore formation is more stochastic and involves fewer peptides. The thickness of the bilayer can be affected by the presence of the peptides, or the membrane itself can be remodelled to form domains rich in anionic lipids surrounding the peptides. In more specific cases, non-bilayer intermediates in the membrane can be induced; peptide adsorption to the membrane can be enhanced by targeting them to oxidised phospholipids; a peptide may couple with small anions across the bilayer, resulting in their efflux; the membrane potential can be dissipated without other noticeable damage; or conversely, in the molecular electroporation model, the accumulation of peptide on the outer leaflet increases the membrane potential above a threshold that renders the membrane transiently permeable to various molecules including the peptides themselves. (From Nguyen, L.T. et al., *Trends Biotechnol.*, 29, 464–472, 2011.)

membrane of the gram-negative cell wall and the fungal cell wall to reach the cytoplasmic membrane remain unclear.

The microbicidal activity of AMPs is considered to result from their ability to disrupt the integrity of the cytoplasmic membrane such that susceptible microbes die as a result of osmotic lysis. However, in addition to the cytoplasmic membrane, AMPs appear to have cellular targets, including DNA and protein synthesis, protein folding, enzymatic activity and cell wall synthesis. The effects of AMPs on fungi have largely been limited to the study of *Candida albicans*, although histatins have been shown to be fungicidal for *C. dubliniensis*, *Cryptococcus neoformans* and *Aspergillus fumigatus* also. AMPs target fungal cell wall synthesis, which may result in cell lysis. Some AMPs such as lactoferricin have been shown to severely damage the candida cell wall. The mechanisms by which AMPs exert antifungal activity include the generation of reactive oxygen species (histatin and lactoferricin), and attack on mitochondria (histatin).

AMPs have been shown to inhibit infection by RNA- and DNA-enveloped viruses and the non-enveloped viruses, adenovirus and echovirus 6. AMPs target the adsorption to, and entry of, viruses into host cells or the viral envelope itself. Heparan sulphate is a highly negatively charged molecule on the surface of mammalian cells that serves as a frequent ligand for virus attachment. Thus, because AMPs have a strong net positive charge, they are able to bind to heparan sulphate and sterically hinder virus attachment. This mechanism has been demonstrated for herpes simplex virus (HSV) and human immunodeficiency virus-1 (HIV-1). Examples of AMPs binding specific cellular receptors have been described. A case in point is the binding of AMPs to the chemokine receptor CXCR4 on T cells, thereby inhibiting HIV-1 entry. In addition, AMPs have been shown to inhibit cell-to-cell spread and the formation of syncytia by inactivating viruses, although the mechanism(s) by which this is accomplished are unclear. Defensins have been shown to directly bind to envelope glycoproteins of HSV-2 and HIV-1 preventing fusion of virus to the cell membrane. That AMPs can bind to and disrupt virus envelopes has been shown for HIV-1, HSV-1, HSV-2, vesicular stomatitis virus VSV and influenza A virus. Some AMPs can cross the cell and nuclear membranes and have been shown to interact directly with virus nucleic acids of HIV-1.

The anti-parasite effects of AMPs appear to be mediated *via* mechanisms akin to those employed against bacteria and fungi, that is, the formation of pores and disruption of the cytoplasmic membrane. AMPs have activity against protozoa such as *Leishmania* and *Trypanosoma*.

The functions of AMPs extend beyond their antimicrobial activity. They are chemotactic for DCs, neutrophils, monocyte/macrophages, mast cells and T lymphocytes; they also play a role in apoptosis, angiogenesis and wound repair and induce the release of cytokines. Deficiencies of AMPs have been correlated with increased susceptibility

condition	changes in AMP[a]
atopic dermatitis	↓LL-37, HBD-2, and dermcidin
atopic eczema	↓HBD-2, HBD-3, and LL-37
thermal injury	Lack of HBD-2 production
infectious diarrhea	↓α–defensin
crohn's disease	↓HBD-2 mRNA
diabetes type 1	SNPs in HBD-1; ↓LL-37
oral bacterial infection in morbus Kostmann	↓LL-37,↓human neutrophil peptides
Chediak-Higashi syndrome	↓AMP in neutrophil granules
HIV-1 infection risk	DEFB1 polymorphism
tuberculosis	↓mBD-3 and mBD-4; progressive disease

Table 8.1 Disease associated with AMP deficiency. [a] ↓, diminished production. (From Rivas-Santiago, B. et al., *Infect. Immun.*, 77, 4690–4695, 2009.)

to some inflammatory and infectious diseases (**Table 8.1**), confirming that they are an essential component of innate immunity. Therefore, it is not surprising that pathogenic microorganisms have evolved several strategies to subvert their activities.

Bacterial evasion of AMPSs

Several mechanisms of evading AMPs have been described in bacteria. Because the bacterial envelope is negatively charged, there is strong electrostatic attraction between it and cationic AMPs. One mechanism of decreasing binding of AMPs to the bacterial envelope is to decrease the net negative charge of the envelope. Another method is to bind AMPs away from the cell wall and degrade them. Some bacteria pump out AMPs from the cytosol and others decrease synthesis of AMPs by host cells. A common approach of gram-negative and gram-positive bacteria in resisting AMP binding is to increase the positive charge of peptidoglycan and lipid A, respectively. For example, the gram-positive bacterium *Staphylococcus aureus* D-alanylates its wall teichoic acid and incorporates L-lysine or L-alanine into the phosphatidylglycerol of the cell membrane. The gram-negative bacterium *Salmonella enterica* serovar Typhimurium modifies its lipid A by addition of 4-deoxy-L-arabinose, phosphoethanolamine, or parmitoyl groups or by acylation. Binding AMPs at the cell wall or away from the cell wall can prevent access to the cytoplasmic membrane. For example, staphylokinase, an enzyme on the surface of *Staphylococcus aureus*, can bind and inactivate several AMPs. Similarly, surface appendages such as the M protein of *Streptococcus pyogenes* and the penicillin-binding protein 1a and PilB, the major pilus protein subunit of *Streptococcus agalactiae*, can bind AMPs and prevent access to the cell wall and thereby to the cell membrane. Bacterial capsules are a barrier that blocks access of AMPs to the cell wall and thence to the cell membrane. Furthermore, some bacteria can shed their capsule, carrying

AMPs with it. Many bacteria are able to excrete AMPs from the cytoplasmic membrane through the cell wall and into the external environment using efflux pumps in a manner akin to the excretion of antibiotics.

Despite a relative resistance of AMPs to proteolysis, they can be degraded and inactivated by proteases from a number of bacteria. Bacteria sense their environment using two-component regulatory systems. These sensors can detect AMPs and initiate increased resistance to them. For example, in gram-negative bacteria, these regulatory systems can direct reduction in fluidity and permeability of the cytoplasmic membrane and increase the positive charge of the outer membrane. Similarly, the accessory gene regulator (agr) of *Staphylococcus aureus* is a global regulator that can direct modifications to teichoic acid to increase resistance to AMPs. Some effectors injected *via* bacterial type 3 secretion systems (T3SSs) are able to suppress the expression of AMPs. For example, *Shigella flexneri* and *S. dysenteriae* employ the MxiE bacterial regulator, which controls a set of effectors injected into host cells that suppress transcription of several genes encoding AMPs. An interesting approach to AMP inhibition is the ability of some bacteria to affect the release of host cell-surface proteoglycans such as dermatin sulphate and syndecan-1, which inhibit AMPs. Mechanisms used by bacteria to evade AMPs are shown in **Figure 8.2**.

Fungal evasion of AMPs

In a manner comparable to the AMP evasion mechanisms described above for bacteria, the dimorphic fungus *C. albicans* is able to secrete the aspartyl proteases SAP9 and SAP10 that degrade histatin 5 and the Msb2 glycoprotein which binds AMPs away from the cell wall. *C. albicans* utilises the polyamine efflux transporter Flu1 to pump out histatin 5 from the cytosol and up-regulates signalling pathways to increase resistance to AMPs. *C. albicans* AMP resistance mechanisms are shown in **Figure 8.3**.

Virus evasion of AMPs

Currently it is not known whether viruses are able to evade AMPs.

Parasite evasion of AMPs

A common strategy of parasites' evasion of AMPs is to degrade them with secreted proteases. For example, the surface-metalloprotease, leishmanolysin of *Leishmania,* and the cysteine proteases of *Entamoeba histolytica* and *Trichomonas vaginalis* all degrade various AMPs. In contrast, *T. gondii* suppresses induction of β-defensin 2 expression.

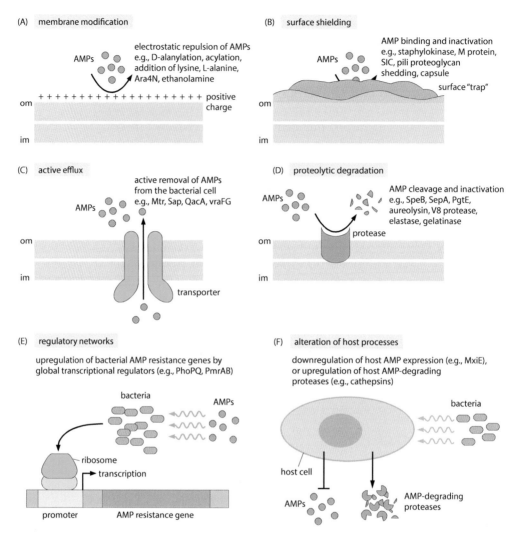

Figure 8.2 Mechanisms developed by bacteria to overcome host antimicrobial peptides. (A) Modification of the bacterial outer membrane. Bacterial resistance to cationic antimicrobial peptides is mediated by alterations in surface charge. Gram-positive bacteria: D-alanine modification of cell wall teichoic acid (*dlt*), L-lysine (*mprf*), or L-alanine modification of phosphatidylglycerol (*mprf*). Gram-negative bacteria: aminoarabinose or acylation modifications of lipid A in lipopolysaccharide (LPS; *pmr, pagP*), or addition of ethanolamine to lipid A (*pmrC, lptA*). The increased positive charge on the bacterial surface repels cationic AMPs. (B) Shielding of the bacterial surface through the trapping and inactivation of AMPs in the extracellular milieu enhances resistance and pathogenicity. Surface-associated capsule traps AMP (e.g., *K. pneumoniae cps* operon), surface protein binds AMP (e.g., GAS M1 protein, Group B *Streptococcus* (GBS) PilB pilus protein), secreted protein binds AMP (e.g., Group A *Streptococcus* (GAS) SIC protein or *S. aureus* staphylokinase), or bacterial proteases release host proteoglycans to block AMP (e.g., *P. aeruginosa* LasA). (C) Membrane efflux pumps function by translocating the AMP out of cell (e.g., *Neisseria* spp. Mtr, *S.* Typhimurium Sap, *S. aureus* QacA, and *Staphylococcus* spp. VraFG). (D) Degradation and inactivation of AMPs by bacterial proteases (e.g., GAS SpeB protease, *S. epidermidis* SepA, *S.* Typhimurium PgtE, *S. aureus* aureolysin and V8 protease, *P. aeruginosa* elastase, and *E. faecalis* gelatinase). (E) Bacterial exposure to AMPs up-regulates the expression of AMP-resistance genes through global gene regulatory networks (e.g., *S.* Typhimurium and *P. aeruginosa* PhoPQ and PmrAB). (F) Alteration of host processes by bacteria, including the down-regulation of host AMP production (e.g., *Shigella* spp. transcriptional factor MxiE), or the up-regulation and activation of host AMP-degrading proteases (e.g., *P. aeruginosa*). im, bacterial inner membrane; om, bacterial outer membrane. (From Cole, J.N. and Nizet, V., *Microbiol. Spectr.*, 4, 2016.)

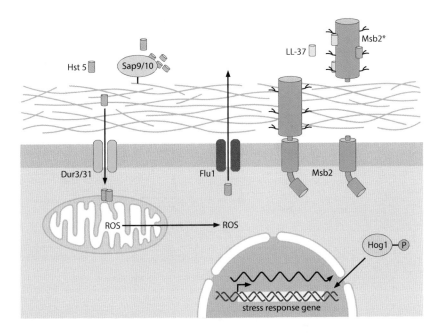

Figure 8.3 *C. albicans* mechanisms to evade AMP responses. Histatin 5 (Hst 5) is taken up by the *C. albicans* influx transporters Dur3 and Dur31, inducing the formation of reactive oxygen species (ROS); in addition, Hst 5 acts by promoting the efflux of ions and ATP. The Hog1 MAPK pathway is activated during AMP stress and up-regulates anti-oxidative and other response mechanisms to overcome AMP activity. The toxicity of Hst 5 is decreased further by its extrusion from fungal cells *via* the polyamine efflux transporter Flu1. The cell wall-anchored protease Sap9 cleaves and inactivates Hst 5 on the outside of fungal cells. In addition, the shed exodomain fragment of the Msb2 membrane sensor (Msb2*) binds several AMPs extracellularly to provide broadrange protection against AMPs. MAPK, mitogen-activated protein kinase; P, phosphate. (From Swidergall, M. and Ernst, J.F., *Eukaryot. Cell*, 13, 950–957, 2014.)

THE COMPLEMENT SYSTEM

The complement cascade consists of a number of component proteins distributed throughout the blood, lymph and tissues. Some of the components are inactive serine proteases, termed zymogens, which are activated by cleavage. There are three complement pathways (**Figure 8.4**). The component proteins of the complement system are designated C1–C9. In the classical pathway, components are activated in the following order: C1, C4, C2, C3, C5, C6, C7, C8 and C9. C1 is a three-component complex consisting of C1q, C1r and C1s. The classical pathway is activated when C1q binds to IgM or IgG antibodies or C-reactive protein bound to microbe surfaces, or when C1q binds directly to molecules on bacterial surfaces such as lipoteichoic acid and phosphocholine. The other two pathways are termed innate pathways because they do not require antibody for activation. These innate pathways are the mannose-binding lectin (MBL) pathway and the alternative pathway. The MBL pathway is homologous to the classical pathway except that it is activated when the mannose-binding lectin or ficolins bind to sugar arrays on microbial surfaces. Like C1, the

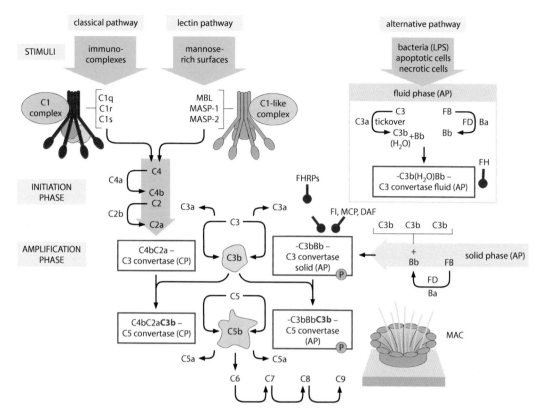

Figure 8.4 The complement system can be activated by the classical, lectin, and alternative pathways, all resulting in the formation of C3 convertases. The C3 convertases continuously cleave C3 in a powerful amplification loop. The terminal complement cascade is initiated by the C5 convertase and ultimately generates the MAC complex that inserts pores into cell membranes to induce cell lysis. AP, alternative pathway; CP, classical pathway; DAF, decay-accelerating factor; FB, factor B; FD, factor D; FH, factor H; FHRPs, factor H–related proteins; FI, factor I; LPS, lipopolysaccharide; MAC, membrane attack complex; MASPs, MBL-associated serine proteases; MBL, mannose-binding lectins; MCP, membrane cofactor protein; P, properdin. Red arrows, anaphylatoxins; white boxes, convertases; red circles, inhibitors. (From Angioi, A. et al., *Kidney Int.*, 89, 278–288, 2016.)

mannose-binding lectin is a three-component complex consisting of the MBL and two mannose-associated serine proteases, MASP-1 and MASP-2. The alternative pathway is activated by the spontaneous hydrolysis of C3, which combines with two proteins, B and D. The spontaneous hydrolysis of C3 is called C3 tick-over. Therefore, the alternative pathway does not use components C1, C4 or C2. C1r, C1s, MASP-2, C2, factors B and D are the serine proteases in the three complement pathways. Serine proteases cleave components C2–C5 and factor B into two fragments, for example, component C2 into C2a and C2b or component B into Ba and Bb. With the exception of C2, where the binding fragment is C2a, b fragments bind to the microbe surface and a fragments are soluble mediators of inflammation. Although the three pathways differ in the way that they are activated, they converge into a single effector pathway at the stage of formation of a C3 convertase. The function of the C3 convertase is to generate a large amount of C3b, which is covalently bound to the microbe. Pathogens coated (opsonised) with C3b are more readily taken up by phagocytes that have membrane receptors for C3b. The addition of C3b to C3 convertase generates a C5 convertase that cleaves C5 into C5a and C5b. C3a and C5a fragments bind to receptors on endothelial cells, mast cells and macrophages, and C5a fragments also bind to receptors on neutrophils and initiate an inflammatory response. Formation of the C5 convertase begins the terminal sequence of the complement cascade, which is the assembly of the membrane attack complex (MAC). One molecule of C6 and C7 bind sequentially to C5b, exposing a hydrophobic site on C7 that inserts into the pathogen cell membrane, as do C8 and C9. C8 initiates the polymerisation of between 10 and 16 molecules of C9 that create a pore in the cell membrane. The formation of many hundreds of such pores results in osmotic lysis of the pathogen (**Figure 8.5**).

The complement system contributes to host immunity in several important ways. Microbes opsonised by C3b are efficiently captured and destroyed by phagocytes bearing C3b receptors. C3b receptors on erythrocytes allow pathogens and pathogen molecules to be cleared from the circulation and degraded in the spleen and liver. Three fragments of the complement system (C5a, C3a and C4a) initiate inflammation and the MAC is microbicidal. In addition, complement plays a role in the positive selection,

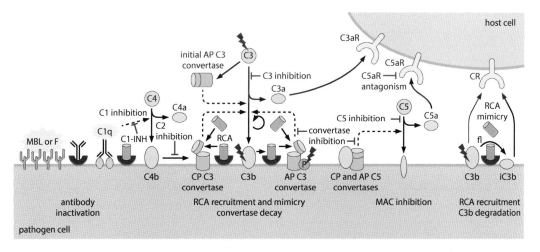

Figure 8.5 Evasion of the complement cascade by microbial pathogens. Suppression of classical pathway (CP) activation can be achieved by trapping endogenous C1 inhibitor (C1-INH) to the surface or by inactivating antibodies through the capture of their Fc regions. Whereas the recruitment of soluble regulators by capturing host proteins is a common strategy to impair downstream complement actions, certain viruses also produce structural mimics of these regulators. In addition, some microbial proteins have activities similar to CD59 in preventing MAC formation. Direct inhibition of C3, the C3 and C5 convertases, C5 or the C5a receptor (C5aR) is a prominent strategy of *Staphylococcus aureus*. Finally, a set of different microbial proteases can degrade many of the crucial components of the complement system. These proteases act directly or by capturing and activating a human protease. Increased and decreased activity is represented by thick and thin arrows, respectively. F, ficolin; fB, factor B; fD, factor D; fI, factor I; MAC, membrane attack complex; MASP, MBL-associated serine protease; MBL, mannose-binding lectin; RCA, regulators of complement activation. (From Lambris, J.D. et al., *Nat. Rev. Microbiol.*, 6, 132–142, 2008.)

activation and expansion of B lymphocytes and in the initiation, effector function and contraction of T lymphocytes.

The complement system can target host cells as well as pathogens, so it is necessary that the cascade be tightly regulated. Control is exerted on the initiation steps of the cascades and on the assembly of the MAC and comprise soluble and cell membrane molecules termed *regulators of complement activation* (RCAs). RCAs function by preventing deposition of complement components on host cell surfaces or inhibiting their activation. A list of the RCAs is shown in **Table 8.2**. The complement cascade is such a powerful component of both innate and adaptive immunity that it is not surprising that pathogenic microorganisms have invested extensively in mechanisms that subvert it. Recruitment of RCAs is the most common strategy used by pathogens, so we will begin by examining those pathogens that use this strategy.

Recruiting and mimicking RCAs

Representatives of all classes of pathogens, bacteria, viruses, fungi and parasites are able to bind the soluble plasma RCAs C4CP, factor H and FHL-1, and this is the most common strategy of circumventing the complement system. Because the different RCAs share the same consensus sequences, termed short consensus repeats (SCRs), pathogens are able to bind various RCAs. It might not be surprising then to find that the ligands that bind

name	ligand/binding factor	action
soluble factors regulating complement		
C1 inhibitor (C1INH)	C1r, C1s (C1q); MASP-2 (MBL)	displaces C1r/s and MASP-2, inhibiting activation of C1q and MBL
C4-binding protein (C4BP)	C4b	displaces C2a; cofactor for C4b cleavage by factor I
CPN1 (Carboxypeptidase N)	C3a, C5a	inactivates C3a and C5a
factor H	C3b	displaces Bb, cofactor for factor I
factor I	C3b, C4b	serine protease, cleaves C3b and C4b
protein S	C5b67 complex	inhibits MAC formation
membrane-bound factors regulating complement		
CRIg	C3b, iC3b, C3c	inhibits activation of alternative pathway
complement receptor 1 (CR1, CD35)	C3b, C4b	cofactor for factor I; displaces Bb from C3b, and C2a from C4b
decay-accelerating factor (DAF, CD55)	C3 convertase	displaces Bb and C2a from C3b and C4b, respectively
membrane-cofactor protein (MCP, CD46)	C3b, C4b	cofactor for factor I
protectin (CD59)	C8	inhibits MAC formation

Table 8.2 Regulatory proteins of the classical and alternative complement pathways. (From *Janeway's Immunology*, 9th ed., Garland Science, 2017.)

RCAs on the surface of various bacterial genera are similar. *Helicobacter pylori* and *Escherichia coli* are able to acquire the host cell surface-bound complement regulator CD59.

Viruses can also acquire RCAs, but some viruses such as variola virus, vaccinia virus, human herpesvirus 8 (HHV-8), and cowpox and monkeypox viruses produce proteins that contain SCRs and mimic RCAs. **Table 8.3** lists examples of various classes of pathogens and shows the mechanisms they use to evade the complement pathways as well as the complement components and accessory molecules they target.

pathogen protein		target	Action
Bacteria			
Actinobacillus spp.			
Omp100	outer membrane protein 100	fH	recruitment of regulators
Bordetella spp.			
FHA	filamentous hemagglutinin	C4BP, (fH, FHL-1)	recruitment of regulators
Borrelia spp.			
CRASP	complement regulator-acquiring surface proteins	fH, FHL-1	recruitment of regulators
Erp	OspE/F-related proteins	fH	recruitment of regulators
n/a	CD59-like protein	C8, C9	prevents MAC formation
Escherichia spp.			
OmpA	outer membrane protein A	C4BP	recruitment of regulators
StcE	secreted protease of C1 esterase inhibitor	C1-INH	recruitment of C1-INH
TraT	TraT outer membrane protein	C5b6	prevents MAC formation
Fusobacterium spp.			
n/a	unknown factor	fH	recruitment of regulators
Haemophilus spp.			
n/a	unknown factor	C4BP, fH	recruitment of regulators
Moraxella spp.			
UspA1/2	ubiquitous surface protein A1/A2	C4BP	recruitment of regulators
Neisseria spp.			
LOS	lipooligosaccharide	fH, FHL-1	recruitment of regulators
GNA1870	genome-derived neisserial antigen 1870	fH, FHL-1	recruitment of regulators
Por	outer membrane porins	C4BP, fH, FHL-1	recruitment of regulators
n/a	Type IV pili	MCP	attachment to epithelial cells
Porphyromonas spp.			
prtH	prtH protease	C3, IgG	degrades C3 and IgG
Pat	Pseudomonas elastase	C1q, C3	degrades C1q and C3
PaAP	pseudomonas alkaline protease	C1q, C3	degrades C1q and C3
Tuf	elongation factor	fH, FHL-1	recruitment of regulators
Serratia spp.			
n/a	56 kDa protease	C5a, C1-INH	degrades C5a, degrades C1-INH
Staphylococcus spp.			
CHIPS	chemotaxis inhibitory protein of *S. aureus*	C5aR	antagonizes C5a

(Continued)

pathogen protein		target	Action
Efb	extracellular fibrinogen-binding protein	C3/C3b/C3d	inhibition of C3 and C3b-containing convertases
Ehp[a]	Efb-homologous protein	C3/C3b/C3d	inhibition of C3 and C3b-containing convertases
SAK	staphylokinase	C3b, IgG (via Plasmin)	cleaves complement proteins, removes C3b from surface
Sbi	S. aureus IgG-binding protein	IgG	inhibits Ig interaction with C1q
SCIN	staphylococcal complement inhibitor	C3 convertases	inhibits C3 activation to C3a/C3b
SpA	S. aureus protein A	Ig's, gC1q-R	inhibits Ig interaction with C1q
SSL-7	staphylococcal superantigen-like protein 7	C5	prevents C5 cleavage
Streptococcus spp.			
Bac	-protein	IgA, fH	recruitment of regulators
Fba	fibronectin-binding protein	fH, FHL-1	recruitment of regulators
Hic[b]	factor H-binding inhibitor of complement	fH	recruitment of regulators
IdeS	IgG-degrading Enzyme of S. pyogenes	IgG	cleaves IgG, no interaction with C1q
M[b]	surface proteins M family (Arp, Sir, etc.)	fH, C4BP, FHL, FHR, MCP	recruitment of regulators
PLY	pneumolysin	IgG, C1q	complement activation/depletion
PspA	pneumococcal surface protein A	unknown	potential impairing of AP and complement receptors
PspC[c]	pneumococcal surface protein C	fH (C3, IgA)	recruitment of regulators, potential degradation of C3/C3b
scpA/B	streptococcal C5a peptidase	C5a	degrades C5a, disrupts signaling
SIC	streptococcal inhibitor of complement	C5b-7, C5b-8	prevention of MAC formation
SPE B	streptococcal pyrogenic exotoxin B	properdin, Ig's	degrades Properdin, Ig's
SpG	streptococcus protein G	Ig's	inhibits Ig interaction with C1q
Yersinia spp.			
YadA	yersinia adhesin A	fH	recruitment of regulators
viruses			
herpes viruses			
gC1/2	transmembrane glycoproteins C1,C2 (HSV)	C3b	binds to C3b, decay acceleration (only AP), Inhibits bind properdin and C5
gE+gI	glycoproteins E+I (HSV)	IgG	Fc-receptor, less activation
gp34,68	glycoproteins 34, 68 (HCMV)	IgG	Fc-receptors, less activation
gpI+gpIV	glycoproteins I+IV (VZV)	IgG	Fc-receptor, less activation
KCP[d]	kaposi's sarkoma-associated complement control protein (KSHV)	C3b	mimics regulators (cofactor/decay acceleration)
retroviruses			
gp41	envelope glycoprotein 41 (HIV)	C1q, fH, CD59	direct CP activation, recruitment of regulators, decreased expression
gp120	envelope glycoprotein 120 (HIV)	MBL, fH	direct LP activation, recruitment of regulators
Tat	transactivator of transcription (HIV)	C1-INH	induces C1-INH expression
poxviruses			
IMP	cowpox control inflammation modulatory protein (cowpox virus)	C3b, convertases	mimics regulators (cofactor/decay acceleration)
MOPICE	monkeypox inhibitor of complement enzymes (monkeypox virus)	C3b	mimics regulators (only cofactor activity)

(*Continued*)

pathogen protein		target	Action
SPICE	smallpox inhibitor of complement enzymes (variola virus)	C3b, convertases	mimics regulators (cofactor/decay acceleration)
VCP	vaccinia virus complement control protein (vaccinia virus)	C3b, convertases	mimics regulators (cofactor/decay acceleration)
filoviruses			
NS1	non-structural protein 1 (West Nile virus)	fH	recruitment of regulators
fungi			
Aspergillus fumigates			
n/a	unknown factor	fH, FHL-1, C4BP	RCA recruitment
Candida albicans			
CRASP-1	complement regulator-acquiring surface protein 1	fH, FHL-1, C4BP	recruitment of regulators
Gpm1p	phosphoglycerate mutase	fH, FHL-1	recruitment of regulators
parasites			
Echinococcus spp.			
n/a	hydatid cyst wall	fH	recruitment of regulators
Ixodes spp.			
IRAC	ixodes ricinus anti-complement protein	AP convertase	decay acceleration
ISAC	ixodes scapularis anti-complement protein	AP convertase	decay acceleration
Onchocerca spp.			
mf	microfilariae	fH	recruitment of regulators
Ornlthodoros spp.			
OmCI	ornithodoros moubata complement inhibitor	C5	binds to C5 (potentially blocks binding to C5 convertase)
Schistosoma spp.			
CRIT	complement C2 receptor trispanning	C2	inhibits CP convertase formation
m28	28kDa membrane serine protease	iC3b	cleaves iC3b, restricts CR3 binding
Pmy[e]	paramyosin	C8, C9, C1q, IgG	prevents MAC formation decreases AP activation
Trypanosoma spp.			
CRIT	complement C2 receptor trispanning	C2	inhibits CP convertase formation
T-DAF	trypanosoma decay-accelerating factor	convertase	destabilizes convertase

Table 8.3 Complement evasion proteins and their targets. [a]Ehp has also been termed extracellular complement-binding protein (Ecb). [b]As Hic is not a member of the classical PspC families PspC-like protein, it is listed as separate protein. [c]Former names: cholin-binding protein (CbpA), *Streptococcus pneumonia* secret binding protein (SpsA) and pneumococcal C3-binding protein A (PbcA). [d]KPC has also been termed kaposica. [e]Paramyosin has been described as Schistosome complement inhibitor protein 1 (SCIP-1). AP, alternative pathway; C1-INH, C1 esterase inhibitor; binding protein; C5aR, C5a receptor; CP, classical pathway; DAF, decay accelerating factor; fH, factor H; FHL, factor H-like protein immunoglobulin; MBL, mannose-binding lectin; MCP, membrane cofactor protein. (From Lambris, J.D. et al., *Nat. Rev. Microbiol.*, 6, 132–142, 2008.)

Destroying complement components

Proteases of many bacteria, fungi, parasites and some viruses destroy complement components. Complement-degrading proteases are utilised by bacterial pathogens such as *Staphylococcus aureus* (C3b-staphylokinase), *Streptococcus pyogenes* (C5a peptidase), *S. agalactiae* (C5a peptidase), *Pseudomonas aeruginosa* (C1q-elastase), to name but a few. Parasites also employ complement-inactivating enzymes. *Shistosoma mansoni* has a 28-kDa protease that degrades iC3b. *Leishmania* species use a protease to release C3b from their cell membrane. In addition, membrane protein kinases phosphorylate C3, C5 and C9. Among fungi, *Candida albicans* contains a family of secreted aspartyl proteases (SAPs), and *Aspergillus fumigatus* an alkaline serine protease (ALp1) serine protease, that cleave C3. Hepatitis C virus (HCV) uses its NS3/4A protease to cleave complement component C4 (Table 8.3).

Microbial envelope/wall components that inhibit complement

The capsules of bacteria and fungi inhibit the deposition of complement components on their surfaces and/or inactivate them. These two events inhibit opsonophagocytosis mediated by the complement cascade. In addition, capsules block access of antibody and C3b to the cell wall located beneath the capsule. For example, the capsule of *Streptococcus pneumoniae* impairs deposition of C3b and decreases conversion of C3b to iC3b. The capsules of *Streptococcus agalactiae* and *Neisseria meningitidis* contain sialic acid that prevents alternative pathway complement activity by creating a non-activating surface and by binding factor H, a plasma complement regulator. *Borrelia burgdorferi* CD59, *Staphylococcus aureus* SCIN, Efb and Ecb and *Streptococcus pyogenes* SIC are bacterial envelope proteins that inhibit formation of the MAC. The transmembrane protein gC1 of herpes simplex virus type 1 and the transmembrane protein gC2 of herpes simplex virus type 2 bind C3b and increase the rate of breakdown of the alternative pathway C3 convertase. Both shistosomes and trypanosomes have cell membrane-associated proteins that inhibit the interaction between C2 and C4 and, thus, prevent the formation of the classical and MBL pathway C3 convertase. In addition, the shistosome protein paramyosin inhibits MAC formation.

Evasion resulting from cell wall structure

The thick peptidoglycan layer of the cell wall of gram-positive bacteria, the thick and hydrophobic cell wall of mycobacteria and the complex cell walls of fungi are resistant to lysis by the MAC. Many blood-borne parasites are similarly resistant to lysis. This observation has lead to the view that the opsonic and inflammatory components of complement may be more important in anti-pathogen immunity than pathogen lysis.

Consuming complement in the fluid phase

The secretion or liberation of cell wall or cell membrane molecules by pathogens can consume immune factors such as complement components and antibodies so that they are unavailable to bind to targets on the surface of the pathogen. For example, *Trypanosoma bruci* blebs off pieces of its membrane carrying away antibodies and complement components and *T. cruzi* releases a glycoprotein (gp160) that binds C3b and blocks C3 and, thus, C5 convertase formation. Several bacteria liberate their capsule so that complement and antibodies bind capsule in the fluid phase rather than on the surface of the organism.

CIRCUMVENTION OF PHAGOCYTOSIS

Phagocytosis is the process by which certain leucocytes ingest and destroy pathogens. The principal phagocytes are neutrophils (also called polymorphonuclear leucocytes), and monocyte/macrophages, and DCs, known as mononuclear phagocytes. Collectively, these cells are termed professional phagocytes. Eosinophils and basophils are also phagocytic, but neutrophils and macrophages are the foremost cells involved in pathogen uptake and killing. The principal role of DCs is antigen presentation to naïve T lymphocytes. Macrophages and DCs are resident in the tissues where they act as surveillance cells, whereas neutrophils are circulating cells that must be recruited into infected tissue from the blood. In humans, neutrophils comprise about one-half to two-thirds of all blood leucocytes, and their number can increase 10-fold during infection. Neutrophils have a lifespan of only a few days. Macrophages begin their life either as progenitor cells that seed tissues during embryogenesis and self-renew throughout life or as monocytes. Monocytes emigrate from the circulation into tissues where they differentiate into macrophages or DCs. Tissue macrophages live from months to years. DCs are distributed throughout the body and their lifespan is generally short, but they may live longer in certain locations. In order to destroy pathogens, phagocytes must first locate them. This is accomplished by recognition of chemoattractant gradients created by interaction of the pathogen with molecules of the innate immune system such as the complement components C3a, C4a and C5a that have been described above and by components of the pathogens themselves that are detected by phagocyte cell membrane receptors. In order to take up pathogens, phagocytes must capture them using a variety of cell membrane endocytic receptors such as C-type lectins (CTLs), complement receptors or Fcγ receptors. Once captured, in a process analogous to receptor-mediated endocytosis phagocytes enclose the pathogen in a vacuole called a phagosome in one of two ways that depend on which type of receptor captures the pathogen. The phagocyte may extend pseudopods that wrap around the pathogen, or the phagocyte membrane may invaginate to form a cup-shaped depression into which the pathogen sinks. In either case, the phagocyte cell membrane flows over the pathogen and seals it in a vacuole termed a phagosome. Phagosomes move towards the centrosome and, during their passage, their interior becomes increasingly acidic, reaching a terminal pH of about 4.5. During traffic along the microtubular transport system, endosomes fuse with the phagosome and some of the cell

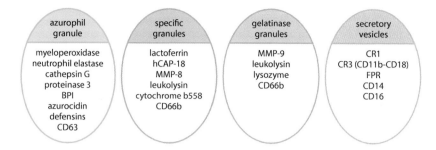

Figure 8.6 Contents of neutrophil granules. (From Chapter 11, New research on the importance of cystic fibrosis transmembrane conductance regulator function for optimal neutrophil activity, in: Wat D (Ed.), *Cystic Fibrosis in the Light of New Research*, IntechOpen.com, https://www.intechopen.com/books/cystic-fibrosis-in-the-light-of-new-research, 2015.)

membrane that makes up the phagosome is recycled back to the cell surface. Vesicles called lysosomes that contain antimicrobial molecules such as cationic antimicrobial peptides and lactoferrin and hydrolytic enzymes fuse with the phagosome to form a phagolysosome. In addition to lysosomes, neutrophils contain several types of granules that contain acidic hydrolases and antimicrobial proteins that fuse with phagosomes. The content of these granules are shown in **Figure 8.6**. Secretory vesicles/recycling endosomes contain cell membrane receptors that become incorporated in the formation of the phagosome and are recycled back to cell membrane of the phagocyte.

In addition, to the fusion of phagosomes and lysosomes a multicomponent NADPH oxidase and nitric oxide synthase assemble in the membrane of the phagolysosome and generate microbicidal reactive oxygen and nitrogen species. Activity of the NADPH oxidase results in a transient increase in oxygen consumption by the phagocyte that is termed the respiratory burst. Steps in phagosome maturation are shown in **Figure 8.7**.

Chemoattraction, attachment, uptake, vacuole formation, and killing are all stages of phagocytosis that are targeted for evasion by pathogens. However, pathogens focus mainly on killing phagocytes, resisting uptake by them and resisting killing in the phagolysosome. The following sections consider mechanisms of phagocyte evasion by pathogens.

Chemoattraction

The professional phagocytes, neutrophils and macrophages, are attracted to the site of infection by gradients of molecules termed chemoattractants. Host-generated chemoattractants include the chemokine CXCL8 and stress proteins released by stressed or damaged host cells and, importantly, the complement components C3a,

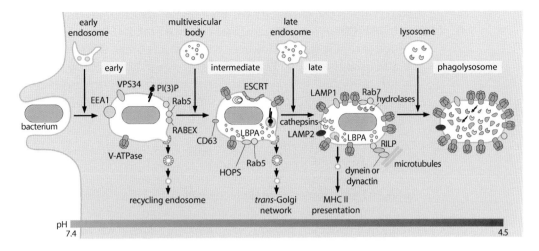

Figure 8.7 Stages of phagosomal maturation. Shortly after pathogen uptake, the phagosome under-goes a series of transformations that result from its sequential interaction with subcompartments of the endocytic pathway. Different stages of maturation are recognised – (a) early, (b) intermediate, and (c) late phagosomes – that culminate with the formation of (d) phagolysosomes. During maturation, the phago-somes acquire various hydrolases and undergo a progressive acidification caused by proton pumping by the V-ATPase. EEA1, early endosome antigen 1; ESCRT, endosomal-sorting complex required for transport; HOPS, homotypic protein sorting; LAMP, lysosomal-associated membrane protein; LBPA, lysobisphos-phatidic acid; MHCII, major histocompatibility complex II; PI(3)P, phosphatidylinositol-3-phosphate; RILP, Rab-interacting lysosomal protein. Multi-vesicular bodies are a specialised subset of endosomes that contain membrane-bound intraluminal vesicles. The content of MVBs can be degraded, *via* fusion with lysosomes, or released into the extracellular space, *via* fusion with the plasma membrane. (From Flanna-gan, R.S. et al., *Nat. Rev. Microbiol.*, 7, 355–366, 2009.)

C5a and C4a. Molecules secreted from live, or liberated from dead or dying, microorganisms also attract phagocytes. A notable example is *N*-formylmethionyl-leucyl-phenylalanine (fMLP). In bacteria, protein synthesis begins with *N*-formylmethionine, whereas in eukaryotic cells it does not. This tri-peptide is a potent chemoattractant molecule but probably not the only one. Phenol-soluble extracts of certain bacte-ria have been shown to contain chemoattractant molecules that have been termed modulins. The binding of fMLP and C5a to their recep-tors initiate the generation of microbicidal reactive oxygen species by neutrophils and macrophages. The surfaces of all classes of pathogen activate the complement cascade, thereby releasing C3a and C5a. The chemoattractant receptors on the phagocyte cell membrane belong to a family of receptors termed G-protein–coupled receptors (GPCRs). Liga-tion of GPCRs activates small GTPases that set in motion downstream signal transduction pathways. G proteins are a frequent target of bacte-rial exotoxins, and they were discussed in Chapter 6. Many pathogenic bacteria, fungi, and parasites secrete proteases that are capable of cleav-ing both pathogen-derived and host-generated chemoattractant factors and GPCRs on the phagocyte cell membrane. For example, the elastase secreted by *Pseudomonas aeruginosa* cleaves both fMLP and the formyl peptide receptor (FPR). However, mechanisms other than proteolysis are also used to subvert chemoattractants. For example, *Staphylococcus aureus* produces two proteins: chemotactic-inhibitory protein of *S. aureus* (CHIPS) that inhibits FPR, and FLPL-1-inhibitory protein that inhibits FL1Pr.

Regulation of chemokines

Pathogens can evade chemokine signalling by either down-regulating chemokine receptors, producing chemokine mimics, producing chemokine receptor mimics or producing chemokine-binding proteins. For example, the bacterium *Staphylococcus aureus* produces a family of proteins related to superantigens termed staphylococcal superantigen-like (SSL) proteins that are able to block chemokine receptors and thus chemotaxis. In contrast, the bacterium *Mycobacterium avium* and the parasite *Toxoplasma gondii* produce proteins that bind to and activate chemokine receptors. In these cases, chemotaxis and the release of cytokines by phagocytes aid in the infectious cycle of these pathogens. Various viruses such as HHV8 and molluscum contagiosum virus (MCV) produce proteins that are chemokine analogues. These proteins bind to chemokine receptors but do not activate them, whereas human cytomegalovirus (HCMV) synthesises a CXCL1 analogue that activates chemotaxis. SSL5 of *S. aureus* binds a glycosaminoglycan-binding site common to all chemokines, which prevents activation of chemokine receptors. The parasite *Shistosoma mansoni* secretes a chemokine-binding protein that blocks binding of chemokines to their receptors. Some viruses express chemokine-binding proteins and/or chemokine receptors. HHV8 is a virus that expresses both chemokine receptor analogues and proteins that bind cytokines, whereas CMV expresses cytokine receptor analogues only. Many pathogenic bacteria, some fungi and parasites produce proteases, which, in addition to cleaving complement components, can degrade chemokines. The bacteria include various species of gram-positive cocci and saccharolytic and non-saccharolytic gram-negative rods. Among fungi, *Candida albicans* contains a family of secreted SAPs and *Aspergillus fumigatus* secretes an ALp1 serine protease. The parasite *Necator americanus* secretes a metalloprotease with activity against chemokines. Glycosaminoglycans (GAGs) facilitate the retention of chemokines on cell surfaces, allowing the accumulation of the high local concentration of cytokines required for cell activation. Accordingly, it has been proposed that pathogens that secrete enzymes such as hyaluronidase, condroitinase and heparinase may be capable of degrading GAGs and thereby releasing cytokines.

For example, *Staphylococcus aureus*, *Streptococcus pyogenes*, and *Clostridium perfringens* produce hyaluronidase, as do parasitic nematodes. **Tables 8.4** and **8.5** list the mechanisms of chemokine evasion for the various classes of pathogen.

mechanism	modulator	species	target
intracellular signaling	pertussis toxin	*B. pertussis*	$G_i\alpha$
	cholera toxin	*V. cholerae*	$G_s\alpha$
receptor binding	CHIPS	*S. aureus*	FPR, C5aR
	FLIPr	*S. aureus*	FPRL-1
	SSL5	*S. aureus*	glycosylated GPCRs
	SSL10	*S. aureus*	CXCR4
	CCL2-like molecule	*M. avium*	CCR5
	cyclophalin-I8	*T. gondii*	CCR5
	chemokine mimics	viruses	
stimulus binding	SSL5	*S. aureus*	all chemokines
	evasin-1	*R. sanguineus*	CCL3, CCL4, CCL18
	evasin-3	*R. sanguineus*	CXCL8, CXCL1
	evasin-4	*R. sanguineus*	CCL5, CCL11
	smCBP	*S. mansoni*	
	receptor mimics	viruses	
	chemokine-binding proteins	viruses	
stimulus and receptor cleavage	SpyCEP, ScpC	*S. pyogenes*	CXCL8, KC, MIP2, CXCL1, CXCL6
	ScpA, ScpB	*S. pyogenes*	C5a
	CepI	*S. iniae*	CXCL8
	elastase	*P. aeruginosa*	fMLP, C3, C5
			CXCL5, CCL2, CCL5
	alkaline protease	*P. aeruginosa*	CXCL5, CCL2, CCL5
	cystein/serine proteases	*P. gingivalis*	CXCL8, CCL2
	gingipains	*P. gingivalis*	C3, C5, C5aR
	metalloprotease	*N. americanus*	CCL11
	56-kDa protein	*S. marcescens*	C5a
	elastase	*E. faecalis*	C3, C3a
inhibition stimulus	Ecb, Efb	*S. aureus*	C3b-containing convertases
generation	SCIN, SCIN-B, SCIN-C	*S. aureus*	C3 convertases
	SSL7	*S. aureus*	C5

Table 8.4 Pathogens that modulate GPCR with their targets and mechanism of action. CHIPS, Chemotaxis inhibitory protein of *S. aureus*; Ecb, extracellular complement-binding protein; Efb, extracellular fibrinogen-binding protein; FLIPr, FPRL-1-inhibitory protein; FPR, formylated peptide receptor; FPRL-1, FPR-like receptor; SCIN, staphylococcal complement inhibitor 1; Scp, streptococcal C5a peptidase; smCBP, *S. mansonii* chemokine-binding protein; SpyCEP, *S. pyogenes* cell envelope protease; SSL, staphylococcal superantigen-like. (From Bestebroer, J. et al., *FEMS Microbiol. Rev.*, 34, 395–414, 2010.)

gene	virus	homolog	features
viral chemokine homologs			
vMIP-I/K6	human herpesvirus 8	MIP-Iα	CCR8 agonist
vMIP-II/K4	human herpesvirus 8	MIP-Iα	broad-spectrum CC, CXC and CX₃C chemokine antagonist
vMIP-III/BCK	human herpesvirus 8	MIP-Iβ	ND
U83	human herpesvirus 6	MIP-Iα	CC chemokine-like agonist
MCK-I/m131	Murine cytomegalovirus	CC chemokines	CC chemokine-like agonist; promotes viral dissemination
vCXC-I/UL146	human cytomegalovirus	IL-8	CXC chemokine-like agonist
vCXC-2/UL147	human cytomegalovirus	IL-8	ND
unmapped	stealth virus	GRO-α/MGSA	ND
unmapped	marek's disease virus	IL-8	ND
vMCC-I/MC148R	molluscum contagiosum	MCP-I	broad-spectrum CC and CXC chemokine antagonist
viral chemokine receptor homologs			
ORF74	human herpesvirus 8	CXCR2	constitutively signaling and agonist-independent receptor
ECRF3IORF74	herpesvirus saimiri	CXCR2	functional CXC chemokine receptor
ORF74	murine γ-herpesvirus 68	CXCR2	ND
ORF74, E1,E6	equine herpesvirus 2	CXCR2/CCRI	ND
U12	human herpesvirus 7	CCR	ND
U12	human herpesvirus 6	CCR	functional CC chemokine receptor
US28	human cytomegalovirus	CCRI	functional CC chemokine receptor; HIV entry coreceptor
UL33	human cytomegalovirus	CCRI	localizes to virus envelope particles
M33	murine cytomegalovirus	CCRI	important for viral dissemination to salivary glands
K2R	swinepox virus	CXCR	ND
Q2/3L	capripox virus	CCR	ND
viral chemokine-binding proteins			
M-T7[a]	myxoma virus	IFN-γ receptor	binds C, CC, CXC chemokines via heparin-binding domains
M-TI[b]	myxoma virus	?	broad-spectrum CC chemokine inhibitor
S-TI[b]	shope fibroma virus	?	broad-spectrum CC chemokine inhibitor
35 kDa[b]	rabbitpox virus	?	broad-spectrum CC chemokine inhibitor

(Continued)

gene	virus	homolog	features
C23L/B29R[b]	vaccinia virus (strain Lister)	?	broad-spectrum CC chemokine inhibitor
G3R[b]	variola (smallpox) virus	?	broad-spectrum CC chemokine inhibitor
DIL/H5R[b]	cowpox virus	?	broad-spectrum CC chemokine inhibitor

Table 8.5 Virus-encoded modulators of chemokines and chemokine receptors. Abbreviations: CCR, CC chemokine receptor; CXCR, CXC chemokine receptor; GRO-α, growth-related oncogene α; IFN-γ, interferon γ; IL-8, interleukin 8; MCP, monocyte chemotactic protein 1; MIP-1α, macrophage inflammatory protein 1α; ND, not determined; ORF, open reading frame. [a]Designated type I vCkBP (vCkBP-I) in text. [b]Designated type II vCkBP (vCkBP-II) in text. (From Lalani, A.S. et al., *Immunol. Today* (now: *Trends in Immunology*), 21, 100–106, 2000.)

CIRCUMVENTING PATTERN RECOGNITION RECEPTORS

The human immune system recognises microbes (in the larger context, non-self) using germ line-encoded receptors termed pattern recognition receptors (PRRs). PRRs are expressed most completely by antigen-presenting cells and professional phagocytes such as DCs, macrophages and neutrophils and the mediator-releasing cells, mast cells and eosinophils. A more limited range of PRRs is displayed by B lymphocytes and by epithelium. PRRs recognise motifs that are parts of evolutionarily conserved, largely structural components of bacteria, fungi, viruses and parasites. These motifs were initially termed pathogen-associated molecular patterns (PAMPs). However, the realisation that these motifs are shared by both pathogens and non-pathogens resulted in changing the acronym from PAMPs to microbe-associated molecular patterns (MAMPs). It has become apparent that cellular damage is detected by PRRs and the endogenous molecules that signal host cell damage are termed alarmins or DAMPs (danger-associated molecular patterns).

The PRRs that recognise pathogens can be divided into the following families: (1) Toll-like receptors (TLRs); (2) calcium-type lectins (CTLs); (3) nucleotide-binding leucine-rich repeat-containing receptors (NLRs) that contain the nucleotide-binding oligomerisation domain (NOD)-like receptors; (4) retinoic acid-inducible gene-1 (RIG-1) receptors (RLRs) and (5) absent-in-melanoma (AIM)-like receptors (ALRs). The TLRs, the best described PRRs, are located both on the cell membrane and in phagosomes/endosomes. There are 10 TLRs in humans (TLR-1 to TLR-10). Each TLR recognises motifs on a group of evolutionarily conserved molecular structures that are not expressed in vertebrates. The range of structures recognised by TLRs and their cellular distribution is shown in **Table 8.6**. TLRs act as homo- or heterodimers. CTLs (DC-SIGN, Dectin 1 Dectin 2,

PRRs	localization	ligand	origin of the ligand
TLR			
TLR1	plasma membrane	triacyl lipoprotein	bacteria
TLR2	plasma membrane	lipoprotein	bacteria, viruses, parasites, self
TLR3	endolysosome	dsRNA	virus
TLR4	plasma membrane	LPS	bacteria, viruses, self
TLR5	plasma membrane	flagellin	bacteria
TLR6	plasma membrane	diacyl lipoprotein	bacteria, viruses
TLR7 (human TLR8)	endolysosome	ssRNA	virus, bacteria, self
TLR9	endolysosome	CpG-DNA	virus, bacteria, protozoa, self
TLR10	endolysosome	unknown	unknown
RLR			
RIG-I	cytoplasm	short dsRNA, 5'triphosphate dsRNA	RNA viruses, DNA virus
MDA5	cytoplasm	long dsRNA	RNA viruses (picornaviridae)
LGP2	cytoplasm	Unknown	RNA viruses
NLR			
NOD1	cytoplasm	iE-DAP	bacteria
NOD2	cytoplasm	MDP	bacteria
CLR			
dectin-1	plasma membrane	β-Glucan	fungi
dectin-2	plasma membrane	β-Glucan	fungi
MINCLE	plasma membrane	SAP130	self, fungi

Table 8.6 Pattern recognition receptors (PRRs) and their ligands. (From Takeuchi, O. et al., *Cell*, 140, 805–820, 2010.)

macrophage C-type lectin (MCL) and macrophage inducible C-type lectin (Mincle) bind mannose, *N*-acetylglucosamine, L-fucose, glucose and galactose. CTLs also recognise carbohydrates, such as β-glucan, and many non-carbohydrate ligands, such as lipids and proteins by mechanisms that are currently unclear. CTLs are located on the cell membrane and function as both signalling and endocytic receptors (**Table 8.6**). Thus, TLRs and CTL are sensors of extracellular pathogens. Similar innate sensors are also located in the cytoplasm to detect cytosolic pathogens such as viruses. A list of cytosolic sensors and their ligands are shown in Table 8.6. An important function of cytosolic PRRs is to activate the inflammasome. The inflammasome is a complex of certain NLRs that associate and recruit the proenzyme caspase-1, which becomes activated, to cleave pro-interleukin-18 (IL-18), and pro-interleukin-1β (IL1β) so that these pro-inflammatory cytokines can be secreted from the host cell. However, it has become clear that the cytoplasmic PRRs have a wide range of activities that extend beyond pathogen recognition and initiating inflammation and the adaptive immune response. They also sense metabolic changes in the cell and play roles in embryonic development and cell death.

PRRs are the means by which the host discriminates between self and dangerous non-self and the ligation of PRRs by MAMPs initiate immune responses. Because different PRRs recognise different MAMPs and are located on the cell membrane, in endosomes or phagosomes and in the cytosol, antigen-presenting cells such as DCs are alerted not only to the type of pathogen but also to its extra- or intracellular location. Based on this information, DCs drive naïve T lymphocytes to develop into the most appropriate effector subset to combat a particular type of pathogen (see next section). The extent of PRR ligation informs the magnitude and the duration of the infection. The ligation of PRRs of DCs initiates a process termed *licencing* in which DCs increase expression of major histocompatibility (MHC) class I and II molecules, co-stimulatory molecules such as CD80 (B7.1), and CD86 (B7.2), cell adhesion molecules, and cytokines, all of which are involved in activating naïve T-cells. In addition to the expression/up-regulation of the aforementioned molecules, DCs express the chemokine receptor CCR7, which allows them to respond to the chemokine CCL21 that is produced constitutively by secondary lymphoid tissue, attracting DCs to the T-cell areas of these tissues. Because of the critical role of PRRs in the initiation of both innate (inflammation) and adaptive immunity, many pathogens attempt to circumvent PRR recognition.

Subversion of PRR crosstalk

TLR–TLR and TLR–CTL interactions are exploited by pathogens to subvert pro-inflammatory signalling cascades effected by specific MAMP–PRR interactions. TLR2 appears to be an important target in pathogen subversion because, although it induces a pro-inflammatory response, it can induce secretion of the immunosuppressive cytokine IL-10. The significance of IL-10 is its ability to drive naïve T cells to become regulatory T cells (T_{REG}) that down-regulate adaptive immunity. For example, *Candida albicans* binds both TLR2 and TLR4 on macrophages with TLR4 conferring protection and TLR2 susceptibility to infection with the fungus. *Mycobacterium tuberculosis* and *M. bovis* MAMPs bind to several PRR on DCs to subvert their maturation. TLR2, TLR4 and DC-SIGN are simultaneously ligated, resulting in down-regulation of NF-κB, which reduces the expression of co-stimulatory molecules and induces production of IL-10. Other pathogens exploit the TLR-DC-SIGN axis. For example, HIV-1 activates the Raf-1 pathway *via* DC-SIGN, which modulates TLR signalling resulting in IL-10 production. Certain pathogens such as HCV and *Toxoplasma gondii* bind TLR2, inducing IL-10, which down-regulates positive signalling through other TLRs. *Helicobacter pylori* induces IL-10 production as a result of fucose-containing Lewis antigens on its LPS being bound by DC-SIGN. This interaction results in the dissociation of the KSR1-CNK-Raf1 complex modifying downstream signal transduction. In the tick-borne zoonosis, Lyme disease, caused by *Borrelia burgdorferi*, bacterial lipoproteins trigger TLR2 activation and a tick salivary protein Salp15 binds to DC-SIGN, leading to attenuation of pro-inflammatory cytokines and enhanced IL-10 production by the human host. Dectin-1, an important PRR for the detection of *Candida albicans*, *Aspergillus fumigatus*, and *Pneumocystis carinii*, collaborates with TLR2 to induce the secretion of pro-inflammatory cytokines in DCs and macrophages. However, when ligated alone, Dectin-1 leads to the secretion of IL-10.

Pathogens may also utilise interactions between TLRs and complement receptors that include the C3b family receptors and the receptors for the chemokines C3a and C5a. Generally, binding of MAMPs, C3b and C3a and C5a by DCs and macrophages deliver activation signals to these immune cells; however, interaction between TLRs and the C5a receptor has been shown to result in down-regulation of IL-12, IL-18, IL-23 and IL-27. These cytokines are important in driving naïve T cells to become T helper 1 (T_H1) effector/memory cells. T_H1 cells are essential to defence against extracellular and facultatively intracellular pathogens by increasing the microbicidal activity of macrophages and in helping B cells make opsonic IgG antibodies. Several pathogens such as HCV, *Listeria monocytogenes*, and *Staphylococcus aureus* bind to the gC1q receptor (gC1qR). The gG1q receptor is a ubiquitous membrane protein that binds to the globular heads of the complement component C1q. However, this receptor appears to be a multifunctional protein with affinity for a variety of ligands. Ligation of this receptor inhibits production of IL-12 that occurs when TLR2 and TLR4 recognise MAMPs. *Histoplasma capsulation* and *Bordetella pertussis* inhibit IL-12 production by cooperatively ligating complement receptor 3 (CR3) and TLRs. IL-12 is essential for activation of natural killer (NK) cells and generation of T_H1 cells, both of which are critical to antimicrobial activity. The effectors of type 3, type 4 and type 6 secretion systems of gram-negative intracellular bacteria (see Chapter 5) can subvert the downstream signal transduction pathways that link PRR activation to the secretion of pro-inflammatory cytokines *via* the transcription factor NF-κB in phosocytes that have taken them up. The parasitic protozoans *Leishmania donovani*, *L. major* and *Toxoplasma gondii* also subvert host signalling pathways by various means. Pathogens such as *Escherichia coli*, *Chlamydia* and *Bacillus anthracis* thwart PRR signalling by directly removing intermediate molecules from the NF-κB and mitogen-activated protein kinase (MAPK) signalling cascades.

Targeting cytosolic PRRs, IPS-1, RIG-I and MDA5

Interferon-beta (IFN-β) promoter stimulator 1 (IPS-1) is an adaptor involved in RIG-I- and MDA5-mediated antiviral immune responses. HCV encodes a protease that cleaves IPS-1 and blocks RIG-I signalling and, thus, type 1 interferon production. A protein produced by paromyxovirus directly binds and inhibits MDA5. Viruses such as HIV and human herpesviruses can degrade the transcription factor IRF-3 to prevent secretion of the type 1 interferons, IFN-α and IFN-β.

Masking microbe-associated molecular patterns

Fungi try to evade CTLs and TLRs that recognise fungal sugar motifs by masking or modifing cell wall carbohydrates, particularly β-1-3-glucan. For example, *Candida albicans* covers β-1-3-glucan with a layer of mannoproteins and *Histoplasma capsulatum* covers β-1-3-glucan with α-1-3-glucan, thus avoiding recognition by Dectin-1, and TLR2:TLR4. In addition, *H. capsulatum* yeast cells secrete an endo-β-1,3-glucanase, Eng1, which trims β-glucans exposed on the fungal cell to reduce recognition by Dectin-1. The fungus *Paracoccidioides brasiliensis* converts β-1-3-glucan to α-1-3-glucan to avoid PRR recognition. The capsule of *Cryptococcus neoformans* effectively masks its MAMPs from recognition by PRRs.

MANIPULATING HOST INHIBITORY SIGNALING

Because inflammation initiated by MAMP-PRR ligation may result in tissue destruction, inhibitory receptors exist to regulate PRR, Fc receptor and other cell membrane receptor signalling. These inhibitory receptors have inhibitory motifs on their cytoplasmic tail, the most common of which is the immunoreceptor tyrosine-based inhibitory motif (ITIM). ITIMs recruit tyrosine phosphatases that contain an SH2 domain (SHP1 and SHP2). By dephosphorylating signalling intermediates on nearby activating receptors that contain immunoreceptor tyrosine-base activation motifs (ITAMs), the activation of ITAMs receptors such as the Fcγ receptor or TREM1, a member of the TREM family, is dampened. TREM1 (triggering receptor expressed on myeloid cells) plays an important role in the amplification of inflammation, crosstalk with other PRRs and activation of antigen-presenting cells. Pathogens use various methods to inhibit host receptor signalling (**Figure 8.8**). Some pathogens such as *Moraxella catarrhalis*, *Neisseria meningitidis*, *Staphylococcus aureus* and *Streptococcus agalactiae* bridge inhibitory and activating receptors as shown in Figure 8.8A. In the case of the closely related bacteria, *M. catarrhalis* and *N. meningitidis*, both bridge the inhibitory receptor CEACAM1 and the activating receptor TLR2. The lipoteichoic acid of *S. aureus* bridges the inhibitory receptor paired immunoglobulin-like receptor B (PIR-B), which has four ITIMs with activating TLRs. *S. agalactiae* binds the inhibitory sialic acid-binding Ig-like receptors (siglec-5 and siglec-9) using surface β-protein and capsular sialic acid to suppress activation of neutrophils. Activation of ITAM-containing receptors requires high-affinity binding between the receptor and its ligand. Low-affinity binding causes ITAMs to function in an inhibitory manner. An ITAM that transduces an inhibitory signal is termed an ITAMi. Low-affinity binding to ITAMs is the strategy used by *Escherichia coli* (Figure 8.8B) to avoid phagocytosis by macrophages. *E coli* binds directly to an Fcγ receptor III (FcγRIII) and bridges it to a macrophage receptor with collagenous structure (MARCO) receptor. MARCO is a scavenger receptor with specificity for low-density lipoprotein

(A)

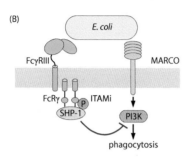

Hijacking ITIM-bearing inhibitory receptors: Various human bacterial pathogens evolved virulence factors that co-ligate inhibitory receptors with recognized activating receptors. This leads to the suppression of several antimicrobial functions and evasion of the host immune response.

(B)

Exploiting inhibitory ITAM signaling: E. coli escapes phagocytosis through low-avidity engagement of FcγRs and the induction of inhibitory ITAM signaling. This serves to resist clearing of the bacterial pathogen by the host.

(C)

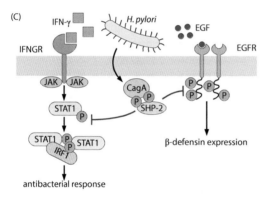

Altering first-line immune functions: After translocation into the host cell cytoplasm, the ITIM-containing effector protein CagA modulates epithelial defense responses, including IFN-γ signaling. This strategy helps to overcome the first-line inflammatory response.

Figure 8.8 Bacterial pathogens evade host defence responses by manipulating inhibitory signalling. (A) M. catarrhalis, N. meningitidis, Group B Streptococcus and Staphylococcus aureus evolved specific virulence factors to engage inhibitory receptors, which co-ligate with and attenuate pattern recognition receptor (PRR) signalling. (B) Escherichia coli escapes macrophage receptor with collagenous structure (MARCO)-dependent killing through hijacking of inhibitory ITAM signalling. Non-opsonised E. coli binds to FcγRIII with low affinity and induces weak phosphorylation of the FcR common γ chain (FcRγ), leading to recruitment of SHP-1. In turn, SHP-1 dephosphorylates PI3K and abrogates MARCO-dependent phagocytosis. (C) Upon infection, Helicobacter pylori translocates the ITIM-dephosphorylation of activated STAT1 and epidermal growth factor receptor (EGFR). This abrogates IFN-γ signaling and human β−defensin 3 (hBD3) synthesis, and enhances bacterial survival. *(Continued)*

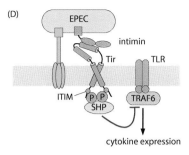

(D)

Dampening TLR signaling:
E. coli inserts the ITIM-bearing virulence factor Tir
into the epithelial cell membrane to attenuate first-line
TLR responses and pro-inflammatory cytokine release.

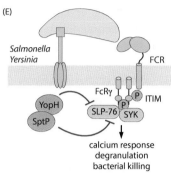

(E)

Mimicking host protein tyrosine phosphatases:
Salmonella and *Yersinia* developed effector proteins
that resemble host phosphatases and target essential
signaling intermediates to dampen inflammatory responses
and increase survival within the host.

Figure 8.8 (Continued)   Bacterial pathogens evade host defence responses by manipulating inhibitory
signalling. (D) During infection with the bacterium enteropathogenic *E. coli* (EPEC), the intimin receptor
(Tir) translocates into the epithelial cell. The intracellular tail of EPEC Tir recruits host cell phosphatases
SHP-1 and SHP-2. As a result, the activation of TRAF6 is inhibited, and EPEC-induced expression of
pro-inflammatory cytokines is suppressed. (E) *Salmonella* and *Yersinia* secrete protein tyrosine phosphatases SptP and YopH, respectively. SptPt targets the protein tyrosine kinase SYK in mast cells and suppresses degranulation. During in vivo infection, YopH targets the signaling adaptor SLP-76 in neutrophils.
This leads to reduced calcium responses and IL-10 production.

that is involved in phagocytosis of pathogens. The FcγRIII ITAMi impairs
phosphorylation of MARCO, which compromises the ability of the receptor to participate in phagocytosis of *E. coli*. The effectors of type 3 and type
4 secretion systems of some gram-negative pathogens contain ITIM-like
motifs that suppress immune cell activation. The *Helicobacter pylori* T4SS
effector CagA contains ITIM-like motifs (Figure 8.4C), as does the enteropathogenic *Escherichia coli* type III secretion effector Tir (Figure 8.8D). Both
reduce secretion of pro-inflammatory cytokines. In contrast, the YopH effector of *Yersinia* and the SptP effector of *Salmonella enterica* serovar Typhimurium mimic host cell protein tyrosine phosphatases.

PATHOGEN SURVIVAL INSIDE HOST CELLS

As we learned above, pathogens enter non-phagocyte host cells by inducing uptake by endocytosis. Cell entry is a step in the local or systemic spread of the pathogen. The majority of pathogens seek to avoid uptake by professional phagocytes, using one or more methods described in this chapter because it usually leads to their death. However, some pathogens have evolved mechanisms to avoid being killed inside professional phagocytes and to use them as a Trojan horse to spread throughout the human body. As you will recall, whether an epithelial cell or a professional phagocyte, the pathogen is enclosed in a vacuole formed by the cell membrane. This vacuole is termed an endosome in non-phagocytic cells or a phagosome in professional phagocytes. Maturation of the endosome/phagosome depends on localising the GTPase Rab5 to the early endosome/phagosome membrane because Rab5 is responsible for recruiting additional effector proteins including Rab7. Replacement of Rab5 by Rab7 is associated with conversion of the early endosome/phagosome to the late endosome/phagosome. The proton pump ATPase (V-ATPase) inserts into the vacuole membrane and the contents of the endosome/phagosome are acidified. The end point of the endosomal and phagosomal pathways is fusion of both types of vacuole with lysosomes to form endolysosomes or phagolysosomes, respectively, both of which are digestive organelles (Figure 8.7). The phagolysosome is, however, a superior microbicidal compartment compared to the endolysosome because of the presence of NADPH oxidase and nitric oxide synthase as well as the contents of lysosomes and specific granules that provide various antimicrobial proteins that were discussed earlier in the chapter (Figure 8.6). The strategies employed by pathogens to avoid destruction in the endolysosomal or phagolysosomal pathways are the topic of the following sections. They include arresting the maturation of the endosome or phagosome, escaping from the endosome or phagosome before fusion with lysosomes, or surviving in the phagolysosome. In instances where pathogens escape from endosomes or phagosomes, they become susceptible to uptake by autophagy, a process used by cells to remove non-functional proteins and clear damaged organelles, such as mitochondria and endoplasmic reticulum (ER). Autophagy consists of enclosing the target molecule, organelle or pathogen in a double membrane that fuses with lysosomes to form a degradative vacuole. We will discuss evasion of autophagy later in the chapter. Intracellular survival strategies used by various bacteria are shown in **Figure 8.9**.

Bacteria

There are several strategies used by intracellular bacterial pathogens to take control of the vacuole in which they are contained to prevent its acidification and fusion with lysosomes. Broadly, these can be divided into arresting the maturation of the endosome/phagosome, diverting the endosomal/phagosomal pathway away from fusion with lysosomes,

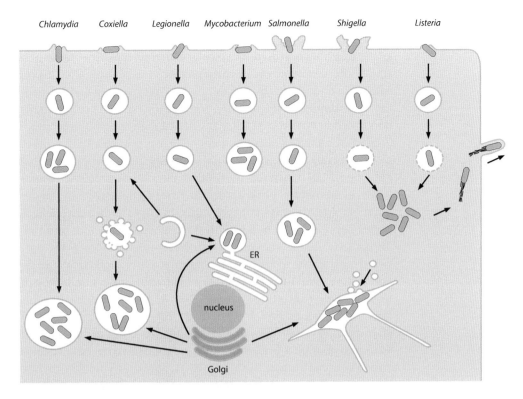

Figure 8.9 Intracellular survival strategies of some bacterial pathogens. The survival mechanisms of these bacteria are described in detail in the text. (From Cain, R.J. and Vazquez-Boland, J.A., *Molecular Medical Microbiology*, 2nd ed, Vol 1, Academic Press, Amsterdam, the Netherlands, pp. 491–515, 2015.)

survival in the endolysosome/phagolysosome, or escape from these vacuoles into the cytosol. These strategies are mediated by altering the phosphoinositide composition of the vacuole membrane, manipulating the cell cytoskeleton, regulating small GTPases, and chemical interference of vacuole–lysosome fusion (Figure 8.9).

As discussed earlier, pathogens that gain entry to a host cell by inducing reorganisation of the actin cytoskeleton have mechanisms that inactivate GTPases, resulting in rapid depolymerisation of actin after the entry process is complete in order to return the cell cytoskeleton to its resting state. These same mechanisms can serve to arrest the maturation of the vacuole containing the pathogen and its fusion with lysosomes. Inactivation of GTPases appears to be a common strategy of intracellular bacteria including *S. enterica* serovar Typhimurium and *Legionella pneumophila*.

Arresting the phagosome/endosome

The major phosphoinositide targeted by intravesicular bacterial pathogens is phosphatidylinositol 3-phosphate (PI3P) because it helps to recruit a range of proteins that are involved in protein trafficking to membranes.

Mycobacterium tuberculosis: *Mycobacterium tuberculosis* is a bacterial pathogen that enters the human host in respiratory droplets and is phagocytosed by alveolar macrophages. The bacterium arrests the development of early phagosomes shortly after fusion with early endosomes by preventing PI3P recruitment and/or facilitating its depletion (Figure 8.7). Some of the bacterial components implicated in this process are mannose-capped lipoarabinomannan (ManLAM), a cell wall component, which suppresses PI3P synthesis in early phagosomes and SapM, a phosphatase that can dephosphorylate PI3P. However, the notion that *M. tuberculosis* resides solely in immature phagosomes has been challenged by longer duration observations of phagocytosis that have shown that phagosomes containing *M. tuberculosis* do fuse with lysosomes and that the bacteria escape from this environment into the cytosol. The proposed mechanism of arresting phagosome maturation is *via* a secretion system, akin to those in gram-negative bacteria, and has been termed a type 7 secretion system (T7SS). *M. tuberculosis* has five different T7SS named ESX-1 to ESX-5. *M. tuberculosis* has a modified gram-positive–type cell wall and, for this reason, it is thought that T7SS may also be a feature of other gram-positive bacteria such as *Staphylococcus aureus*. The T7SS, ESX-1, appears to mediate escape from the phagosome into the cytosol and macrophage apoptosis by secretion of two proteins, ESAT-6 and CFP-10, but the mechanisms by which these processes are accomplished remain to be elucidated (**Figure 8.10**).

***Salmonella enterica* serovar Typhimurium**: The ability of *Salmonella enterica* serovar Typhimurium to survive inside a modified phagosome known as a *Salmonella*-containing vacuole (SCV) is mediated by effectors from two type 3 secretion systems (T3SS1 and T3SS2) encoded by *Salmonella* pathogenicity islands 1 and 2 (SPI-1 and SPI-2) that are used to transfer bacterial effector proteins into the host cell. Much remains to be understood about the intracellular lifestyle of *Salmonella*, but the SCV resembles a late endosome/phagosome based on the presence of markers such as the late endosomal/lysosomal glycoprotein (lgp), LAMP1 and Rab7 and an acidic intra-SCV pH. However, there is little hydrolase activity in the SCV. Among the more than 30 effectors transferred across the SCV membrane by *Salmonella*, SifA is important for the intracellular survival of *Salmonella*. Expression of the *Salmonella* T3SS-2 delivers additional effectors that further modify the SCV. Under the influence of SifA, the SCV forms *Salmonella*-induced filaments (SIFs) that are long tubular structures that extend from the SCV membrane along microtubules. Sifs are most evident in epithelial cells. Because phagosomes, endosomes and other vesicles travel along the microtubular system and because the actin cytoskeleton is centrally involved in the formation of endosomes and phagosomes, it is not surprising that intracellular pathogens attempt to interfere with these components

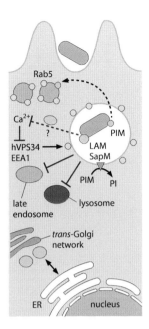

Figure 8.10 Strategies used by *Mycobacterium tuberculosis* to modulate phago-some maturation. After internalisation, the bacterium uses an array of effector molecules, including the lipids phosphatidylinositol mannoside (PIM) and lipoarabinomannan (LAM), and the phosphatidylinositol-3-phosphate (PI(3)P) phosphatase SapM to arrest phagosome maturation at an early stage. Rab5, a regulatory guanosine triphosphatase, regulates entry of the bacterium from the plasma membrane to the early endosomes and generation of phosphotidy-linositol-3-phosphate lipid; Rab5A acts using multiple effectors, including the p150–hvPS34 complex, early endosome antigen 1 (EEA1); and SapM eliminates PI3P from the phagosomal membrane by catalysing its hydrolysis, and thus contributes to inhibition of phagosome maturation. (From Flannagan, R.S. et al., *Nat. Rev. Microbiol.*, 7, 355–366, 2009.)

of the cytoskeleton. As described above, under certain conditions the SCV of *Salmonella enterica* serovar Typhimurium forms long tubes termed SIFs (**Figure 8.11**). The effectors SifA and PipB2, from T3SS-2, indirectly allow the membrane of the SCV to bind to the microtubular motor protein, kinesin-1, and bring about filamenta-tion of the SCV so that it can contact late endosomes and lysosomes. How SIF formation benefits the intra-vacuolar bacteria is not fully understood. The T3SS effector *Salmonella* outer protein (SoPE) of *S. enterica* serovar Typhimurium is an activator of small GTPases and also activates Rac1 and Cdc42, which play a role in the cell entry process of this bacterium. Therefore, SopE is involved not only in cell entry but also in the generation and rerouting of the intracellular vacuole in which the bacterium is contained. It appears that T3SS effectors imitating guanine nucleotide exchange factors (GEFs) may be common among intracellular gram-negative pathogens.

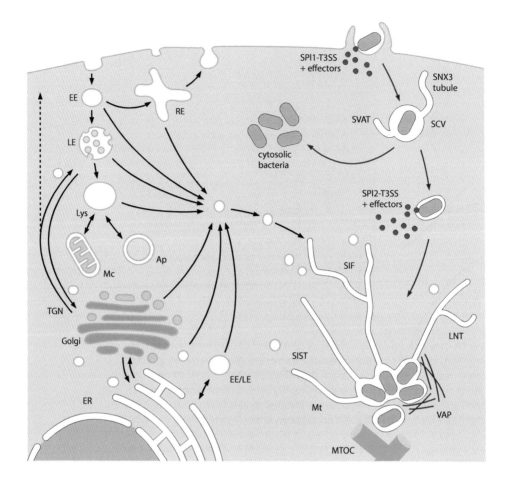

Figure 8.11 Model of *Salmonella* intracellular lifestyle and interactions with host membrane compartments. After SPI1-T3SS-mediated entry of host cells or phagocytic uptake, internalised *Salmonella* are located in a host-derived membrane compartment, termed a *Salmonella*-containing vacuole (SCV). A subpopulation of *Salmonella* escapes from the SCV, replicates in the cytosol and is targeted by macro-autophagy. Generally, bacteria remain in an early SCV from which *Salmonella*-induced spacious vacuole-associated tubules (SVATs) and sorting nexin 3 tubules (SNX3s) extend. In parallel, the SCV moves along microtubules towards the microtubular organising centre (MTOC). By means of the SPI2-T3SS, *Salmonella* translocates another set of effector proteins. This activity induces various types of *Salmonella*-induced tubules (SITs) such as SIFs, *Salmonella*-induced SCAMP3 tubules (SISTs) and LAMP1-negative tubules (LNTs). SITs are formed through fusion processes with recruited vesicles derived from different host cell membrane compartments and move along microtubules. SIT membranes contain proteins from various cellular origins. Actin polymerises around the SCV vacuole. EE, early endosome; ER, endoplasmic reticulum; LE, late endosome; Lys, lysosome; Mc, mitochondrion; RE, recycling endosome; TGN, *trans*-Golgi network. (From Liss, V. and Hensel, M., *Cell. Microbiol.*, 17, 639–647, 2015.)

Diverting the endosomal/phagosomal pathways

Chlamydia and **Chlamydophila**: *Chlamydia trachomatis* and *Chlamydophila psittaci* and *Chlamydophila pneumoniae* are human gram-negative pathogens that have a unique life cycle that comprises a metabolically inactive, but infectious, elementary body (EB) and a metabolically active, non-infectious reticulate body (RB) also called the *Chlamydia/Chlamydophila*-containing vacuole (CCV). Entry into mucosal epithelia is facilitated by a T3SS effector named translocated actin-recruiting phosphoprotein (TarP) that nucleates actin. Immediately after entry into epithelial cells, endosomes containing EBs are rapidly diverted by the microtubule transport system from the endosomal pathway to the Golgi apparatus where they form a replicative organelle termed an inclusion body. Inclusion bodies are surrounded by a scaffold consisting of actin and intermediate filaments. No early or late endosomal markers are present in the inclusion membrane and the inclusion is not acidified nor do lysosomes fuse with it. Thus, the bacteria are not subject to destruction. Fusion with the Golgi and the late-endosomal multivesicular body (MVB) compartment provides nutrients to the CCV. Fusion is mediated by recruiting RAB GTPases and their effectors, phosphoinositide lipid kinases and soluble *N*-ethylmaleimide-sensitive factor attachment protein receptor (SNARE) proteins (**Figure 8.12**).

Legionella pneumophila: *Legionella pneumophila* is a respiratory pathogen that is transmitted by aerosols and phagocytosed by alveolar macrophages, but the bacterium can be taken up by non-phagocytic cells as well. Shortly after phagocytosis, the pathogen employs a T4SS termed Dot/Icm to divert the phagosome from its maturation pathway to fuse with vesicles derived from the ER and with mitochondria. This replicative entity is termed a *Legionella*-containing vacuole (LCV) that eventually resembles the rough ER and is studded with ribosomes. The LCV is characterised by the presence of several small GTPases that are central to its generation. Some of the Dot/Icm effector proteins, for example RalF and DrrA, are guanine nucleotide exchange factors (GEFs) involved in the activation of small GTPases. When the ER-derived vesicles fuse with the LCV, *L. pneumophila* disrupts microtubule-dependent transport of the host cell by means of a phosphocholine transferase AnkX (**Figure 8.13**).

Survival in the endolysosome/phagolysosome

Coxiella burnetii: *C. burnetii* is a gram-negative, obligate intracellular pathogen that has a biphasic life cycle consisting of an infectious small-cell variant and a large-cell replicative variant. Similar to *L. pneumophila*, *C. burnetii* encodes a T4SS that is similar to the Dot/Icm secretion system. *C. burnetii* replicates intracellularly in monocyte/macrophages in a phagolysosome-like compartment that is acidic and contains several antimicrobial factors. After *C. burnetii* is uptaken into a phagosome, the bacterium causes the vacuole to take on characteristics of an autophagosome by recruiting the autophagic microtubule-associated protein light-chain 3 (LC3), Rab24 and the Rab1GTPase to the phagosome. The presence of LC3 delays fusion of the phagosome with lysosomes, allowing the bacterium time to transition to the large-cell replicative variant. The *C. burnetii*-containing vacuole progresses along the phagolysosomal pathway and acquires the normal maturation markers,

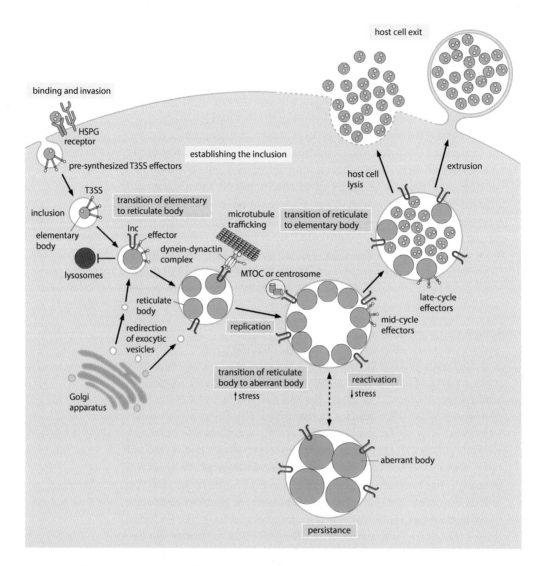

Figure 8.12 The lifecycle of *Chlamydia trachomatis*. The binding of elementary bodies to host cells is initiated by the formation of a trimolecular bridge between bacterial adhesins, host receptors and host heparan sulfate proteoglycans (HSPGs). Next, pre-synthesized type III secretion system (T3SS) effectors are injected into the host cell, some of which initiate cytoskeletal rearrangements to facilitate internalization and/or initiate mitogenic signaling to establish an anti-apoptotic state. The elementary body is endo-cytosed into a membrane-bound compartment, known as the inclusion, which rapidly dissociates from the canonical endolysosomal pathway. Bacterial protein synthesis begins, elementary bodies convert to reticulate bodies and newly secreted inclusion membrane proteins (Incs) promote nutrient acquisition by redirecting exocytic vesicles that are in transit from the Golgi apparatus to the plasma membrane. The nascent inclusion is transported, probably by an Inc, along microtubules to the microtubule-organizing centre (MTOC) or centrosome. During mid-cycle, the reticulate bodies replicate exponentially and secrete additional effectors that modulate processes in the host cell. Under conditions of stress, the reticulate bodies enter a persistent state and transition to enlarged aberrant bodies. The bacteria can be reactivated upon the removal of the stress. During the late stages of infection, reticulate bodies secrete late-cycle effectors and synthesize elementary-body-specific effectors before differentiating back to elementary bodies. Elementary bodies exit the host through lysis or extrusion.

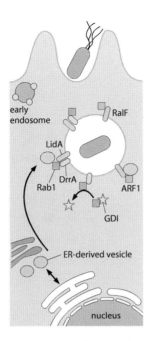

Figure 8.13 Strategies used by Legionella pneumophila to subvert phagosome maturation. *L. pneumophila* impairs fusion of the *Legionella*-containing vacuole (LCV) with endolysosomal compartments and, instead, promotes fusion with endo-plasmic reticulum (ER)-derived membranes. DrrA, protein that recruits the small guanosine triphosphatase (GTPase) Rab1 to the cytosolic face of LCV; GDI, guanine nucleotide-dissociation inhibitor; LidA, a Rab effector; Rab1, GTPase; RalF, recruits ADP-ribosylation factor (ARF1) to *Legionella*-LCVs in the establishment of a replica-tive organelle. (From Flannagan, R.S. et al., *Nat. Rev. Microbiol.*, 7, 355–366, 2009.)

Rab5 and Rab7 GTPases; the V-type H1 ATPase; lysosomal markers LAMP1, LAMP2 and LAMP3; and some lysosomal enzymes. The terminal pH of the vacuole is approximately 5.0. The *C. burnetii*-containing vacuole is known as a parasitophorous vacuole (PV) and, although it appears to be a large phagoly-sosome, it is a specialised compartment that has been actively modified by the pathogen. It is clear that *C. burnetii* has adapted to the acidic pH of the PV such that it requires a low pH for certain metabolic activities and must have mechanisms to evade antimicrobial systems inside the PV (**Figure 8.14**).

Fungi

Histoplasma capsulatum is an intracellular fungal pathogen of macro-phages and DCs and opsonised spores or mycelial fragments are taken up by phagocytosis into phagolysosomes. However, the pH of phagolysosomes containing *H. capsulatum* is less acidic (~pH 6.5), perhaps because the host vacuolar membrane ATPase/proton-pump is partially excluded. Non-opsonised spores may enter phagocytes through different receptor–ligand

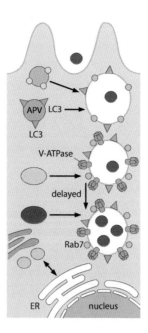

Figure 8.14 Strategies used by *Coxiella burnetti* to subvert phagosome maturation. Phagosomes containing this *C. burnetti* undergo delayed maturation as they fuse with autophagocytic vesicles (APVs) bearing LC3. The delay enables *C. burnetii* to acquire features that allow it to replicate in a membrane-bound compartment that resembles phagolysosomes. (From Flannagan, R.S. et al., *Nat. Rev. Microbiol.*, 7, 355–366, 2009.)

pathways For example, the fungal receptor used to enter macrophages is heat shock protein (HSP) 60, whereas a separate fungal adhesin is used to enter DCs. Fungi that enter by routes other than opsonophagocytosis reside in phagosomes that do not fuse with lysosomes and therefore survive. The mechanism(s) by which the fungus modulates phagosome maturation remain to be determined. The yeast form of the fungal pathogen *Cryptococcus neoformans* enters macrophages in phagosomes that rapidly fuse with lysosomes. However, capsular polysaccharide elaborated by the yeast fills the phagosome and is extruded into lysosomes that become polysaccharide-containing vacuoles. In addition, although the yeast remains in the phagolysosome, the membrane loses continuity, possibly allowing nutrients to be acquired by the pathogen. *Candida albicans* may interfere with phagosome-lysosome fusion. The consensus is that *C. albicans* resides in a membrane-bound compartment that is not fully acidified and contains only a subset of lysosomal markers. *Candida glabrata* also appears to inhibit phagosome–lysosome fusion and replicates in a non-acidic compartment. Fungal replication eventually destroys infected macrophages.

Parasites

Microsporidian infections in humans are most frequently caused by *Encephalitozoon cuniculi* and occur in immunocompromised individuals. *E. cuniculi* infects a wide range of cells, including macrophages.

Most microsporidians develop directly within the cytosol of their host cells following puncture of the host cell plasma membrane and injection of their sporoplasm. Other microsporidia, such as *E. cuniculi*, also may locate within a PV, where they replicate by binary fission and produce new spores. Spores that are phagocytosed proceed through the phagosomal pathway to fuse with lysosomes and are destroyed. However, spores may hatch inside the phagosome and deposit sporoplasm in the cytosol of the phagocyte. When the sporoplasm of *E. cuniculi* is injected into the host, it is contained in a vacuole derived from the host cell membrane. The creation of this vacuole does not require actin. This non-phagocytic vacuole does not acquire endocytic markers such as the transferrin receptor, EEA1 or LAMP1, so it does not fuse with lysosomes.

Leishmania donovani and *L. major* are taken up by monocyte/macrophages and enter the early phagosome, but phagosome maturation is prevented by the synthesis of a family of glycoconjugates termed phosphoglycans by the parasite. One of these, the membrane-anchored lipophosphoglycan (LPG), is inserted into the phagosome membrane. LPG retains the molecular machinery required for the assembly of F-actin, particularly the Rho family small GTPases, CDdc42, and F-actin that accumulates on the phagosome membrane. LPG may activate Cdc42-specific RhoGEFs or inhibit Cdc42 GTP release by inactivating RhoGAPs. LPG on *Leishmania* promastigotes also prevents acidification of the phagosome by interfering with the V-ATPase pump, which allows promastigotes to differentiate into resistant amastigotes. *Leishmania* amastigotes scavenge host sphingolipids to counteract the acidic environment of the phagolysosome.

Once *Toxoplasma gondii* crosses the intestinal epithelium, it encounters DCs, macrophages, and intraepithelial lymphocytes. *T. gondii* is capable of invading various types of nucleated cells. The DC is considered to act as a Trojan horse for the spread of the parasite. *T. cruzi* trypomastigotes do not enter phagocytic or non-phagocytic cells by the process of phagocytosis or endocytosis. Reorganisation of the actin cytoskeleton is not involved in uptake of trypomastigotes. However, uptake of amastigotes requires actin remodelling and the small GTPase Rac1. *Toxoplasma* takes up residence in a PV surrounded by the ER, from which lipids are obtained. The membrane of the PV is connected to the ER and mitochondria by finger-like extensions. The *Toxoplasma* PV membrane does not intersect with the endosomal pathway given that it lacks the ligands required for docking and fusion of endosomes (**Table 8.7**).

	encephalitozoon	*toxoplasma*	*leishmania*
invasive stages	spore/ sporoplasm	tachyzoite	promastigote
		bradyzoite	
		sporozoite	
intracellular stages	Meront	tachyzoite	amastigote
		bradyzoite	
cell types	various, including macrophages	all nucleated, including macrophages	primarily macrophages
host receptors	GAGs, mannose receptors	GAGs, sialic acid	C1, C3, mannose scavenging
entry mechanism	injection-induced invagination	active penetration	phagocytosis
host markers in the vacuole	excludes endocytic and exocytic markers	excludes endocytic and exocytic markers	EEA1, Rab5, Rab7, LAMP I, LAMP2
niche	non-fusigenic vacuole	non-fusigenic vacuole	delayed maturation, phagolysosome
secreted-surface virulence factors	polar tube proteins	rhoptry proteins ROPs, RONs Microneme proteins MICs	lysophosphoglycan LPG
nutrient acquisition	pores in vacuole membrane	pores in vacuole membrane	induced autophagy– phagolysosome degradation

Table 8.7 Survival strategies of intracellular parasites. (From Sibley, L.D., *Immunol. Rev.*, 240, 72–91, 2011.)

ESCAPE TO THE CYTOSOL

Instead of attempting to control the environment and the maturation of the vacuole to avoid destruction in the endolysosome or phagolysosome, certain pathogens escape from the vacuole into the cytosol almost immediately after uptake by the host cell. Escape from the vacuole is typically mediated by enzymes that lyse the vacuole membrane. Some pathogens escape from vacuoles later along the endosomal/phagosomal maturation pathway. An example of enzymatic lysis is the production of listeriolysin and the phosphatidylinositol-specific phospholipase C, PlcA and PlcB, by *Listeria monocytogenes.* Listeriolysin is a pore-forming exotoxin that inserts into the vacuole membrane by binding to cholesterol. The creation of pores perturbs ion gradients across the vacuolar membrane and prevents vacuole maturation and fusion with lysosomes. Interestingly, acidification of the vacuole to pH 5.5 is optimum for the activity of listerolysin, as is the presence of IFN-γ-induced lysosomal thiol reductase (GILT). GILT is an enzyme that facilitates the complete unfolding of proteins destined for lysosomal degradation by disrupting disulphide bonds and has an important role in MHC class II-restricted antigen processing in antigen-presenting cells such as DCs and macrophages (**Figure 8.15**). *Francisella tularensis* appears similar to *Listeria* in its uptake and escape into the cytosol from late endosomes. The *Francisella* pathogenicity island (FPI) protein IglC and its regulator MglA and a lipoprotein termed FTT1103 that have been implicated in escape from the endosome, as have two effectors of a type 6 secretion system, VgrG and IglI (**Figure 8.16**). *Rickettsia* species are intracellular parasites of macrophages and endothelial cells. These

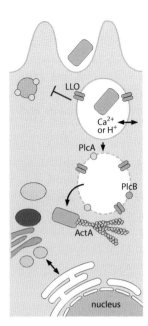

Figure 8.15 Strategies used by *Listeria monocytogenes* to subvert phago-some maturation. *Listeria monocytogenes* escape into the cytosol. ActA, actin assembly-inducing protein; LLO, listeriolysin; PlcA and PlcB, phospholipases. (From Flannagan, R.S. et al., *Nat. Rev. Microbiol.*, 7, 355–366, 2009.)

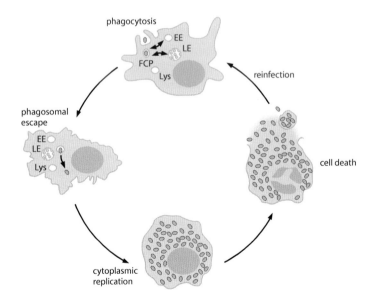

Figure 8.16 Model of the *Francisella* intracellular cycle in phagocytes. Upon phagocytosis, bacteria reside in an early phagosome (FCP) that interacts with early (EE) and late (LE) endocytic compartments but not lysosomes (Lys). Bacteria rapidly disrupt the FCP membrane and reach the cytosol where they undergo extensive replication, a process followed by cell death, bacterial release, and subsequent infection. (From Celli, J. and Zahrt, C. Cold Spring Harbor Perspective in Medicine, Cold Spring Harb Prospect Med, 2013;3:010314.) (From Celli, J. and Zahrt, C., *Cold Spring Harb. Prospect. Med.*, 3, 010314, 2013.)

bacteria gain access to the host cell cytoplasm using several phospholi-pases and haemolysins. *Shigella flexneri* may use the same pore-forming effectors, IpaB-IpaC, of its T3SS that it uses for facilitated uptake into cells to escape from the vacuole. IpaB and IpaC bind cholesterol and assemble to form a pore with the aid of IpaD, However, escape from the vacuole may be mediated by other bacterial products as well.

The trypomastigote stage of the protozoan parasite, *Trypanosoma cruzi* degrades the PV using a trypsin-sensitive enzyme (TcTox) and *trans*-sialidase/neuraminidase. The removal of sialic acid from the PV mem-brane makes it susceptible to the action of TcTox. Low pH PV is required for optimal enzyme activity. *Babesia* spp. parasitise red blood cells (RBCs) by invaginating the RBC membrane to form a PV. The vacuole membrane gradually disintegrates and the parasite rapidly escapes into the cytosol, but how this is accomplished remains to be determined.

VIRUS INTERACTIONS WITH INTRACELLULAR VACUOLES

In order to cause infection, viruses must deliver their nucleic acid to the cytosol (RNA viruses) or to the nucleus (DNA viruses). This necessitates crossing the plasma membrane and intracellular membranes. In Chapter 4, the mechanisms of adhesion of viruses were discussed, and we saw that for some viruses it is not possible to separate adhesion from entry because they are part of the same continuum. Here we consider what happens to enveloped and non-enveloped viruses after they have attached to the host cell plasma membrane. Enveloped viruses are enclosed by a lipid bilayer, whereas non-enveloped viruses are not. For enveloped viruses, multi-func-tional spike glycoproteins are the virus adhesins that bind host cell plasma membrane receptors, whereas for non-enveloped viruses, projections or indentations of the capsid serve this purpose. Some enveloped viruses fuse their envelope with the host cell plasma membrane, releasing the virus contents into the cytosol. In almost all cases, viruses enter permissive cells through the endocytic pathway and are taken by professional phagocytes through phagocytosis (**Figure 8.17**). Viruses take advantage of all of the various forms of endocytosis that were discussed previously. The uptake of viruses through the endocytic pathway is similar to that of other classes of intracellular pathogens that ultimately escape into, and replicate in, the cyto-sol and, in the case of viruses, also in the nucleus. The endosome provides a means by which the virus can travel through the cortical cytoskeleton and the highly structured cytoplasm. In addition, the changing environ-ment of the endosome, particularly increased acidity, may facilitate escape from the vacuole, as is the case with other classes of pathogen. For exam-ple, non-enveloped (double-stranded DNA [dsDNA]) adenoviruses enter host cells *via* the endocytic pathway using clathrin coated pits and, as the endosome acidifies, a capsid component undergoes a conformational

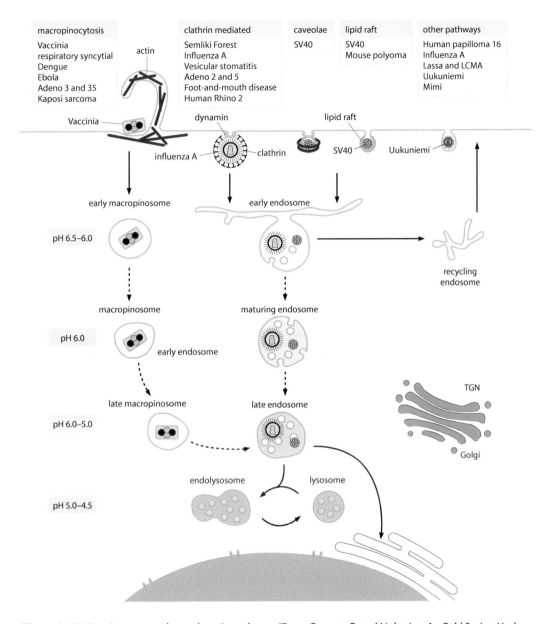

Figure 8.17 Virus in entry to the endocytic pathway. (From Cossart, P. and Helenius, A., *Cold Spring Harb. Perspect. Biol.*, 6, a016972, 2014.)

change, resulting in endosome lysis and escape of the virion into the cytosol. Similarly, the enveloped (negative-strand RNA) influenza virus is taken up by clathrin-coated pits into the endosomal pathway where low pH activates the integral membrane protein (M2) ion channel, allowing the acidification of the interior of the capsid. This results in fusion between the viral envelope haemagglutinin and the endosome membrane and the virus nucleoprotein core is released into the cytoplasm. The non-enveloped (dsDNA) SV40 virus, as well as other polyomaviruses, are taken up in caveolae and enter the endosomal pathway that delivers the virus to the ER. Within the ER lumen, the capsid undergoes partial disassembly that is completed after the virus exits the ER into the cytosol. In addition to the use of endosomes for intracellular transport to the nucleus or specific intracellular locations, capsids can employ the microtubular transport system or, in the case of vaccinia virus, polymerised actin attached to one pole of the capsid acts as a motor. The acquisition of polymerised actin for propulsion is a strategy used by certain bacteria such as *Shigella flexneri*, *Listeria monocytogenes*, *Burkholderia pseudomallei* and *Rickettsia* species that we will be discussed later in this chapter (**Table 8.8**).

virus	family	penetration
clathrin-mediated endocytosis		
semliki forest virus	togaviridae enveloped ss(+)RNA	pH < 6.2 early endosome
vesicular stomatitis virus	rhabdoviridae enveloped ss(−)RNA	pH < 6.4 early endosome
influenza A virus	myxoviridae enveloped ss(−)RNA	pH < 5.4 early endosome
foot-and-mouth disease virus	picornaviridae nonenveloped ss(+)RNA	early endosome recycling endosome
rhinovirus (HRV2) minor group	picornaviridae nonenveloped ss(+)RNA	pH < 5.5 endosome
adenovirus 2	adenoviridae nonenveloped dsDNA	endosome
macropinocytosis		
vaccinia virus	poxviridae	low pH for MV macropinosome
mature particle (MV)	enveloped	
enveloped particle (EV)	dsDNA	
respiratory syncytial virus	paramyxovirida enveloped ss(−)RNA	pH-independent macropinosome
ebolavirus	filoviridae enveloped ss(−)RNA	cathepsins B and C macropinosome
kaposi sarcoma virus	herpesviridae enveloped dsDNA	macropinosomes
caveolar/lipid-mediated endocytosis		
simian virus 40	polyomaviridae nonenveloped dsDNA	endoplasmic reticulum
polyomavirus (mouse)	polyomaviridae nonenveloped dsDNA	endoplasmic reticulum
other mechanisms		
papillomavirus 16 (human) (HPV16)	papillomavirida nonenveloped dsDNA	late endosome macropinosome
polio virus	picornavirusss(+)RNA	pH-independent early endosome
lymphocytic choriomeningitis virus (LCMV)	lassaviridae ss(−)RNA	pH < 6.3
rhinovirus (HRV14)	picornaviridae	low pH
major group	nonenveloped ss(+)RNA	macropinosome?

Table 8.8 Virus entry into cells by various endocyte pathways. Examples of endocytosis of viruses. (From Cossart, P. and Helenius, A., *Cold Spring Harb. Perspect. Biol.*, 6, a016972, 2014; Extensively modified by Michael Cole.)

CYTOSOLIC MOTILITY OF INTRACELLULAR PATHOGENS

Certain intracellular bacteria and viruses assemble host cell actin at one of their poles and use it as a motor to propel them through the cell and/ or move from one cell to the next. In *Listeria monocytogenes, Burkholderia pseudomallei, Shigella flexneri* and *Mycobacterium marinum*, the actin is composed of short cross-linked filaments, whereas the actin tail of *Rickettsia* spp. consists of long, unbranched filaments. Actin assembly begins with the recruitment by the pathogen of the actin-related complex 2/3 (Arp 2/3), which is central to the process (**Figure 8.18**). Some pathogens (*Listeria monocytogenes, Burkhoderia pseudomallei,* and *Rickettsia* spp.) possess a molecule that serves as a nucleation-promoting factor (NPF) and others (*Shigella flexneri*) acquire such from the host cytosol. The NPF of *L. monocytogenes* is the actin assembly-inducing protein (ActA), for *B. pseudomallei*, BimA, and for *Rickettsia*, RickA. *S. flexneri* employs the outer membrane protein IcsA, which functions like Cdc42 to recruit the Wiscott-Aldrich syndrome protein (WASP), which directly activates Arp 2/3. Vaccinia virus, monkeypox virus, and variola virus use actin-based motility to facilitate spread of the virus between cells, although the micro-tubule system has a role in intracellular virus movement too. Vaccinia virus replicates into two different forms in the cytosol: the intracellular mature virus (IMV) and the intracellular enveloped virus (IEV). IEVs form

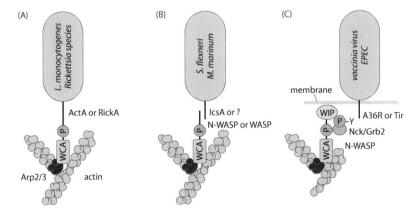

Figure 8.18 Three types of Arp2/3 activation by various pathogens. (A) *Listeria monocytogenes* and *Rickettsia* species; (B) *Shigella flexneri* and *Mycobacterium marinum*; and (C) and vaccinia virus. The surface proteins ActA of *L. monocytogenes* and RickA of *Rickettsia* species, directly activate the Arp2/3 complex. IcsA of *Shigella* recruits N-WASP, which then activates the Arp2/3 complex. *M. marinum* recruits WASP and by doing so probably activates the Arp2/3 complex. Vaccinia virus triggers the recruitment of N-WASP and thus induces the activation of the Arp2/3 complex. In this case the pathogen is separated from the actin cytoskeleton machinery by the plasma membrane. (From Gouin, E. et al., *Curr. Opin. Microbiol.*, 8, 35–45, 2005.)

when some of the IMVs become wrapped in a double membrane derived from the *trans*-Golgi network or endosomal cisternae. IEV form actin tails because they acquire the Golgi protein A36R that is required for virus actin assembly. Images of actin comet tails in Vaccinia virus and *L. monocytogenes* are shown in **Figures 8.19** and **8.20**.

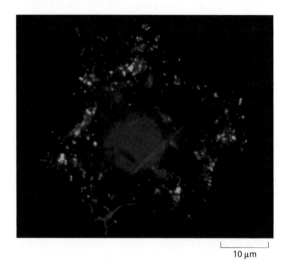

10 µm

Figure 8.19 Vaccinia virus particles (envelope protein labelled in green) exit infected HeLa cells on actin tails (visualised by phalloidin in red) at 8 hpi. The nucleus is stained with DAPI (in blue). Scale bar: 10 µm. (From Jockusch, B.M., *The Actin Cytoskeleton*, Series of *Handbook of Experimental Pharmacology 235*, Springer, Cham, Switzerland, 2016.)

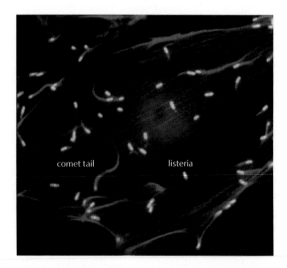

comet tail   listeria

Figure 8.20 *Listeria monocytogenes* comet tails. *L. monocytogenes* in green, polymerised actin in red. (Modified from Pollard, T.D. and Cooper, J.A., *Science*, 326, 1208–1212, 2009.)

ESCAPE OF INTRACELLULAR PATHOGENS FROM HOST CELLS

Exit from the host cell by intracellular pathogens occurs in several ways. The crudest method is to lyse the cell, which is generally attributed to the replication load of the pathogen exerting stress on the cytoplasmic membrane or perturbing the cellular functions of the cell and causing it to burst. However, evidence indicates that exit is, for the most part, a pathogen-facilitated process. Just as pathogens enter host cells without damage by taking advantage of host cell machinery, they can do the same when they exit. Furthermore, they may exit into the extracellular environment or exit one cell and enter an adjacent cell. The strategies employed by intracellular pathogens to exit host cells are illustrated in **Figure 8.21**.

Cytolysis

Cytolysis is a mechanism employed by various classes of intracellular pathogens. It is generally mediated by hydrolytic enzymes, usually proteases. In *Chlamydia* spp., the vacuole is lysed by cysteine proteases, which may, in turn, result in the calcium-dependent lysis of the cell membrane. Cysteine proteases are also responsible for the escape of *Plasmodium falciparum* from both the PV and the host cell. *Leishmania amazonensis* and *L. guyanensis* produce pore-forming cytolysins, termed leishporins, which are optimally active at a pH of between 5.0 and 5.5. Leishporins are non-covalently associated with an inhibitor and are activated by proteases in the low pH environment of PVs. Following activation by proteases, leishporins have increased ability to bind to phospholipids and acquire pore-forming ability, leading to disruption of both vacuoles and plasma membranes with the release of amastigotes. The replication of *T. gondii* tachyzoites inside the PV results in its acidification, which acts as a signal that initiates parasite motility and the secretion of perforin-like protein 1 (PLP1), a pore-forming protein, by the microneme. Damage to the membrane of the PV by secreted PLP1 allows leakage of potassium ions from the PV that activate a phospholipase C (PLC). Both PLP1 and PLC are likely involved in the focal destruction of the host cell membrane. Non-enveloped viruses use cytolysis to escape from the host cell, a process known as the burst. Use of the autophages pathway may allow, non-enveloped viruses to escape that do not result in host cells without cell lysis.

Actin-mediated cell-to-cell spread

As we have seen, various intracellular pathogens use host cell actin to enable them to move through the cytosol of the host cell. The polar actin comet tails can also provide the motive force to propel the pathogen from one cell into the next. Using this mechanism, the agent can spread without

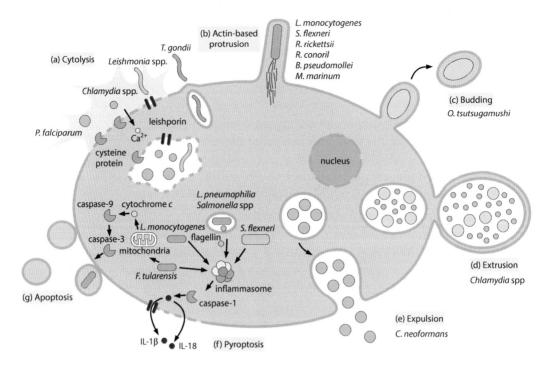

Figure 8.21 Strategies and mechanisms used by intracellular pathogens to exit host cells. (a) The cytolysis, and destructive and sequential rupture, of the vacuole and cell membranes. Putative mechanisms include proteases (*Plasmodium falciparum* and *Chlamydia* spp.), pore-forming proteins (PFPs) (*Leishmania* spp.) and the unique mechanism of *Toxoplasma gondii*. (b) Actin-based protrusion, which is exploited by *Listeria monocytogenes*, *Shigella flexneri*, *Rickettsia rickettsii*, *Rickettsia conorii*, *Burkholderia pseudomallei* and *Mycobacterium marinum*, results in a single bacterium that uses the force that is generated by actin polymerisation to protrude from the cell membrane and force engulfment into a neighbouring cell. (c) The budding of *Orientia tsutsugamushi*, in which a single bacterium is encased by plasma membrane. (d) The extrusion of *Chlamydia* spp., in which the large *Chlamydia*-containing vacuole pinches off and extrudes out of the cell; the extruded vacuole is encased by cytosol and plasma membrane. (e) The phagosomal expulsion of *Cryptococcus neoformans*, in which the large vacuole fuses with the plasma membrane by an undefined exocytic process. (f) Pro-inflammatory pyroptosis is defined by the sensing of bacterial molecules (flagellin *of Legionella pneumophila* and *Salmonella* spp. and unknown molecules of *S. flexneri*, *L. monocytogenes* and *Francisella tularensis*) through the host inflammasome. The inflammasome proteolytically activates caspase-1, which leads to IL-1β and IL-18 activation and secretion. Cytokine secretion occurs initially through a caspase-1-dependent pore, and is then released upon necrotic cell lysis. (g) Apoptosis is induced by *F. tularensis* using the intrinsic pathway of activation–cytochrome c release from mitochondria and activation of the initiator caspase-9 and the effector caspase-3. The bacterial molecule (or molecules) that is(are) responsible for apoptotic induction is unknown. (From Hybiske, K. and Stephens, R.S., *Nat. Rev. Microbiol.*, 6, p. 103, 2008.)

exiting the cell and, thus, avoid exposure to the host immune system. However, this mechanism requires that the pathogen must breach two plasma membranes. An analogy is wearing a latex glove and pushing a finger into the surface of an under-inflated balloon so that the finger creates a depression in the balloon. Similarly, the bacterium generates a protrusion in the first cell and a depression in the second. *Listeria monocytogenes* generates a protrusion by employing the T3SS effector InIC to release the tension of apical cell junctions. Tension in the apical cell junction is created by a host signalling pathway that consists of the scaffolding protein Tuba and its effectors N-WASP and Cdc42. InlC interacts with a domain in Tuba to displace N-WASP and inactivate this pathway. Another protein termed ezrin appears to stabilise *L. monocytogenes* protrusions by giving purchase to the actin comet tail. Ezrin is a member of the ERM protein family that comprises three closely related proteins: ezrin, radixin and moesin. ERM proteins cross-link actin filaments with the plasma membrane. The mechanism by which the *Listeria* protrusion is engulfed by the uninfected adjacent cell remain to be determined. *Shigella* protrusion appears to involve the T3SS and the formin-mediated actin polymerisation pathway. E-cadherin appears to be involved in the engulfment of *Shigella* protrusions that appear to occur preferentially at tri-cellular tight junctions. It seems that the adjacent cells that are recipients of the protrusions are actively involved in receiving them, perhaps by a type of endocytosis.

Protrusion into the extracellular environment (extrusion)

Extrusion is akin to exocytosis and resembles the budding of enveloped viruses from the plasma membrane. Exocytosis, like endocytosis, is actin polymerisation-dependent and so is the extrusion process. The net effect of extrusion is that the pathogens are released into the extracellular environment, the plasma membrane reseals, and the host cell remains viable. Intracellular pathogens that employ this method do so in slightly different ways. Budding is a form of extrusion in which the vacuole containing either a single pathogen cell, or many cells, fuses with the plasma membrane of the host cell and then is pinched off. Extrusion is one of the exit strategies employed by *Chlamydia* spp. and the scrub typhus agent *Orientia tsutsugamichi* (formally classified in the genus *Rickettsia*). In a variation of this theme, phagosomes containing individual cells of the fungal pathogen *Cryptococcus neoformans* coalesce to form a large vacuole surrounded by polymerised actin. Plb1, an enzyme secreted by *C. neoformans*, compromises the integrity of the stabilised phagosome membrane, which then fuses with the plasma membrane of the phagocyte and the fungal cells are expelled. Although non-enveloped viruses exit the cell by cytolysis, it has been proposed that the virions of positive-strand RNA viruses, or their RNA genomes, might escape the host cell enclosed in membranes. Extracellular virus-containing vesicles have been observed for polivirus, hepatitis A and Coxsackie B3 virus. The origin of these vesicles is thought to be the autophagosome pathway and the multivesicular body pathway.

Induction of programmed cell death

Rather than lyse the host cell using hydrolytic enzymes or pore-forming exotoxins, some intracellular pathogens induce the programmed cell death pathway termed pyroptosis to escape from their host cell. While apoptosis is a form of physiological cell death in which a programmed sequence of events leads to the elimination of cells without inducing inflammation, pyroptosis is a highly inflammatory form of programmed cell death that is dependent on caspase-1, inflammasomes, and the activation of IL-1β and IL-18. Various gram-negative intracellular bacteria, such as *Shigella flexneri*, *Salmonella* spp. *Legionella pneumophila*, and *Franciscella tularensis,* employ pyroptosis to escape from professional phagocytes, notably the macrophage. In these bacteria, pyroptosis is caused by effectors of either T3SS or T4SS. *S. flexneri* and *Salmonella* spp. escape from the cytosol of macrophages using the T3SS effectors IpaB, and SPI-1, respectively, to activate caspase-1, whereas *L. pneumophila* employs the Dot-Icm T4SS. The pyroptosis pathway(s) utilised by *F. tularensis* are currently unknown.

Preventing programmed cell death

The disadvantage of killing the host cell is that it releases intracellular pathogens into the extracellular environment where they are subject to attack by the complement cascade, antibodies and other components of the immune system. Intracellular pathogens arrest apoptosis in non-phagocytic cells by various means. The *Shigella* effector IpgD inhibits cytochrome c release mediated by caspase-3, whereas *Chlamydia* uses a protease (Chlamydia protease/proteasome-like activity factor [CPAF]) to destroy, and T3SS effectors to sequester, pro-apoptosis proteins. *Legionella* and *Coxiella* both target the Bcl-2 family members that are important pro- and anti-apoptotic proteins. *Legionella* blocks pro-apoptotic Bcl-2 activity and *Coxiella* regulates the anti-apoptotic Bcl-2. Some intracellular pathogens target general cell survival pathways. For example, *Salmonella* targets the PI3K/Act pathway using its T3SS effector, SopB. *Rickettsia*, *Legionella* and *Listeria* target the NF-κB pathway.

Interference with the host cell cycle

Some intracellular bacteria can arrest the cell cycle at the G2/M stage, and this may prolong their residence in the host cell. One mechanism by which bacteria arrest the cell cycle is by secreted toxins called cytolethal distending toxins (CDTs). CDTs are produced by several pathogenic bacteria including *Escherichia coli*, *Shigella dysenteriae*, *Campylobacter* spp., *Salmonella typhi* and *Haemophilus ducreyi*. Certain effectors of T3SS may also be involved in cell cycle arrest. For example, the effector IpaB of *Shigella* can traffic to the nucleus where it targets Mad2L2, an anaphase-promoting complex/cyclosome (APC) inhibitor. Cycle inhibiting factor (Cif) is T3SS effector cyclomodulin produced by enteropathogenic and enterohaemorrhagic *Escherichia coli* that can arrest the cell cycle.

Reprogramming the host cell

It may not be surprising that intracellular bacteria affect gene expression of the host cell to optimise intracellular survival and replication. *Salmonella enterica* serovar Typhimurium infection of epithelium results in transcriptional reprogramming of the host cell. Such reprogramming is affected by the *Salmonella* SPI-1 T3SS effector proteins SopE, SopE2 and SopB that stimulate unique signal transduction pathways that activate STAT3. The resulting changes in gene expression of the host cell are necessary to create the specialised vesicle that is necessary for intracellular replication. Several intracellular pathogens reprogram the host cell by epigenetic mechanisms. Histone modification appears important in this respect. For example, *Mycobacterium tuberculosis* down-regulates IFN-γ production in macrophages by up-regulating deacetylase activity on certain promotors. Upon infection, the *L. monocytogenes* effector InlB induces the host deacetylase, SIRT2, to translocate from the cytosol to the nucleus and deacetylate histone H3. *Shigella* uses the phosphatase activity of its T3SS effector OspF to inhibit MAPK signalling in the nucleus. This results in reduced phosphorylation of histone H3 in NF-κB regulated promotors, leading to the down-regulation of genes involved in immune responses. The T3SS effector, NUE, of *Chlamydia trachomatis*, enters the nucleus of infected cells and targets chromatin. NUE is a histone methyltransferase that targets histones H2B, H3 and H4.

EVADING AUTOPHAGY

Autophagy is a process by which cellular components varying in size and complexity – for example, protein aggregates and mitochondria – are degraded and recycled. In macroautophagy, targeted cytoplasmic components are enclosed by a double-membrane vesicle called an autophagosome that fuses with lysosomes to create a digestive vacuole similar to an endolysosome or a phagolysosome. It is now recognised that, in addition to its important role in cytosolic housekeeping, autophagy has several other functions, one of which involves eliminating intracellular pathogens. The autophagy-related 5 protein (ATG5) is a key to the extension of the phagophore double-membrane segments to form an autophagic vesicle. A diagram of autophagocytosis and the mechanisms by which intracellular pathogens evade it is shown in **Figure 8.22**. As was discussed earlier, in section "Pathogen Survival Inside Host Cells" intracellular pathogens evade destruction in the endolysosome or phagolysosome either by arresting fusion of endosomes or phagosomes with lysosomes or by escaping to the cytosol. However, the cytosol is not without its defences, one of which is autophagy. Therefore, it is not surprising that cytosolic bacteria modulate

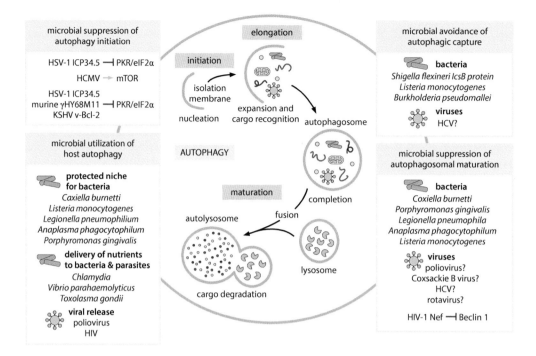

Figure 8.22 Adaptations of microbial pathogens to host autophagy. (From Deretic, V. and Levine, B., *Cell Host Microbe*, 18, 527–549, 2009.)

autophagy in order to survive in the cytosolic environment. Microbial strategies to evade autophagy include (1) preventing the induction of autophagy, (2) preventing the maturation of the autophagosome into an autolysosome, (3) avoiding pathogen recognition by the autophagic machinery, and (4) utilising functions or components of autophagy to enhance intracellular survival, replication, or extracellular release. The recognition of MAMPs displayed by microbes in the cytosol by cytosolic PRRs likely initiates autophagy of pathogens. However, it is thought unlikely that cytosolic pathogens target PRRs but, rather, more general signalling pathways that positively or negatively regulate autophagy.

Preventing the induction of autophagy

Viruses appear to be the only class of pathogen that suppress initiation of the autophagy pathway. Most information about viral mechanisms used to inhibit induction of autophagy come from the study of herpesviruses. Herpesviruses block autophagy by (1) inhibiting signalling by the stimulatory PKR/eIF2a kinase, (2) blocking the autophagy function of Beclin 1 or (3) activating signalling of the autophagy-inhibitor mTOR. Whether other virus families inhibit autophagy induction remains to be determined, but appears likely that they do.

Preventing the maturation of the autophagosome into an autolysosome

Just as intracellular pathogens are relatively safe in an endosome or phagosome that does not fuse with lysosomes, autophagosomes that do fuse with lysosomes provide a similar protected habitat. In the case of *Legionella pneumophila* soluble bacterial type 4 secretion products are sufficient to induce autophagy and the bacterial replication vacuoles have autophagy markers early after infection, It is likely that the bacteria delay autophagosome maturation, allowing time for them to differentiate into an acid-tolerant form. A similar situation pertains for *Coxiella burnetii* replicative vacuoles and *Anaplasma phagocytophilum* bacterial replicative inclusions. Both the replicative vacuole and the inclusion have autophagosomal markers, but not lysosomal markers. This observation is consistent with a mechanism that delays or, perhaps, blocks fusion of the autophagosome with lysosomes.

Several families of viruses may also block autophagosomal maturation. Poliovirus, coxsackie B virus, HCV, and rotavirus induce early stages of autophagy; however, autolysosomes are not formed. HIV-1 Nef inhibits autophagic maturation in macrophages through its interaction with the autophagy protein, Beclin 1.

Avoiding pathogen capture by the autophagosome

Listeria monocytogenes autophagy is activated by listeriolysin (LLO) immediately after the bacterium enters the cytosol. The phospholipases PI-PLC and PC-PLC and ActA mediate escape from autophagy. Cytosolic *L. monocytogenes* can be destroyed in the autolysosome, escape from the autophagosome or form spacious *Listeria*-containing phagosomes (SLAPs). SLAPs require autophagy to form and allow slow replication of *Listeria*. *L. monocytogenes* replicates in SLAPs when LLO activity is not sufficient to drive escape into the cytosol, but is sufficient to block lysosomal fusion. The *S. flexneri* T3SS effector, IcsA, induces autophagy by binding ATG5. However, the T3SS effector IcsB competitively inhibits IcsA and prevents its association with ATG5 blocking autophagy. T3SS effector BopA of *B. pseudomallei* is involved in escape from autophagy perhaps by blocking ATG5 binding to other of its proteins. Several viruses such as HCV, poliovirus, coxsackie B virus, rotavirus, and HIV trigger autophagy, but the autophagosome does not proceed to the formation of the autolysosome (**Figure 8.23**).

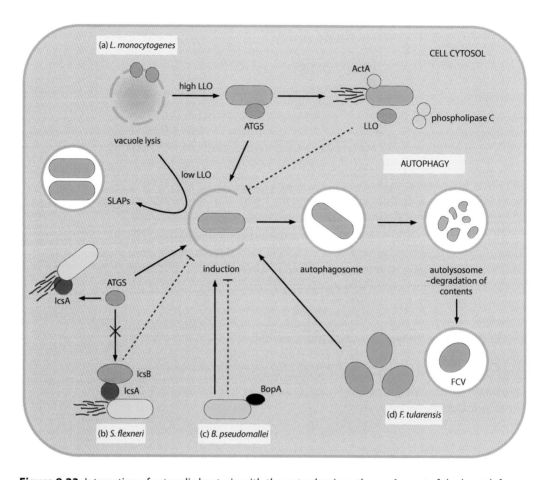

Figure 8.23 Interaction of cytosolic bacteria with the autophagic pathway. As part of the host defence against pathogens, autophagy is used to target intracellular bacteria for degradation. (a) After invasion of cells, *Listeria monocytogenes* escapes the vacuole by secreting a high level of listeriolysin O (LLO) to disrupt the vacuole membrane. The autophagy protein ATG5 is required for recognition of cytosolic *L. monocytogenes* and induction of the autophagic pathway. *L. monocytogenes* is able to evade auto-phagy using multiple mechanisms; notably, LLO, phospholipase and ActA expression are all required for successful evasion. Low-level expression of LLO damages the primary *Listeria*-containing vacuoles, which are recognised by the autophagy system of the cell and induce autophagosome formation around the damaged vacuole. *L. monocytogenes* is then able to block maturation of the phagosome, resulting in the formation of spacious *Listeria*-containing phagosomes (SLAPs), which support slow bacterial growth. (b) *Shigella flexneri* induces autophagy through recognition of the bacterial protein IcsA by ATG5. However, it also blocks the autophagic response by secreting IcsB, which competitively binds IcsA, thereby camouflaging the protein from ATG5 recognition. (c) Intracellular *Burkholderia pseudomallei* induces autophagy, but blocks autophagy induction, using the bacterial protein BopA, which is a homologue of *S. flexneri* IcsB. (d) *Francisella tularensis* induces autophagy following replica-tion of bacteria within the cytosol. It then re-enters the endocytic pathway and subsequently resides in a *Francisella*-containing vacuole (FCV), which displays features of an autolysosome. (From Ray, K. et al., *Nat. Rev. Microbiol.*, 7, 333–340, 2009.)

Utilising the autophagosome as a habitat for survival, replication, or escape from the host cell

It is presumed that some intracellular pathogens employ autophagy for their own benefit for protection, nutrition and replication. For example, poliovirus may use the autophagosome for non-lytic virus release, and this may be the case for HIV as well. Rotavirus may use the autophagosome to form viroplasms. Viroplasms are inclusion bodies in a host cell in which viral replication and assembly occur.

Autophagosomes provide a protected habitat for several intracellular bacterial pathogens, for example, *L. monocytogenes*, *C. burnetii*, and *L. pneumophila*. Thus, in parallel with suppression of autophagosomal fusion with lysosomes and/or acidification of pathogen-containing compartments, the bacterial autophagosomal-like compartments may enable the bacteria to persist (and potentially multiply intracellularly) in a non-acidic compartment. **Tables 8.9** through **8.12** compare and contrast the mechanisms by which various RNA and DNA viruses, fungi and protozoa manipulate autophagy.

The role of autophagy in eukaryotic pathogens

The lower eukaryotes, fungi and parasites, have their own autophagosomal systems that are thought to play a role in surviving nutrient stress and in transitioning to forms that infect humans. Autophagy plays a role in the survival of the fungus *Cryptococcus neoformans* in macrophage phagosomes that are nutrient limited. In addition, autophagy is essential for *Entamoeba* to undergo the developmental transition from trophozoite to cyst stage, for *Leishmania* to differentiate into metacyclic promastigotes and for *Trypanosoma cruzi* to differentiate into infective metacyclic trypomastigotes.

EVADING NATURAL KILLER CELLS

NK cells are innate lymphoid cells that are the counterpart of the adaptive CD8$^+$ cytotoxic T cell. Both cells target host cells that have cytosolic infections or have became malignant. NK cells kill target cells using the same mechanism as CD8$^+$ cytotoxic T cells, that is, the induction of apoptosis. Apoptosis results from the production of perforin, granzymes and granulysin or by the engagement of Fas with FasL. The principle difference between a cytotoxic CD8$^+$ T cell and an NK cell is that the former expresses a highly diverse T-cell receptor (TCR) that recognises MHC class I:non-self peptide, whereas the latter does not express

microbe	adaptation	effects on host-pathogen interactions	microbial virulence factor/mechanism
VIRUSES			
		RNA viruses	
Picornaviridae			
poliovirus	infection induces formation of LC3-positive double membranes	proposed mechanism for nonlytic virus egress	poliovirus 2BC and 3 proteins induce GFP-LC3 colocalization with LAMP1
coxsackievirus B3, B4 (CVB3, CVB4)	CVB3 infection induces early but not late stages of autophagy; CVB4 infection induces autophagy in neurons	may increase viral replication or yields	calpain-dependent (CVB4)
Flaviviridae			
dengue virus	infection induces autophagy	increases viral yields; colocalization of LC3, viral dsRNA, and endosomal marker may indicate association of viral replication complex with amphisomes	unknown
hepatitis C virus	infection induces early but not late stages of autophagy in hepatocyte cell lines	increases HCV replication (without colocalization of viral proteins and autophagosomes)	activation of unfolded protein response
Orthomyxoviridae			
influenza A virus	infection increases autophagy and autophagic flux	may increase viral replication or yields	unknown
Reoviridae			
rotavirus	viral nonstructural protein NSP4 localizes with LC3 but not LAMP1	postulated to play a role in viral morphogenesis, possibly by creating lipid membrane scaffold for formation of viroplasms	unknown
Lentiviridae			
HIV-1	infection inhibits autophagy in primary CD4+ lymphocytes and in macrophage cell lines	proposed mechanism of viral evasion of innate immunity	unknown
	virus may utilize Atg proteins for replication in HeLa cells	proposed mechanism of viral utilization of autophagic machinery for replication	unknown
	infection induces autophagy gene-dependent cell death in bystander cells	proposed mechanism of CD4+ T cell depletion	HIV envelope protein triggers autophagy in bystander lymphocytes by binding to CD4 and CXCR4 through receptor signaling-independent mechanisms thought to involve fusogenic activity of gp41
	infection induces early stages of autophagy and inhibits autophagosomal maturation in macrophages	proposed mechanism for increasing HIV yields	HIV Gag interacts with the LC3 autophagy protein to augment Gag processing; HIV Nef binds to Beclin 1 and inhibits autophagosome maturation

Table 8.9 Adaptations of RNA viruses to evade or use autophagy to promote survival or replication. (Modified from Deretic, V. and Levine, B., *Cell Host Microbe*, 5, 527–549, 2009.)

microbe	adaptation	effects on host-pathogen interactions	microbial virulence factor/mechanism
VIRUSES			
DNA viruses			
Herpesviridae			
herpes simplex virus 1 (HSV-1)	infection inhibits autophagy in neurons	confers neurovirulence	HSV-1 protein ICP34.5 inhibits PKR signaling and binds to Beclin 1 to block autophagy
bovine herpesvirus type 1 (BHV-1)	BHV-1 WT virus induces apoptosis in MDCK cells, whereas BHV-1 bICP0 mutant virus induces nonapoptotic cell death with autophagy	proposed mechanism of regulating of cell death	BHV-1 bICP0 may inhibit autophagy
human cytomegalovirus (HCMV)	infection inhibits autophagy in primary fibroblasts		activates mTOR pathway and rapamycin-insensitive signals
γ-herpesviruses			
KSHV, murine γ-HV68	viral Bcl-2-like proteins inhibit autophagy	unknown	KSHV vBcl-2 and γ-HV68 M11 inhibit autophagy by binding to Beclin 1
Epstein-Barr virus (EBV)	EBV LMP1 protein induces autophagy	proposed mechanism to decrease LMP1 levels and block its cytostatic effects on B cells	unknown
Hepadnaviridae			
hepatitis B virus	HBV X protein transfection enhances autophagy in hepatocytes	unknown	HBV X protein increases Beclin 1 promoter activity

Table 8.10 Adaptations of DNA viruses to evade or use autophagy to promote survival or replication. (Modified from Deretic, V. and Levine, B., *Cell Host Microbe*, 5, 527–549, 2009.)

microbe	adaptation	effects on host-pathogen interactions	microbial virulence factor/mechanism
fungi			
Aspergillus fumigatus (Af)	fungal autophagy protein functions in metal ion homeostasis		*Af* Atg1 required for metal ion homeostasis
Cryptococcus rieoformans (Cn)	fungal autophagy activated during infection in mammalian cells	increases fungal multiplication and lethal infection in mouse model	*Cn* Class III PI3K/Vps34 and Atg8 required for fungal autophagy and virulence

Table 8.11 Adaptations of fungi to evade or use autophagy to promote survival or replication. (Modified from Deretic, V. and Levine, B., *Cell Host Microbe*, 5, 527–549, 2009.)

microbe	adaptation	effects on host-pathogen interactions	microbial virulence factor/mechanism
PROTOZOA			
Entamoeba histolytica (Eh) *Entamoeba invadens* (Ei)	parasite autophagy occurs during proliferation and encystation (Ei)	may facilitate growth of trophozoites and encystation (Ei)	*Eh* and *Ei* possess Atg8 but not Atg5-Atg12 autophagy protein conjugation systems
Leishmania amazonensis (La) *Leishmania major* (Lma) *Leishmania mexicana* (Lme)	parasite may exploit IFN-γ induced host autophagy response (La)	increased intracellular *La* (but not *Lma*) parasite load in macrophages during starvation or IFN-γ-induced autophagy in mouse strain-specific manner	mechanism of increased parasite load with host autophagy induction unknown
	parasite autophagy promotes differentiation and survival during starvation (*Lma*, *Lme*)	parasite autophagy important for transformation to mammalian infective form and parasite virulence	*Lma* Vps and Atg proteins function in endosome sorting (Vps4), autophagy (Vps4, Atg4, Atg8 homologs, Atg12), and differentiation (Vps4, Atg4); *Lme* lysosomal cysteine peptidases (CPA, CPB) required for autophagy and differentiation
Toxoplasma gondii	induces autophagy	autophagy promotes parasite intracellular proliferation in nutrient-limiting conditions *in vitro*	Calcium-, Atg5-, and Beclin 1-dependent but Tor-independent
Trypanosoma cruzi	uses host LC3-positive membranes for cellular entry	host autophagy enhances parasite invasion	mechanism of host autophagic membrane recruitment to parasite unknown
	parasite autophagy promotes differentiation/development and survival during starvation	parasite autophagy important for parasite maintenance and survival	parasite autophagy mediated by conserved Atg proteins (Atg8) but not Atg5-Atg12 protein conjugation system and TOR inhibition

Table 8.12 Adaptations of protozoa to evade or use autophagy to promote survival or replication. (Modified from Deretic, V. and Levine, B., *Cell Host Microbe*, 5, 527–549, 2009.)

a TCR but instead has several activating and inhibitory receptors that are triggered when changes in expression of various surface proteins on the target cell are detected. A potent activator of NK cells is the loss of surface expression of MHC class I, which occurs when certain viruses infect cells. Other activating ligands are those expressed on the target cell membrane as the result of stress resulting from infection or malignancy. Thus, the balance between inhibitory and activating signals determines whether a host cell is killed by an NK cell. Receptors that regulate activity of NK cells are members of the killer cell immunoglobulin-like receptor family (KIRs) and members of the killer cell lectin-like receptor family (KLRs). If the cytoplasmic tail of the receptor contains ITIMs, then it is inhibitory, whereas if the cytoplasmic tail contains ITAMs, it is an activating receptor. ITAMs and ITIMs have been discussed earlier in this chapter. Natural cytotoxicity receptors (NCRs) are another type of activating receptor that recognise ligands such as MHC class I chain-related protein A (MICA) and MHC class I chain-related protein B (MICB) and ULBPs that are a family of human cell-surface molecules distantly related to classic MHC class I molecules. These ligands are expressed as the result of infection, malignancy and physical and chemical damage. Viruses have several ways of avoiding destruction of the cells that they infect by modulating activating receptors.

Evasion of the natural killer group 2D receptor

The natural killer group 2D (NKG2D) receptor is the most important activating receptor for virally infected cells where it binds to MICA, MICB and ULBP1-6. Some viral proteins block the expression of NKG2D ligands on the cell surface by various means. Some DNA viruses employ microRNAs (miRNA) that are small non-coding RNAs that play a role in RNA silencing post-transcriptional regulation of gene expression. Viral miRNAs can down-regulate the expression of NKG2D ligands. HIV can drive the secretion of MICA, MICB and ULB2 that bind NKG2D away from the virus infected target cell surface and block the interaction of the NK cell with the virus-infected cell. Certain viruses, such as HCMV, drive the production of types 1 and 2 interferons and IL-12 by infected mononuclear cells. These cytokines prevent surface expression of NKG2D by NK cells. Prostaglandin E_2 and other small molecules can inhibit NK function. **Table 8.13** summarises viral strategies to evade NKG2D.

virus	viral product	mechanisms
HCMV	pp65	inhibits the dissociation of NKp30 and CD3ζ chain
poxvirus	HA	inhibits NKp30-triggered activation
influenza virus	HA	inhibits NKp46 through lysosomal degradation of CD3ζ chains
	NA	inhibits NKp44 and NKp46 recognition via the removal of sialic acid residues
KSHV	ORF54/dUTPase	inhibits the NKp44 ligand by interfering with intracellular trafficking
HIV	Nef	inhibits the NKp44 ligand through intracellular retention
HCV	?	downregulates NKp30 expression in NK cells

Table 8.13 Viral evasion of NKG2D in NK cell. AICL, activation-induced C-type lectin; BKV, BK virus; EBV, Epstein-Barr virus; ER, endoplasmic reticulum; HBV, hepatitis B virus; HCMV, human cytomegalovirus; HCV, hepatitis C virus; HHV-7, human herpesvirus 7; HIV, human immunodeficiency virus; HSV, herpes simplex virus; IFN, interferon; IL, interleukin; JCV, John Cunningham virus; KSHV, Kaposi's sarcoma-associated herpesvirus; MICA, MHC class I polypeptide-related chain A; MICB, MHC class I polypeptide-related chain B; miRNA, micro RNA; OMPC, orthopoxvirus MHC class I-like protein; TGF-β, transforming growth factor-beta; ULBP, UL16 binding protein; VSV, vesicular stomatitis virus; VZV, varicella-zoster virus. (Modified from Ma, Y. et al., *Viruses*, 8, 95, 2016.)

Evasion of natural cytotoxicity receptors

The NPRs – NKp30 and NKp46 – are expressed by all NK cells. The CD3 zeta chain is important in transduction of these NCRs, and one mechanism of virus evasion is to dissociate CD3 zeta chain from the NCR. NCRs are also disrupted directly by viral ligands such as the influenza virus hemagglutinin **Table 8.14**.

virus	viral product	mechanisms
viral proteins		
HSV	?	decreases MICA, ULBP2, ULBP3 and ULBP1 on the cell surface
VZV	?	reduces ULBP2 and ULBP3 on the cell surface
HCMV	UL16	retains ULBP1, ULBP2, ULBP6 and MICB in the ER/*cis*-Golgi
	UL142	retains ULBP3 and MICA in the *cis*-Golgi apparatus
	US9	induces *MICA*008* proteasomal degradation
	US18, US20	induces MICA lysosomal degradation
HHV-7	U21	redirects ULBP1 to lysosomal degradation
		downregulates expression of MICA and MICB
EBV	LMP2A	reduces the expression of MICA and ULBP4
KSHV	K5	redistributes MICA to an intracellular compartment
		induces AICL endolysosomal degradation
adenovirus	E3/19K	retains MICA and MICB in the ER
HBV	HBsAg	downregulates MICA and MICB by inducing human miRNAs
HIV	Nef	downregulates the cell surface abundance of MICA, ULBP1 and ULBP2
	Vpu, Nef	downregulates the expression of NTB-A and PVR
HCV	NS2, NS5B	downregulates MICA and MICB expression
	?	downregulates NKG2D expression via cell-to-cell interaction
VSV	?	suppresses MICA, MICB and ULBP2 expression
cytokines and secretory molecules		
HCMV	?	inhibits NKG2D/DAP10 expression through type I IFN and IL-12
HCV	NS5A	downregulates NKG2D expression through inducing IL-10-TGF-β
HBV	?	reduces NKG2D/DAP10 and 2B4/SAP expression through TGF-β
KSHV	?	downregulates NKG2D expression through PGE2
viral miRNA		
HCMV	miR-UL112	inhibits MICB mRNA translation
EBV	miR-BART2-5p	inhibits MICB mRNA translation
KSHV	miR-k12-7	inhibits MICB mRNA translation
JCV, BKV	3p* miRNA	inhibits ULBP3 mRNA translation
soluble receptor and ligands		
zoonotic orthopoxviruses	OMCP	secretes soluble NKG2D ligand
HIV	?	releases soluble NKG2D ligands via proteolytic shedding

Table 8.14 Virus evasion of NCRs in NK cells. HSV, herpes simplex virus; MICA, MHC class I polypeptide-related chain A; MICB, MHC class I polypeptide-related chain B; ULBP, UL16 binding protein; VZV, varicella-zoster virus; HCMV, human cytomegalovirus; ER, endoplasmic reticulum; HHV-7, human herpesvirus 7; EBV, Epstein-Barr virus; KSHV, Kaposi's sarcoma-associated herpesvirus; AICL, activation-induced C-type lectin; HBV, hepatitis B virus; miRNA, micro RNA; HIV, human immunodeficiency virus; HCV, hepatitis C virus; VSV, vesicular stomatitis virus; IFN, interferon; IL: interleukin; TGF-β, transforming growth factor beta; JCV, John Cunningham virus; BKV, BK virus; OMPC, orthopoxvirus MHC class I-like protein; ?, viral product unknown. (Modified from Ma, Y. et al., *Viruses*, 8, 95, 2016.)

KEY CONCEPTS

- Pathogens have mechanisms by which to evade almost every component of innate immunity.

- These mechanisms are common to all classes of pathogens.

- Individual pathogens lack a full complement of evasion mechanisms.

- Redundancy in the innate immune system allows it to control pathogens.

BIBLIOGRAPHY

Aktories K. 2011. Bacterial protein toxins that modify host regulatory GTPases. *Nature Reviews Microbiology*, 9: 487–498.

Alcami A. 2003. Viral mimicry of cytokines, chemokines and their receptors. *Nature Reviews Immunology*, 3: 37–50.

Angioi A et al. 2016. Diagnosis of complement alternative pathway disorders. *Kidney International*, 89: 278–288.

Bestebroer J et al. 2010. How microorganisms avoid phagocyte attraction. *FEMS Microbiology Reviews*, 34(3): 395–414.

Bliska JB, Casadevall A. 2009. Intracellular pathogenic bacteria and fungi – A case of convergent evolution? *Nature Reviews Microbiology*, 7: 165–171.

Bowie AG, Unterholzner L. 2008. Viral evasion and subversion of pattern-recognition receptor signalling. *Nature Reviews Immunology*, 8: 911–922.

Brakhage AA et al. 2010. Interaction of phagocytes with filamentous fungi. *Current Opinion in Microbiology*, 13: 409–415.

Bratton DL, Henson PM. 2011. Neutrophil clearance: When the party is over, clean-up begins. *Trends in Immunology*, 32(8): 350–357.

Brinkmann V, Zychlinsky A. 2007. Beneficial suicide: Why neutrophils die to make NETs. *Nature Reviews Microbiology*, 5: 577–582.

Brodsky IE, Medzhitov R. 2009. Targeting of immune signalling networks by bacterial pathogens. *Nature Cell Biology*, 11(5): 521–526.

Brogden KA. 2005. Antimicrobial peptides: Pore formers or metabolic inhibitors in bacteria? *Nature Reviews Microbiology*, 3: 239–250.

Cain RJ, Vázquez-Boland JA. 2015. Chapter 28, Survival strategies of intracellular bacterial pathogens. In: Yi-Wei T. and Andrew S. (Eds.), *Molecular Medical Microbiology*, 2nd ed., Vol. 1, pp. 491–515, London, UK, Academic Press.

Celli J, Finlay BB. 2002. Bacterial avoidance of phagocytosis. *Trends in Microbiology*, 10(5): 232–237.

Celli J, Zahrt C. 2013. *Cold Spring Harbor Perspectives in Medicine*, 3: 010314.

Cole JN, Nizet V. 2016. Bacterial evasion of host antimicrobial peptide defenses. *Microbiology Spectrum*, 4(1). doi:10.1128/microbiolspec.VMBF-0006-2015.

Collette JR, Lorenz MC. 2011. Mechanisms of immune evasion in fungal pathogens. *Current Opinion in Microbiology*, 14: 668–675.

Cooper D, Eleftherianos I. 2016. Parasitic nematode immunomodulatory strategies: Recent advances and perspectives. *Pathogens*, 5: 58. doi:10.3390/pathogens5030058.

Cossart P, Helenius A. 2014. Endocytosis of viruses and bacteria. *Cold Spring Harbor Perspectives in Biology*, 6(8): a016972.

Deretic V, Levine B. 2009. Autophagy, immunity, and microbial adaptations. *Cell Host & Microbe*, 5: 527–549.

Deretic V. 2011. Autophagy in immunity and cell autonomous defense against intracellular microbes. *Immunological Reviews*, 240: 92–104.

Dorn BR et al. 2002. Bacterial interactions with the autophagic pathway. *Cellular Microbiology*, 4(1): 1–10.

Elwell C et al. 2016. *Chlamydia* cell biology and pathogenesis. *Nature Reviews Microbiology*, 14(6): 385–400.

Erwig LP, Gow NAR. 2016. Interactions of fungal pathogens with phagocytes. *Nature Reviews Microbiology*, 14: 163–176.

Fairn GD, Grinstein S. 2012. How nascent phagosomes mature to become phagolysosomes. *Trends in Immunology*, 33(8): 397–405.

Flannagan RS et al. 2009. Antimicrobial mechanisms of phagocytes and bacterial evasion strategies. *Nature Reviews Microbiology*, 7: 355–366.

Freeman SA, Grinstein S. 2014. Phagocytosis: Receptors, signal integration, and the cytoskeleton. *Immunological Reviews*, 262: 193–215.

Gordon S. 2016. Phagocytosis: An immunobiologic process. *Immunity*, 44: 463–475.

Gouin E et al. 2005. Actin-based motility of intracellular pathogens. *Current Opinion in Microbiology*, 8(1): 35–45.

Hajishengallis G, Lambris JD. 2011. Microbial manipulation of receptor crosstalk in innate immunity. *Nature Reviews Immunology*, 11(3): 187–200.

Hallstrom T, Riesbeck K. 2010. *Haemophilus influenzae* and the complement system. *Trends in Microbiology*, 18: 258–265.

Ham H et al. 2011. Manipulation of host membranes by bacterial effectors. *Nature Reviews Microbiology*, 9: 635–646.

Hornef MW et al. 2002. Bacterial strategies for overcoming host innate and adaptive immune responses. *Nature Immunology*, 3(11): 1033–1040.

Huang J, Brumell JH. 2014. Bacteria–autophagy interplay: A battle for survival. *Nature Reviews Microbiology*, 12(2): 101–114.

Hybiske K, Stephens RS. 2008. Exit strategies of intracellular pathogens. *Nature Reviews Microbiology*, 6: 99–110.

Isberg RR et al. 2008. The *Legionella pneumophila* replication vacuole: Making a cosy niche inside host cells. *Nature Reviews Microbiology*, 7: 13–24.

Jockusch BM. 2016. *The Actin Cytoskeleton*, Series of *Handbook of Experimental Pharmacology 235*. Springer, Cham, Switzerland.

Jordan TX, Randell G. 2012. Manipulation or capitulation: Virus interactions with autophagy. *Microbes and Infection*, 14: 126–139.

Justice SS et al. 2008. Morphological plasticity as a bacterial survival strategy. *Nature Reviews Immunology*, 6: 162–168.

Kanayama M, Shinohara ML. 2016. Roles of autophagy and autophagy-related proteins in antifungal immunity. *Frontiers in Immunology*, 7: 47.

Katze MG, Korth MJ, Law GL, Nathanson N. 2016. *Viral Pathogenesis: From Basics to Systems Biology*. 3rd ed. Academic Press, Elsevier, London, UK.

Kaufmann SHE, Dorhoi A. 2016. Molecular determinants in phagocyte-bacteria interactions. *Immunity*, 44: 476–491.

Kirkegaard K et al. 2004. Cellular autophagy: Surrender, avoidance and subversion by microorganisms. *Nature Reviews Microbiology*, 2: 301–314.

Lalani AS et al. 2000. Modulating chemokines: More lessons from viruses. *Trends in Immunology*, 21(2): 100–106.

Lambris JD et al. 2008. Complement evasion by human pathogens. *Nature Reviews Microbiology*, 6: 132–142.

Ling YM et al. 2006. Vacuolar and plasma membrane stripping and autophagic elimination of *Toxoplasma gondii* in primed effector macrophages. *Journal of Experimental Medicine*, 203: 2063–2071.

Liss V, Hensel M. 2015. Take the tube: Remodelling of the endosomal system by intracellular *Salmonella enterica*. *Cellular Microbiology*, 17(5): 639–647.

Loker ES, Hofkin BV. 2015. *Parasitology: A Conceptual Approach*. Garland Science, New York.

Lüder CGK et al. 2009. Intracellular survival of apicomplexan parasites and host cell modification. *International Journal for Parasitology*, 39: 163–173.

Ma Y et al. 2016. Viral evasion of natural killer cell activation. *Viruses*, 8(4): 95.

Medzhitov R. *Innate Immunity and Inflammation*. Cold Spring Harbor Laboratory Press, Cold Spring Harbor, NY.

Mercer J, Greber UF. 2013. Virus interactions with endocytic pathways in macrophages and dendritic cells. *Trends in Microbiology*, 21(8): 380–388.

Mitchell G et al. 2016. Strategies used by bacteria to grow in macrophages. *Microbiology Spectrum*, 4(3).

Monie TP. 2017. *The Innate Immune System: A compositional and Functional Perspective*. Academic Press, Elsevier, London, UK.

Nash AA, Dalziel RG, Fitzgerald JR. 2015. *Mim's Pathogenesis of Infectious Disease*. 6th ed. Elsevier, London, UK.

Nathanson N. 2007. *Viral Pathogenesis and Immunity*. 2nd ed. Academic Press, Elsevier, London, UK.

Nguyen LT et al. 2011. The expanding scope of antimicrobial peptide structures and their modes of action. *Trends in Biotechnology*, 29(9): 464–472.

Nourshargh S et al. 2016. Reverse migration of neutrophils: Where, when, how, and why? *Trends in Immunology*, 37(5): 273–286.

Orvedahl A, Levine B. 2008. Viral evasion of autophagy. *Autophagy*, 4(3): 280–285.

Orvedahl A, Levine B. 2009. Eating the enemy within: Autophagy in infectious diseases. *Cell Death and Differentiation*, 16(1): 57–69.

Ozanic M et al. 2015. The divergent Iintracellular lifestyle of *Francisella tularensis* in evolutionarily distinct host cells. *PLoS Pathogens*, 11(12): e1005208. doi:10.1371/journal.ppat.1005208.

Pauwels A-M et al. 2017. Patterns, receptors, and signals: Regulation of phagosome maturation. *Trends in Immunology*, 38(6): 407–422.

Pollard TD, Cooper JA. 2009. Actin, a central player in cell shape and movement. *Science*, 326(5957): 1208–1212.

Ray K et al. 2009. Life on the inside: The intracellular lifestyle of cytosolic bacteria. *Nature Reviews Microbiology*, 7: 333–340.

Rivas-Santiago B et al. 2009. Susceptibility to infectious diseases based on antimicrobial peptide production. *Infection and Immunity*, 77(11): 4690–4695.

Sarantis H, Grinstein S. 2012. Subversion of phagocytosis for pathogen survival. *Cell Host & Microbe*, 12(October 18): 419–431.

Serruto D et al. 2010. Molecular mechanisms of complement evasion: Learning from staphylococci and meningococci. *Nature Reviews Microbiology*, 8: 393–399.

Sibley LD. 2011. Invasion and intracellular survival by protozoan parasites. *Immunology Reviews*, 240(1): 72–91.

Singh B et al. 2010. Vitronectin in bacterial pathogenesis: A host protein used in complement escape and cellular invasion. *Molecular Microbiology*, 78(3): 545–560.

Sperandioa B et al. 2015. Mucosal physical and chemical innate barriers: Lessons from microbial evasion strategies. *Seminars in Immunology*, 27: 111–118.

Swidergall M, Ernst JF. 2014. Interplay between *Candida albicans* and the antimicrobial peptide armory. *Eukaryotic Cell*, 13(8): 950–957.

Takeuchi O et al. 2010. Pattern recognition receptors and inflammation. *Cell*, 140(6): 805–820.

Tang D et al. 2012. PAMPs and DAMPs: Signal 0s that spur autophagy and immunity. *Immunology Reviews*, 249(1): 158–175.

Van Avondt K et al. 2015. Bacterial immune evasion through manipulation of host inhibitory immune signaling. *PLoS Pathogens*, 11(3): e1004644. doi:10.1371/journal.ppat.1004644.

White MW et al. 2015. Chapter 11, New research on the importance of cystic fibrosis transmembrane conductance regulator function for optimal neutrophil activity. In: Wat D (Ed.), *Cystic Fibrosis in the Light of New Research*. London, UK, Intech Open.

Wileman T. 2013. Autophagy as a defence against intracellular pathogens. *Essays in Biochemistry*, 55: 153–163.

Wilson BA, Salyers AA, Whitt DD, Winkler ME. 2011. *Bacterial Pathogenesis: A Molecular Approach*. 3rd ed. ASM Press, Washington, DC.

Wilson M, McNab R, Henderson B. 2002. *Bacterial Disease Mechanisms: An Introduction to Cellular Microbiology*. Cambridge University Press, Cambridge, UK.

Yap GS et al. 2007. Autophagic elimination of intracellular parasites convergent induction by IFN-γ and CD40 ligation? *Autophagy*, 3(2): 163–165.

Chapter 9: Evasion of the Human Adaptive Immune System

INTRODUCTION

In order to initiate an adaptive immune response a naïve thymus-derived lymphocyte (T lymphocyte or T cell) must enter into a receptor–ligand interaction, which is termed a cognate interaction, with a licenced dendritic cell (DC) in the T-cell area of secondary lymphoid tissue. This interaction is required to elevate a naïve T cell to effector/memory status. Before exploring how pathogens evade adaptive immunity, we will begin with a brief review of antigen presentation that leads to the production of various types of effector T cells, one of which, the follicular helper T (T_{FH}) cell, helps B cells to become antibody-secreting plasma cells. More detailed descriptions of antigen presentation can be found in standard immunology textbooks.

ANTIGEN PRESENTATION

There are three types of professional antigen-presenting cells (APCs): the DC, the bone-marrow-derived lymphocyte (B lymphocyte [or B cell]) and the macrophage (MO). The role of APCs is to present/display on their surface peptides derived from pathogen proteins. DCs presents antigenic peptides to naïve T cells to activate them to become effector/memory T cells, whereas B cells and MOs present antigenic peptides to effector, helper T cells, so that these T cells can help B cells make high-affinity IgG, IgA, or IgE antibodies and help MOs increase their microbicidal ability. APCs take up extracellular pathogens or secreted pathogen molecules using pinocytosis and receptor-mediated endocytosis, which were described in Chapter 5, *Facilitated Cell Entry*. In addition, DCs and MOs, take up extracellular pathogen antigens by phagocytosis. The peptides generated by the degradation of pathogens by endocytosis or phagocytosis are loaded into and displayed in the groove of major histocompatibility complex (MHC) class II molecules. While only APCs express MHC class II molecules, all nucleated cells express a different type of MHC molecule called MHC class I. MHC class I molecules are loaded with peptides derived from pathogens such as viruses and other classes of intracellular pathogens that take up residence in the cytosol (see Chapter 8, *Evasion of the Human Innate Immune System*). The mechanism by which pathogens are degraded in the cytosol remains unclear, but it is possible they are coated with polyubiquitin chains that recruit proteasomes. Proteasomes are enzyme factories that degrade ubiquitin-labelled proteins into short peptides. Under the influence of

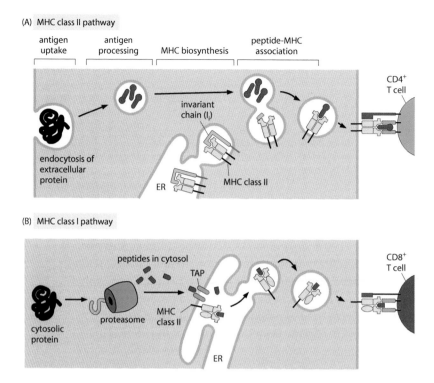

Figure 9.1 The MHC class I (cytosolic) and the MHC class II (endocytic) antigen-processing pathways. (http://slideplayer.com/slide/6086039/.) The top panel (A) shows the MHC class II pathway also known as the endocytic pathway. Endocytosed/phagocytosed antigen is broken down into peptides in an endolysosome or phagolysosome. MHC class II molecules are assembled in the endoplasmic reticulum (ER). Note that the peptide binding groove of MHC class II molecules is blocked by the protein invariant chain (I_i) so that they cannot be loaded with peptide in the ER. Vesicles containing MHC class II:I_i traffic to and fuse with endolysosomes or phagolysosomes, where I_i is removed and the MHC class II binding groove is loaded with pathogen peptides. MHC class II:peptide complexes then traffic to the cell membrane. The bottom panel (B) shows the MHC class I, or cytosolic pathway. Pathogen proteins are broken down into peptides by the proteasome, and these peptides enter the ER *via* a pore called the transporter associated with antigen processing (TAP) and where they are inserted into the peptide-binding groove of nascent MHC class I molecules. The MHC class I:peptide then traffics to the cell membrane in a vesicle.

interferons, proteasomes are modified to become immunoproteasomes designed to produce peptides suitable for binding in the groove of MHC class I molecules. A schematic of the MHC class I and class II pathways of antigen presentation are shown in **Figure 9.1**.

Linking sensing of MAMPs by pattern-recognition receptors with antigen processing

In Chapter 8, *Evasion of the Human Innate Immune System*, we discussed the essential role of sentinel receptors in sensing conserved motifs called microbe-associated molecular patterns (MAMPs). Ligation of MAMPs by pattern-recognition receptors (PRRs) activates several signal transduction pathways that result in the release of cytokines, chemokines and the up-regulation of cell membrane receptors. In DCs, this process is referred to as licencing, and it is required before DCs are able to activate naïve T cells in the secondary lymphoid tissues. Secondary lymphoid tissues are lymph nodes the spleen, tonsils, Peyer's patches, the appendix and other mucosa-associated lymphoid tissues.

The DC licencing process, that is, the ligation of MAMPs by PRRs, results in increased expression of MHC class I and class II molecules, and B7.1 and B7.2 which are co-stimulatory molecules that are involved in the activation of naïve T-cells, cell adhesion molecules, and cytokines. In addition, the chemokine receptor CCR7 is expressed on the DC cell membrane, allowing it to respond to the chemokine CCL21 that is produced constitutively by secondary lymphoid tissues. Simply put, secondary lymphoid tissues consist of an area dominated by naïve T cells in which sit islands of naïve B cells and antigen experienced B cells. These islands are termed B-cell follicles. Licensed DCs are attracted to the T-cell area of the secondary lymphoid tissues by chemokines. Here licenced DCs interact with naïve T cells that have arrived from the blood, attracted by chemokines (**Figure 9.2**).

Activating naïve T cells by licenced dendritic cells

In order for a naïve T cell to be activated to effector/memory status, it must receive three signals from a licenced DC. Signal 1 is delivered by the ligation of MHC class I:peptide or MHC class II:peptide displayed on the cell membrane of the DC by the T-cell antigen receptor (TCR), which is a complex consisting of the TCR itself, CD3, zeta chain and the T-cell co-receptor (CD8 in the case of MHC class I or CD4 in the case of MHC class II). Signal 2 is delivered by the ligation of co-stimulatory molecules such as B7.1/B7.2 on the DC by CD28 on the T-cell plasma membrane, and signal 3 results from the secretion of various cytokines by the DC. The type of cytokines secreted by DCs reflect the type of MAMPs ligated by DC PRRs. In this way, signal 3 dictates the type of effector/memory T cell that the naïve T cell becomes. **Figure 9.3** shows the three signals provided by a licenced DC to a $CD4^+$ naïve T cell. **Figure 9.4** shows subsets of effector $CD4^+$ T cells that develop from naïve T cells under the influence of various cytokines provided by licenced DCs.

If a licenced DC interacts with a naïve $CD8^+$ T cell in the T-cell area of secondary lymphoid tissue, then the naïve T cell becomes an effector or a memory cytotoxic T cell. The role of cytotoxic $CD8^+$ T cells is to kill virus-infected or malignant cells.

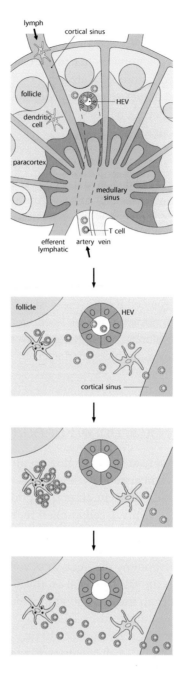

Figure 9.2 Naive T cells, attracted by chemokines (not shown), enter a lymph node from the blood and take up residence in the paracortical areas where the T cells encounter mature dendritic cells (DCs) (first panel). T cells shown in blue encounter their specific antigen on the surface of licenced DCs; they lose their ability to exit from the node and become activated to proliferate and to differentiate into effector T cells. T cells shown in green do not encounter their specific antigen and leave the lymph node *via* the lymphatics to return to the circulation (second and third panels). After several days, some antigen-specific effector T cells exit the node *via* the efferent lymphatics and enter the circulation in greatly increased numbers (fourth panel). Other effector T cells remain in the lymph node and become follicular helper T cells (T_{FH}) that help B cells make antibodies. (From Murphy, K. and Weaver, C., *Janeway's Immunobiology*, 9th ed., Garland Science, New York, 2016.)

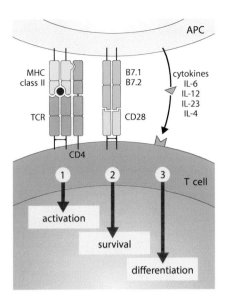

Figure 9.3 Three signals are required from a licenced dendritic cell (DC) to activate a naïve CD4+ T cell. (From Murphy, K. and Weaver, C., *Janeway's Immunobiology*, 9th ed., Garland Science, New York, 2016.)

Follicular helper CD4+ T cells help B cells make high-affinity, class-switched antibodies

After cognate interaction with licenced DCs in the T-cell zone of secondary lymphoid tissues, helper T cells destined to help B cells become antibody-secreting plasma/memory B cells move towards the junction between B-cell follicles and the T-cell zone. These helper T cells are called follicular helper T cells (T_{FH}) and are defined by expression of the transcription factor Bcl6, the chemokine receptor CXCR5, the immune regulatory molecule PD1 and the co-stimulatory molecule ICOS. Concurrently, using their highly specific B-cell receptors (BCRs), which are monomeric immunoglobulin (Ig) M and IgD, naïve B cells capture intact pathogens or pathogen proteins brought to the B-cell follicles *via* the lymph. The pathogen or its proteins captured by the BCRs of naïve B cells are taken up by endocytosis, degraded to peptides in endolysosomes and loaded into MHC class II molecules. The MHC class II:peptide complexes then traffic to the B-cell plasma membrane. The uptake of antigen by naïve B cells activates them, and they move towards the junction with the T-cell zone. Here they present pathogen-derived peptides in MHC class II molecules to T_{FH} cells whose TCR for antigen is specific for the MHC class II:peptide complex displayed by the B cell. Following interaction with T_{FH} cells, a subset of the antigen-selected B cells moves to the medullary cords of the lymph nodes or to the border between the T-cell zone and red pulp in the spleen, where they differentiate into short-lived plasmablasts that secrete IgM antibodies that can immediately engage the invading pathogen. Only those B cells with the highest affinity for pathogen antigens re-enter the follicle with their cognate

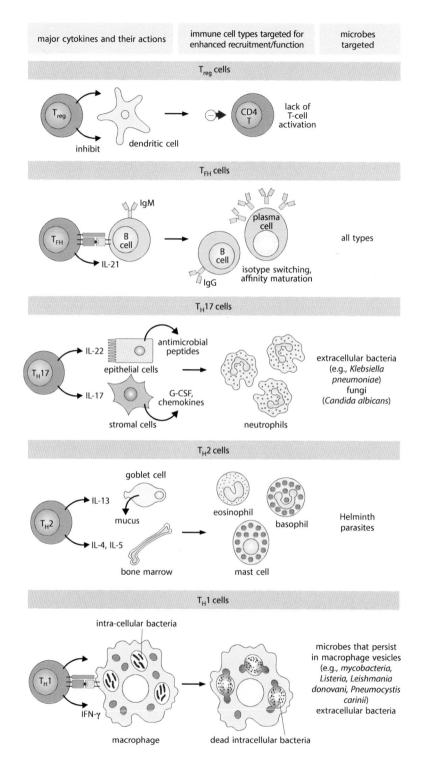

Figure 9.4   Subsets of CD4⁺ T cells that result from different types of cytokines (signal 3) provided by licenced dendritic cells (DCs). The types of cytokines released by the DC depends on which pattern-recognition receptors (PRRs) were ligated when the DC initially encountered the pathogen. (From Murphy, K. and Weaver, C., *Janeway's Immunobiology*, 9th ed., Garland Science, New York, 2016.)

partner T_{FH} cell and initiate formation of a germinal centre. Two important events occur in germinal centres: somatic hypermutation (which increases the affinity of an antibody for its antigen) and class switching (which gives rise to antibodies of different isotypes that have different effector functions). These germinal centre B cells become long-lived plasma cells or memory B cells (**Figure 9.5**).

The expansion of T and B lymphocytes is ultimately controlled by the presence and concentration of antigen. When antigen is removed by the immune system, DCs can no longer be licenced because there are no MAMPs to ligate PRRs and no antigen to cross-link BCRs. However, both T and B lymphocytes have intrinsic regulatory systems in the form of inhibitory receptors such as CTLA4, and programmed death-1 (PD-1) on T cells and FcγRIIB1 (CD32) on B cells. In addition, both the T- and B-cell

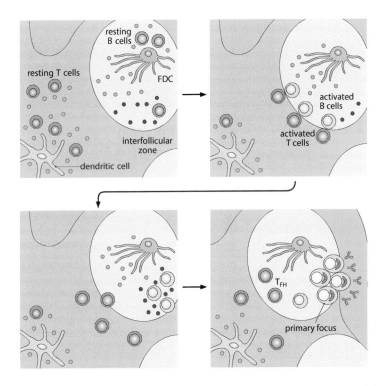

Figure 9.5 Naïve T cells in the T-cell zone are elevated to effector status after interaction with licenced dendritic cells (DCs). Naïve B cells in the B-cell follicles acquire antigen directly or from a follicular dendritic cell (FDC) or a macrophage (first panel). Activated T cells and B cells that have captured antigen migrate to the junction between the T-cell zone and the B-cell follicle (second panel). After 2–3 days, some B cells migrate to the outer follicle and interfollicular regions (spleen) or to the medulla (lymph node) (third panel). After another day or so, some B cells cluster in the interfollicular regions near the red pulp (spleen) or the medulla (lymph node), proliferate, and differentiate into plasmablasts, forming a primary focus with terminal differentiation into antibody-secreting plasma cells. Other T cells enter the follicle and induce Bcl-6 expression to become follicular helper T cells (T_{FH}) that participate with B cells there to form a germinal centre reaction (fourth panel). (From Murphy, K. and Weaver, C., *Janeway's Immunobiology*, 9th ed., Garland Science, New York, 2016.)

compartments contain regulatory cells (T_{REGS} and B_{REGS}, respectively) that secrete interleukin-10 (IL-10), transforming growth factor-beta (TGF-β) and IL-35, which are immunoregulatory cytokines.

INHIBITION OF ANTIGEN PRESENTATION BY MHC CLASS I AND CLASS II PATHWAYS

As we have seen, the presentation of pathogen peptides in the groove of MHC class I molecules by nucleated cells is a prerequisite to generate pathogen-specific cytotoxic $CD8^+$ T cells that can eliminate cells containing cytosolic pathogens. Similarly, the presentation of MHC class II:pathogen peptide complexes by DCs is necessary to generate various types of $CD4^+$ helper T cells that help B cells make IgG, IgA and IgE antibodies; increase the antimicrobial activity of MOs; repair epithelium; and increase synthesis of antimicrobial peptides. Thus, disruption of either of these antigen-processing pathways can impair both cellular and humoral immunity.

Viral subversion of the MHC class I antigen-processing pathway

Viruses are obligate intracellular pathogens, and herpesviruses, in particular, are adept at subverting every step in the cytosolic antigen-processing pathway. Some of these strategies are shown in **Figure 9.6**. These strategies include blocking entry of the peptide into the endoplasmic reticulum (ER), blocking peptide loading into nascent MHC class 1 molecules, blocking release of MHC class I from the ER, targeting MHC class I molecules to the proteasome by polyubiquitinisation, interfering with the trafficking of MHC class I to the cell membrane and increasing degradation of cell surface MHC class I:peptide complexes.

In addition to DNA viruses, RNA viruses such as human immunodeficiency virus-1 (HIV-1) subvert the MHC class I pathway. The HIV Nef protein targets the endocytic sorting machinery to misdirect MHC class I molecules away from the cell surface.

Bacterial subversion of the MHC class I antigen-processing pathway

Chlamydia depresses the expression of cell surface MHC class I molecules by attacking the host transcription factors RFX5 and UAF-1. Invasive *Salmonella* and *Yersinia* are able to force alternative splicing of the pre-mRNA of MHC class I human lymphocyte antigen (HLA)-B27 molecules,

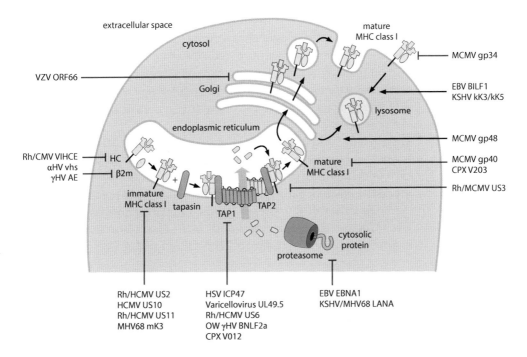

Figure 9.6 Herpesvirus interference with MHC class I antigen presentation. Herpesviruses encode a large variety of proteins that interfere with MHC class I antigen presentation to prevent elimination of the infected cell by CTLs. EBV EBNA1 and KSHV/MHV68 LANA avoid their own proteasomal degradation, thus limiting the generation of antigenic peptides. The VIHCE protein of RhCMV inhibits the translation of MHC class I HC in a signal-peptide dependent manner. Viral host shut-off proteins, such as the UL41 homologues of alpha herpesviruses a(HV) like HSV, BHV-1 and PRV, and the alkaline exonucleases of the gamma herpesviruses g(HV) EBV (BGLF5), KSHV (SOX) and MHV68 (ORF37), and degrade host mRNAs, thereby reducing protein levels of HC and b2m. HCMV US10, RhCMV/HCMV US2 and US11 and MHV68 mK3 direct immature MHC class I molecules to the cytosol where they are degraded. Transport of peptides through the transporter associated with antigen processing (TAP) is inhibited by HSV ICP47, varicellovirus UL49.5 homologues, RhCMV/HCMV US6, Old World primate gamma herpesvirus (OW gHV) BNLF2a homologues and CPX V012. RhCMV/HCMV US3 interferes with the function of tapasin, thereby preventing proper peptide loading of MHC class I molecules. MCMV gp40 and CPX V203 inhibit transport of mature MHC class I molecules out of the ER. VZV ORF66 retains mature MHC class I complexes in the *cis*/medial Golgi. MCMV gp48 interferes with MHC class I molecule migration by directing these molecules from the Golgi to lysosomes. MCMV gp34 prevents cytotoxic T cell–induced lysis by associating with MHC class I molecules at the cell surface. Finally, EBV BILF1 and KSHV kK3 and kK5 increase the endocytosis and lysosomal degradation of cell surface MHC class I molecules. (From Horst, D. et al., *Curr. Opin. Immunol.*, 23, 96–103, 2011.)

which leads to misfolding and degradation of the molecules. These bacteria may also be able to retain MHC class I molecules in the host cytoplasm and alter the repertoire of peptides presented by HLA-B27.

Viral subversion of the MHC class II pathway antigen-presenting pathway

Viruses use two major strategies to inhibit the expression of MHC class II molecules by APCs. The first involves blocking signal transduction pathways that are initiated when interferon-gamma (IFN-γ) binds its receptor on the APC membrane. IFN-γ is essential in up-regulating MHC class II expression *via* activation of the MHC class II transactivator (CIITA). Because IFN-γ is centrally involved in the antiviral activity of cells, it is likely that the IFN-γ signal transduction pathway is inhibited by almost all viruses. Therefore, inhibition of MHC class II expression may simply be one consequence of viral suppression of IFN-γ signalling. **Table 9.1** lists viruses that inhibit MHC class II expression.

The second strategy used by viruses to inhibit the MHC class II pathway is to target the stability or sorting of MHC class II molecules. In the case of cytolytic viruses, it is difficult to determine whether effects on the MHC class II pathway are specific or simply part of the general cytotoxic effect of the viruses. Human cytomegalovirus (HCMV) US2 glycoprotein reduces expression of MHC class II proteins by causing degradation of the α-chains of HLA-DR and HLA-DM, both of which

virus	protein	mechanism
poxviruses		
several species	e.g. T7, B8-R	soluble homologs of IFN-γR that sequester immune IFN-γ
vaccinia	VH1	dephosphorylation of activated STAT1
herpesviruses		
herpes simplex virus 1	unknown	phosphorylation of Jaks and STAT1 affected
human cytomegalovirus	unknown	loss of Jak1; inhibition subsequent to nuclear translocation of STAT1 dimer
murine cytomegalovirus	unknown	inhibition subsequent to nuclear translocation of STAT1 dimer
Epstein–Barr virus	BZLF-1	reduction in transcription of IFN-γR1
varicella zoster virus	unknown	reduction in Jak2 and STAT1 levels
paramyxoviruses		
simian virus 5	V	proteasome-mediated degradation of STAT1
mumps virus	V?	destabilization of STAT1
sendai virus	C	reduced synthesis and phosphorylation of STAT1; destabilization of STAT1
human parainfluenzavirus 3	unknown	STAT1 phosphorylation affected?
nipah virus	V	complex formation with STAT1 and sequestration in the cytoplasm
adenoviruses	E1A or E1A-dependent events	reduction in IFN-γR2 levels; inhibition of function of STAT1 when bound by E1A; decrease in STAT1 levels; inhibition of general transcription
other viruses		
HIV	unknown	reduced transcription of NF-YA
HIV	Tat	binds cyclin T1 to inhibit CIITA-pTEFb interactions
hepatitis C virus	core	decrease in STAT1 expression
murine polyoma virus	large T	binds to Jak1
ebola virus	unknown	inhibition before STAT1 dimer formation
hepatitis B virus	polymerase?	effects on transactivation function of STAT1?

Table 9.1 Viral exploitation of the MHC class II antigen-processing pathway. Viral inhibition of IFNγ–Jak-STAT signal transduction, expression of CIITA, or induction of MHC class II gene expression. Abbreviations: CIITA, class II transactivator; IFN-γR, interferon-γ receptor; Jak, Janus kinase; NF-YA, nuclear factor-Y subunit A; pTEFb, positive transcription elongation factor-b; STAT1, signal transducer and activator of transcription 1. (From Hegde, N.R. et al., *Trends Immunol.*, 24, 278–285, 2003.)

are involved in peptide loading in the MHC class II pathway. Another HCMV glycoprotein, US3, binds nascent MHC class II α:β heterodimers in the ER and prevents the invariant chain (I$_i$) from occupying the peptide-binding groove, impairing traffic of MHC class II molecules to the peptide-loading compartment. The HIV Nef protein reduces MHC class II trafficking to the cell membrane by affecting its intracellular sorting. The envelope protein of HIV-1 may redirect MHC class II away from endosomes and phagolysosomes. The benefit to viruses of inhibiting the MHC class II pathway when, as cytosolic pathogens, their peptides are processed through the MHC class I pathway, may lie with the process of cross-presentation. Cross-presentation is a process by which DCs can switch pathogen antigens from the MHC class II pathway to the class I pathway and from the class I pathway to the class II pathway.

Bacterial subversion of the MHC class II antigen-processing pathway

Several intracellular bacteria such as *Mycobacterium tuberculosis*, *M. avium*, *M. leprae*, *Salmonella typhi*, and *Helicobacter pylori* that infect MOs and DCs subvert the MHC class II pathway. In the case of intravacuolar organisms, interference with the MHC class II pathway is largely a consequence of evading destruction in phagolysosomes of APCs because MHC class II molecules are loaded with peptides in this compartment. We examined microbial evasion of phagocytosis in Chapter 8, *Evasion of the Human Innate Immune System*. The MHC class II pathway is subverted at several points along its pathway from endocytosis/phagocytosis to expression of MHC class II:peptide complexes expressed on the APC cell membrane. These include arresting phagosome–lysosome fusion, assembly of MHC class II, trafficking of MHC class II and cell surface display. *Mycobacterium*, *Chlamydia*, *Salmonella* and *Legionella* actively inhibit phagosome–lysosome fusion. *Escherichia coli*, *Chlamydia* and *M. tuberculosis* may inhibit antigen processing and presentation by targeting CIITA.

Parasite evasion of the MHC class II antigen-processing pathway

The protozoan parasites *Leishmania* species and *Toxoplasma gondii* subvert the MHC class II pathway. They accomplish this by antigen sequestration or by interference with the loading of antigens onto MHC class II molecules. *Leishmania donovani* increases the fluidity of lipid rafts in MOs, which results in defective antigen presentation. Megasomes are large lysosomes found in the amastigote stage of *Leishmania* species. MHC class II molecules within megasomes are endocytosed and degraded by cysteine proteases. *Toxoplasma gondii* also down-regulates expression of MHC class II. However, how this is accomplished is unclear. **Table 9.2** lists human pathogens that subvert MHC class I and class II.

pathogen	gene product	MHC class affected	function/phenotype
HIV[a]	Nef	class I	downregulation of class I
		class II	downregulation of CD4
			proton pump binding
	Vpu	class I	degradation
		class II	blocks CD4 expression
Adenovirus[a]	E3/19K	class I	ER retention
			TAP binding/tapasin inhibitor
HSV	ICP47	class I	TAP binding
			blocks TAP peptide binding
	unknown	class II	blocks neuronal expression
CMV[a]	pp65	class I	antigen phosphorylation
			inhibition of proteolysis
	US2	class I	reverse translocation
			degradation
	US3	class I	ER retention
	US6	class I	TAP binding
			blocks peptide import
	US11	class I	reverse translocation
			degradation
	US18	class I	NK-cell decoy
	ND	class II	blocks γ-interferon signaling
EBV	EBNA1	class I	Gly-Ala repeat blocks its own
			proteasome processing
VZV[a]	ND	class I	unknown
HPV	ND	class I	undetectable viral protein level
	E5	class II	proton pump binding[c]
	E6	class II	AP1 binding[c]
measles virus	ND	class II	mixed effects
C. trachomatis	ND	class II	downregulation
M. tuberculosis[a]	ND	class II	mixed effects
			blocks endosome acidification
			blocks lysosome fusion
	ND	CD1b[b]	downregulation
H. pylori	Vac A	class II	disrupts late endosomes
C. burnetii	ND	class II	distorts loading compartment
S. typhimurium	SPI-II	class II	phagosome alteration
Yersinia	Yops	class II	phagocytosis inhibition
E. coli	LT	class II	inhibits phagocytic processing
V. cholera	CT	class II	inhibits phagocytic processing
L. amazonensis	unknown	class II	degradation

Table 9.2 Human pathogens that interfere with antigen presentation. Abbreviations: HIV, human immunodeficiency virus; CMV, cytomegalovirus; HSV, herpes simplex virus; EBV, Epstein–Barr virus; HPV, human papilloma virus; VZV, varicella-zoster virus; ND, no data; LT, heat labile enterotoxin; CT, cholera toxin. [a]These pathogens have mechanisms for interfering with immune cell interactions, in addition to their effects on antigen presentation. For example, adenovirus inhibits apoptosis, CMV synthesizes chemokine homologs, HIV causes helper T-cell destruction, VZV infects T cells and M. tuberculosis inhibits macrophage activation by γ-interferon. [b]This molecule is a non-classical class I molecule, encoded outside the MHC, that presents glycolipid mycobacterial antigen to T cells following transport through the endocytic pathway, along the route trafficked by classical class II MHC molecules. (From Brodsky, F.M. et al., *Immunol. Rev.*, 168, 199–215, 1999.)

Manipulation of co-stimulatory molecules

Co-stimulatory molecules (signal 2) and their ligands play an essential role in the survival, proliferation and regulation of lymphocytes and activation of APCs. Accordingly, they are targets for intracellular pathogens of all classes that aim to subvert the host adaptive immune response. *M. tuberculosis, Yersinia pseudotuberculosis, Helicobacter pylori* and *Bordetella pertussis* inhibit the expression of CD80 (B7.1), CD86 (B7.2) and CD40 on APCs and T cells, as do HIV, measles, herpesviruses, and hepatitis C virus and the parasites *Leishmania donovoni, L. chagasi* (*L. infantum*), *Toxoplasma gondii, T. cruzi,* and *Plasmodium falciparum.* **Table 9.3** lists human pathogens that down-regulate co-stimulatory molecules.

intracellular pathogens	costimulatory molecules	loss of function
bacteria		
M. tuberculosis	CD80 ↓, CD86 ↓, CD40 ↓, PDL-1/PDL-2 ↑, PD-1 ↑	hampers effective T-cell activation. Induces anergy or apoptosis in T cells, paralyzes IL-2 and chemokines secretion, and inhibits NK cell function.
M. leprae	CD80 ↓, CD28 ↓	blockade of IL-12 secretion, defective T-cell response
S. typhimurium	ICAM-1 ↓	impedes antigen uptake ability of APCs
B. anthracis	CD40 ↓, CD80 ↓, CD86 ↓	impairment of antigen specific B-cell and T-cell immunity, suppresses the function of DCs
H. pylori	PDL-1 ↑, CTLA-4 ↑	exhaustion of DCs, obstructs cytokines secretion, induces anergy in T cells
B. pertussis	CD40 ↓, ICAM ↓	promotes differentiation of Tregs
viruses		
HIV	PD-1 ↑, PDL-1 ↑, CTLA-4 ↑, CD80 ↓, CD86 ↓, CD33 ↓, CD40L ↓, 4-1BB ↓, OX40 ↓	blocks IL-2 but augments IL-10 secretion, induces exhaustion of T cells, defective CTLs response, and hampers antigen uptake ability of APCs
HBV	PDL-1 ↑, PD-1 ↑	induces IL-10 secretion, enhances apoptosis and anergy in T cells
HCV	CD83 ↓, CD86 ↓	reduces stimulatory capacity of DCs
measles	CD40 ↓, CD80 ↓, CD86 ↓, CD25 ↓, CD83 ↓, CD69 ↓	abnormal DCs differentiation, improper CD8 T-cell proliferation
herpes simplex virus	ICAM-1 ↓	blocks APCs T-cell communication
protozoans		
L. donaovani	CD80 ↓	inefficient T-cell response
T. gonodii	CD80 ↓	inhibits T-cell stimulatory activity

Table 9.3 Exploitation of costimulatory molecules by intracellular pathogens. ↑, upregulation; ↓, downregulation. (From Khan, N. et al., *PLoS Pathog.*, 8, e1002676, 2012.)

Manipulation of regulatory receptors and ligands

In addition to stimulatory molecules, lymphocytes express inhibitory receptors that are important in controlling the magnitude of the immune response. CTLA-4 (CD512) and PD-1 are such receptors. The ligand for CTLA-4 is CD28 and the ligands for PD-1 are programmed death ligand-1 (PD-L1) and PD-L2. Many cell types constitutively express PD-L1, but PD-L2 is only expressed by APCs during inflammation. By up-regulating these regulatory molecules, pathogens can suppress immunity. Table 9.3 lists pathogens that up-regulate regulatory molecules.

Up-regulation of IL-10

IL-10 is a regulatory, anti-inflammatory cytokine produced by DCs, MOs, T cells and B cells. IL-10 inhibits the activity of CD4$^+$ T cells, CD8$^+$ T cells, natural killer (NK) cells, and MOs. IL-10 inhibits MHC class II and co-stimulatory molecule expression by monocytes and MOs and limits the production of pro-inflammatory cytokines and several chemokines. The regulation of T cells and NK cells by IL-10 is mediated largely by the effect of the cytokine on monocyte/MOs. IL-10 produced by DCs can act back on them in an autocrine fashion to inhibit chemokine production and trafficking to lymph nodes. Failure of DCs to travel to T-cell areas of secondary lymphoid tissues means they are unable to drive naïve CD4$^+$ T cells to become effector T$_H$1 cells. However, IL-10 can also act directly on T$_H$1, T$_H$2 and T$_H$17 CD4$^+$ T cells to limit their activation and differentiation in secondary lymphoid tissues. In addition, IL-10 can suppress tissue inflammation. Thus, up-regulation of IL-10 can impair immune control of pathogens, while at the same time controlling immune-mediated tissue injury and *vice versa*. Members of all classes of pathogen up-regulate IL-10 to damp down the immune response, and examples are listed in **Table 9.4**. The mechanisms employed by pathogens to up-regulate IL-10 involves receptor crosstalk, which was described in the section *Manipulation of pattern-recognition receptors* in Chapter 8, *Evasion of the Human Innate Immune System*. Crosstalk occurs between a C-type lectin receptor, usually DC-SIGN, specific for mannose and fucose moieties, and TLRs on DCs. The bacteria *Mycobacterium tuberculosis*, *M. leprae* and *Borrelia burgdorferi*; the fungus *Candida albicans*; and the measles and HIV-1 viruses bridge these receptors on DCs. In the case of *B. burgdorferi*, DC-SIGN and TLR2 are bridged by *B. burgdorferi* outer membrane lipoproteins and a tick salivary protein, Salp15, which is captured by outer surface protein C (OspC). Receptor bridging used by *Helicobacter pylori* involves DC-SIGN and fucose moieties on the Lewis X blood group antigen motif on the *O*-antigen of *H. pylori* lipopolysaccharide (LPS).

> **Inhibition of IL-12:** IL-12 is essential to the differentiation and activation of T$_H$1 CD4$^+$ T lymphocytes. Crosstalk between complement receptors such as C5aR, gC1qR, CR3 and CD46 and TLR2 and TLR4 can suppress IL-12 production by human monocyte/MOs. The protozoan parasite

HIV-1 measles virus mycobacteria C. albicans	mannose-containing ligands (e.g., mycobacterial ManLAM, fungal mannan, and gp120) bind DC-SIGN and activate RAF1-dependent signalling	DC-SIGN	TLR3 TLR4 TLR5	DC	increased IL-10, IL-12, and IL-6; unbiased T_H1 cell differentiation. Impaired or intermediate-stage DC maturation also reported
H. pylori	fucose-containing LPS Le antigens bind DC-SIGN and activate RAF1-independent signalling	DC-SIGN	TLR2 TLR4	DC	upregulation of IL-10, downregulation of IL-6 and IL-12; inhibition of T_H1 cell polarization
B. burgdorferi	Salp15 in tick saliva binds DC-SIGN and activates Raf-1 and MEK signalling. Salp15 is bound by B. burgdorferi OspC	DC-SIGN	TLR2	DC	promotion of Il6 and Tnf mRNA decay, impaired nucleosome remodeling at the Il12a promoter, enhanced IL-10 production, inhibition of TLR-dependent DC maturation and function
mycobacteria	ligand unknown	C-type lectin receptor (Clec5A)	TLR2 (MyD88 pathway)	neutrophils	synergistic induction of IL-10, reduction of lung inflammation but persistence of high mycobacterial burden
P. gingivalis	HRgpA & RgpB convert C5 to C5a to activate C5aR, which co-associates with TLR2	C5aR	TLR2	monocytes macrophages	synergistic elevation of cAMP and inhibition of killing in vitro and in vivo
P. gingivalis	fimbriae bind and activate CXCR4, which co-associates with TLR2	CXCR4	TLR2	monocytes macrophages	synergistic elevation of cAMP and inhibition of killing in vitro and in vivo
S. aureus B. anthracis	AdsA converts AMP to adenosine	adenosine receptors	TLR?	neutrophils, whole blood cells	adenosine receptor-mediated immunosuppressive signalling deactivates phagocytes and inhibits immune clearance

Table 9.4 Synergistic induction of immunosuppressive mediators. (From Hajishengallis, G. and Lambris, J.D., *Nat. Rev. Immunol.*, 11, 187–200, 2011.)

Leishmania major inhibits induction of T_H1 CD4$^+$ T by down-regulating IL-12 produced by the MOs it infects. This results in induction of T_H2 CD4$^+$ T that are unable to activate MOs to facilitate clearance of the protozoan. *Plasmodium falciparum* inhibits production of IL-12 *via* interaction of its erythrocyte membrane protein 1 (PfEMP1) with CD36, which is a scavenger receptor. PfEMP1 also drives DCs to secrete high levels of IL-10 by a mechanism unrelated to IL-12 suppression. Viruses, similarly, use complement receptors to suppress IL-12. The measles virus, human herpesvirus 6 and adenovirus (groups B and D) suppress IL-12 production by monocyte/MOs and DCs by crosstalk between CD46 and TLR4. The measles virus also inhibits IL-12 production by DCs *via* crosstalk between CD150 (SLAM) and TLR4. The hepatitis C virus binds gC1qR and TLR4 on MOs to inhibit IL-12 production. Bacterial pathogens can also cross signal between complement receptors and other endocytic receptors and TLR4 to down-regulate IL-12. For example, mannosylated lipo-arabinomannan of *Mycobacterium* species can inhibit IL-12 production *via* crosstalk between the mannose receptor and TLR4, and the filamentous hemagglutinin of *Bordetella pertussis* can down-regulate IL-12 *via* crosstalk between CR3 and TLR4. The fungus *Histoplasma capsulatum* inhibits IL-12 production also *via* CR3–TLR4 crosstalk. Pathogens and their antigens involved in IL-12 suppression are listed in **Table 9.5**.

Inside-out signalling: Inside-out signalling refers to the process by which intracellular signalling activates the ligand binding function of an integrin receptor. Conversely, outside-in signalling is the process by which the act of the ligand binding to the integrin

measles virus	hemagglutinin binds CD46	CD46	TLR4	monocytes	selective inhibition of IL-12 and T_H1 immunity
measles virus	hemagglutinin binds CD150 (SLAM)	CD150 (SLAM)	TLR4	DC	selective inhibition of IL-12 and T_H1 immunity
human herpesvirus 6	glycoprotein H binds CD46	CD46	TLR4	macrophages	selective inhibition of IL-12 and T_H1 immunity
adenovirus (Groups B and D)	fiber protein binds CD46	CD46	TLR4	peripheral blood mononuclear cells	inhibition of IL-12 and other proinflammatory cytokines (IL-1, IL-6)
hepatitis C virus	core protein binds gC1qR	gC1qR	TLR4	macrophages DC	selective inhibition of IL-12 and TH1 differentiation
P. gingivalis	microbial C5 convertase-like enzymes generate C5a	C5aR	TLR2	macrophages	selective inhibition of IL-12 and upregulation of IL-1β, IL-6, and TNF-a *in vivo*; promotion of pathogen survival *in vivo*
L. major	C5a generation via complement activation	C5aR	TLR4	macrophages	inhibition of IL-12 and T_H1 immunity leading to increased pathogen survival
P. gingivalis	fimbriae bind CR3	CR3	TLR2	monocytes macrophages	inhibition of IL-12/IFN-γ-dependent clearance *in vivo*
H. capsulatum	Hsp60 binds CR3	CR3	TLR4	monocytes	selective inhibition of IL-12
B. pertussis	FHA binds CR3	CR3	TLR4	macrophages	selective inhibition of IL-12
Mycobacteria	ManLAM binds mannose receptor	mannose receptor	TLR4	DC	inhibition of IL-12 and other proinflammatory cytokines
P. falciparum	PfEMP-1 (on infected erythrocytes) binds CD36	CD36	TLR4	DC	inhibition of IL-12 and suppression of DC maturation and T cell activation

Table 9.5 Inhibition of IL-12 and T-cell immunity *via* subversive receptor crosstalk. (From Hajishengallis, G. and Lambris, J.D., *Nat. Rev. Immunol.*, 11, 187–200, 2011.)

receptor initiates signal transduction pathways that lead to, for example, cell spreading, retraction, migration, or proliferation. Integrin signalling involves both heterotrimeric G proteins and monomeric small G proteins. Complement receptor 3 (CR3) is an integrin that binds iC3b but also binds fibrinogen, intercellular adhesion molecule (ICAM)-1 and a number of other molecules, including glycans. The affinity of the CR3 receptor for its ligand is significantly increased by the chemokines C3a and C5a binding to their receptors and by TLRs binding to their MAMPs. It has been suggested that pathogens target CR3 because binding to this receptor does not activate the phagocyte oxygen-dependent anti-microbial system and may not lead to fusion of phagosomes with lysosomes. However, whether or not phagosome–lysosome fusion and the respiratory burst occurs likely depends on the particular ligand and the site on CR3 to which it binds. The lipoarabinomannan of *M. tuberculosis* and the BclA glycoprotein of *B. anthracis* exosporium bridge CD14 and TLR-2 such that TLR-2 transactivates CR3. Transactivation is the process by which a gene at one locus is activated by a gene at another locus. Transactivation of CR3 subverts phagocytosis. A surface-bound glycoprotein, aggregation substance, of *Enterococcus faecalis* and the filamentous hemagglutinin of *Bordetella pertussis* bind the $\alpha v \beta 3$ integrin that partners with the integrin-associated protein (IAP). This complex transactivates CR3. The outcome is a failure of the phagocyte respiratory burst following CR3-mediated phagocytosis (**Table 9.6**).

Down-regulation of cell adhesion molecules: Another strategy employed by pathogens is to inhibit the expression of cell adhesion molecules that stabilise the immunological synapse between APCs and lymphocytes. *Salmonella typhi, Bordetella pertussis, B. bronchiseptica* and herpes virus suppress ICAM-1, resulting in reduced antigen uptake by APCs and inadequate T-cell response. Reducing

B. anthracis	BclA induces CD14/TLR2 inside-out signalling	CD14/TLR2	CR3	monocytes macrophages	activation of CR3-mediated internalization of spores leading to increased infection and host mortality
Mycobacteria	LAM induces CD14/TLR2 inside-out signalling	CD14/TLR2	CR3	monocytes macrophages	activation of CR3-mediated internalization, coronin-1-dependent inhibition of lysosomal delivery
B. pertussis	FHA induces avβ3/CD47 inside-out signalling	$\alpha v \beta 3$/CD47	CR3	monocytes macrophages neutrophils	activation of CR3-mediated internalization, suppression of pathogen clearance *in vivo*
E. faecalis	'aggregation substance' glycoprotein induces avβ3/CD47 inside-out signalling	$\alpha v \beta 3$/CD47	CR3	macrophages neutrophils	activation of CR3 entry, inhibition of oxidative burst and killing

Table 9.6 Exploitation of inside-out signalling for safe internalisation by phagocytes. (Modified from Hajishengallis, G. and Lambris, J.D., *Nat. Rev. Immunol.*, 11, 187–200, 2011.)

ICAM-1 on DCs impairs the formation of the immunological synapse with T cells, rendering DCs tolerogenic such that DCs drive naïve CD4$^+$ T cells to become T$_{REG}$ cells that secrete IL-10 rather than effector T$_H$1 T cells that secrete IL-12.

EVASION OF ANTIBODY

Pathogens that exist extracellularly for all, or part, of their life cycle are targeted by antibodies that bind to their cell surface or secreted antigens. Antibodies can bind to pathogen adhesins at barrier epithelia, particularly mucosal epithelia, and prevent pathogens from adhering to host surfaces – a process termed immune exclusion – or they can bind to viral receptors and block virus entry into permissive host cells – a process called neutralisation. Antibodies can also bind exotoxins and prevent them from binding to their cell membrane receptors, a process also termed neutralisation. IgM and IgG antibodies bound to pathogens activate the classical complement pathway resulting in complement-mediated lysis of the pathogen. In addition, IgG, principally, opsonises pathogens to assist phagocytes in taking up and destroying them. The importance of antibody in host defence is shown by increased susceptibility to extracellular encapsulated bacteria in individuals with immunoglobulin deficiency. Such is the importance of antibody in host defence that pathogens have evolved mechanisms to evade it.

Antibody enzymatic degradation: Many pathogens produce proteases and glycosidases that can degrade antibody molecules, and these have been discussed in Chapter 7, *Hydrolytic Enzymes*. All classes of antibody are susceptible to degradation, but secretory immunoglobulin A (SIgA) is most resistant, perhaps not surprisingly, because it functions at mucosal surfaces in the presence of microbiotas that have high enzymatic activity. Resistance of SIgA to proteolysis is thought to be because the highly glycosylated secretory component obscures the susceptible hinge regions of the immunoglobulin dimer.

Immunoglobulin Fc receptors: Many cells in the immune system have cell membrane receptors that capture immunoglobulins by their Fc region. Most of these Fc receptors are specific for IgG and are found in the cell membrane of neutrophils and MOs, which are potent phagocytes. The purpose of these receptors is to capture pathogens coated (opsonised) by antibody. Similarly, some pathogens have Fc receptors on their cell surface to capture IgG, IgA and IgM antibodies. By capturing antibody in this fashion, pathogens render antibody unable to act as an opsonin or to activate the classical complement pathway. In addition, pathogens that have captured immunoglobulins are camouflaged from the immune system because they appear to be self. Some pathogens such as *Staphylococcus aureus* capture host molecules such as fibrin to the same end. *Staphylococcus aureus*, the Lancefield group A streptococcus, *S. pyogenes*, and other Lancefield group streptococci have Fc receptors. Several viral proteins are IgG Fc receptors. For example, Herpes simplex virus type-1 (HSV-1) expresses an Fc receptor called gE-gI that is found on the surface of virions and infected cells. Parasites are also able to bind immunoglobulins and these include *Shistosoma mansoni, Taenia crassiceps, Taenia pisiformis, Taenia solium, Echinococcus multilocularis, Echinococcus granulosus, Heligmosoides polygyrus, Toxoplasma gondii* tachyzoites and pathogenic species of *Trypanosomatidae*.

Molecular mimicry: The term molecular mimicry is generally reserved to describe sequence similarities between foreign and self epitopes that are sufficiently great to result in the cross-activation of autoreactive T or B cells by pathogen-derived peptides. However, pathogen molecules that resemble self can serve to camouflage the pathogen from the immune system. Examples of this strategy are the capsule of *S. pyogenes*, which is composed of hyaluronic acid which is also a human extracellular matrix molecule, and the polysialic acid capsules of *Neisseria meningitidis* serogroup B and *Escherichia coli* K1. Sialic acid is the terminal sugar on human cell surface glycoconjugates. The structure of the *E. coli* K5 capsule is very similar to host heparin. In all these cases, the similarity of the structure of the capsules to host molecules prevents the immune system of the host from recognising them as sufficiently foreign to produce a vigorous antibody response.

Pathogen inaccessibility to antibodies: Cytosolic intracellular pathogens are inaccessible to antibodies, but opsonised pathogens taken up by phagocytosis may still be subject to antibody and complement in the phagosome. One situation where antibodies cannot access pathogens is in the biofilm matrix, and many bacterial and fungal pathogens form biofilms, as we saw in Chapter 3, *Biofilms*.

ANTIGEN MODULATION

Pathogens have several mechanisms by which to modify their surface antigens. They can change the three-dimensional structure of the antigen in a process called epitope masking, replace an antigen with one displaying different epitopes in a process known as antigenic variation, and turn off expression of the antigen in a process termed phase variation. We will consider these mechanisms below.

Epitope masking: We will begin by reviewing what parts of a pathogen that antibodies and T cells see. Antibodies and T cells do not see whole pathogen proteins; rather, they see small pieces of proteins known as epitopes. An epitope, also called an antigenic determinant, is the part of an antigen recognised by the paratopes of antibodies or BCR for antigen on the B cell membrane or by the TCR for antigen of T lymphocytes. The size of protein epitopes recognised by antibodies is generally between 15 and 22 amino acids in length, whereas epitopes recognised by T-cell receptors are between 9 and 20 amino acids in length. Antibodies recognise both linear stretches of amino acids but also stretches of amino acids brought together by the folding of peptide chains. These epitopes are called conformational or discontinuous epitopes. In contrast, TCRs can only recognise linear epitopes in the context of MHC molecules. A comparison of epitopes recognised by antibodies and TCRs is shown in **Figure 9.7**.

Epitope masking is the term given to the effect of intrinsically disordered proteins (IDPs) and IDP regions on the binding of antibodies to pathogen antigens. Although IDPs are active biologically, they fail to form a fixed or ordered three-dimensional structure and this property may allow escape from antibodies. IDPs are commonly found in viruses and apicomplexan parasites and also in prokaryotes. In viruses, IDPs are structural, non-structural, regulatory and accessory proteins. It is suggested that intrinsic disorder allows economic usage of virus genetic material, enables them to tolerate the high mutation rates to which their genomes are subject, and facilitates their adaptation to fluctuating environmental conditions by retaining their ability to utilise host cellular machinery. Importantly, IDPs may assist in evading host defence mechanisms. For example, HIV-1 transactivator of transcription (TaT) undergoes a limited unstructured-to-structured transition under hydrophobic conditions or in the presence of TaT-specific antibodies. Similarly,

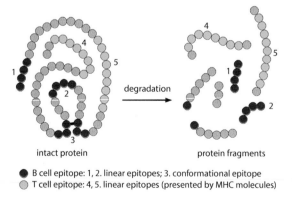

intact protein   degradation →   protein fragments

● B cell epitope: 1, 2. linear epitopes; 3. conformational epitope
○ T cell epitope: 4, 5. linear epitopes (presented by MHC molecules)

Figure 9.7 A comparison of epitopes recognised by antibodies and T-cell receptors. (From http://slideplayer.com/slide/9019630/; Slide 11.)

the mechanism by which mutants escape neutralising antibodies to the conserved V3 loop of the HIV-1 envelope glycoprotein, gp120, may not simply reflect changes in the primary amino acid sequence, but perhaps disorder in this region of the glycoprotein. The orthopoxvirus C10L proteins are highly disordered and, likely, contribute to immune evasion of vaccinia virus. Several IDPs are found in the genus *Plasmodium,* examples being the pre-erythrocytic circumsporozoite protein (CSP) and the merozoite surface protein 2 (MSP2) of *P. falciparum.* Thus, although antibodies to MSP2 can affect a strain-specific immunity in *P. falciparum,* many of them exhibit low-affinity binding to native antigen on the parasite surface. The CSP forms the basis of RTS,S a recombinant protein-based malaria vaccine, but antibodies induced by the antigen appear directed to unstructured repeats, which may, in part, explain its low efficacy (26%–50% in infants and young children). Some of the protein effectors of type-3 secretion systems (T3SSs) of gram-negative bacteria may be IDPs.

Cryptic epitopes (cryptotopes): Some pathogens conceal antigens from the immune response, exposing them only immediately before cell entry. Such is the case with HIV. Immediately before HIV-1 binds and enters a CD4$^+$ T cell, a cryptic epitope is exposed on the viral envelope. Antibodies that recognise this epitope can block viral entry. It is likely that cryptic epitopes are important in pathogens other than HIV-1.

Consuming antibodies away from the pathogen surface and releasing immune complexes from the pathogen surface: Pathogens can shed surface antigens to mop up antibodies before they can bind to the pathogen surface. Several bacteria shed their capsule and many gram-negative bacteria release membrane vesicles (MVs). These MVs contain LPS and other surface antigens and several periplasmic components. The oral bacterium *Streptococcus mutans* is capable of liberating antibodies bound to its cell surface in the form of an antigen:antibody immune complex, and this mechanism is probably used by other pathogens.

ANTIGENIC AND PHASE VARIATION

Adaptive immunity responds to infections with extracellular pathogens by producing high affinity largely IgG antibodies directed against surface antigens of the pathogen. As we have learned, these antibodies activate complement and serve as opsonins to facilitate phagocytosis of the pathogen. The antibody response exerts pressure on pathogens to modify their surface antigens to escape antibody recognition. One strategy employed by pathogens is to change the antigenic structure of molecules on the cell surface targeted by antibodies. Another strategy is to switch off surface display of the targeted antigens.

Antigenic variation: Antigenic variation is the periodic expression of homologous but antigenically distinct surface molecules within a clonal population, and it is a strategy employed by pathogens of all classes to evade antibody recognition. The antigens that undergo antigenic variation are predominantly those located on the cell surface because they are exposed directly to the host immune system. For example, in bacteria these appendages or molecules include capsule, LPS and lipooligosaccharide, fimbriae, pili, flagella and others such as transporters, porins, receptors, and enzymes. Antigens that undergo antigenic variation are encoded by large families of non-allelic genes. These gene families may number as many as 1,000 and vary in their genetic organisation. Pathogens that undergo antigenic variation include the apicomplexans *Trypanosoma brucei*, *Plasmodium* species, *Babesia microti*, and *B. divergens*; the prokaryotes *Anaplasma phagocytophilum*, *Borrelia recurrentis*, *B. Henselae*, *B. bergdorferi*, *Neisseria gonorrhoeae*, *Campylobacter fetus*, and *Treponema pallidum*; the fungi *Pneumocystis carinii*, *Candida glabrata, and Cryptococcus neoformans*; and the protozoan *Giardia lamblia*. Four genetic mechanisms are used by all classes of pathogens to effect antigenic variation. Transcriptional control is a mechanism in which the expression of the gene encoding an expressed antigen at one locus is silenced while the expression of a new gene at another locus that encodes a variant antigen is activated. Transcriptional control is used by the parasites *T. brucei*, *P. falciparum*, and *Giardia lamblia*. In gene conversion (replacement), the expressed gene is replaced by recombination with an archived gene located at another site in the genome. This mechanism is employed by *T. brucei*, *P. carinii* and *B. hermsii*. A related mechanism in which segments of silent genes are exchanged for segments of the active gene by homologous recombination is employed by *T. brucei*, *B. bovis*, *A. marginalis*, and *N. gonorrhoeae*. This process has been termed segmental gene conversion. In *M. bovis* and *C. fetus*, promoter rearrangement occurs by DNA inversion in which a promoter sequence located upstream of the expressed gene encoding the variable antigen is relocated in front of an open reading frame of a silent gene encoding the variable surface antigen.

Phase variation: Phase variation is defined as a reversible switch between expression and non-expression of one or more molecules in a clonal population and which is heritable by a genetic or epigenetic mechanism. In phase variation, pathogen-surface exposed molecules are turned on and off frequently in response to environmental triggers such as an increase in temperature to 37°C and changes in pH, the level of iron and other trace elements, oxygen tension, and gradients of antimicrobial factors. In bacteria and fungi, phase variation can be observed *in vitro* by changes in the colour, shape, and texture of their colonies. For example, in nature, the dimorphic fungi *Coccidioides* species, *Histoplasma capsulatum*, and

Blastomyces dermatitidis exist as moulds, but in the lungs, they convert to yeast forms. In the cases of the commensal dimorphic fungi *Candida albicans* and the environmental pathogen *Aspergillus fumigatus*, the invasive phase is hyphae.

Capsule: Capsule production can be switched on or off, or the amount of capsule produced may vary. Capsule production by both bacterial and fungal pathogens is generally switched on *in vivo* to avoid phagocytosis and, in the case of *Cryptococcus*, to allow them to cross the blood-brain barrier. The bacterium *Bacteroides fragilis* can produce eight types of capsule and the production of each can be switched off independently.

Fimbriae (pili): Some bacterial species express several types of pilus, some of which are capable of independent phase variation.

Flagella: Flagella undergo phase variation in a number of enteric pathogens such as *Campylobacter jejuni* and *C. coli*, *Helicobacter pylori*, and in the respiratory pathogen *Bordetella pertussis*.

Other phase variable moieties: In addition to the surface appendages discussed above, *viz.*, capsule, fimbriae (pili) and flagella, cell wall molecules of both gram-negative and gram-positive bacteria may undergo phase variation. Examples of such molecules are the hyaluronic acid capsule, M protein, C5a peptidase, IgG Fc receptor and pyrogenic exotoxin of the gram-positive coccus *Streptococcus pyogenes* and opacity proteins, porins, siderophores and haemoglobin receptors of the pathogenic *Neisseria* species. The lipoproteins of the pathogenic borelliae, *B. recurrentis* and *B. burgdorferi* undergo both antigenic and phase variation. Examples of bacteria that undergo phase and/or antigenic variation are shown in **Table 9.7**.

Antigenic shift and drift: Antigenic shift is a feature of the negative-sense, single-stranded, segmented RNA virus influenza A in which two different strains of the virus exchange RNA segments. This results in the generation of a new virus subtype with a mixture of the surface antigens of the two original strains. The two important determinants of pathogenesis, hemagglutinin (HA) and neuraminidase (NA) in the new subtype are now unique and escape existing protective antibodies. Pandemic influenza occurs when antigenic shift generates a virus to which humans are susceptible but lack protective antibodies. Antigenic drift, on the other hand results from natural mutation over time that lead to antigenic changes in HA and NA. HA and NA can drift independently, and drift does not necessarily lead to a change in the antigen. However, these genetic changes can accumulate over time and result in viruses that are antigenically different, leading to a loss of immunity or to vaccine mismatch.

bacterial species	affected moiety or phenotype[b]	gene(s) or operon regulated	variation of regulated gene(s)[c]	class(es) of regulated gene or operon	molecular mechanism
Bordetella pertussis	fimbriae	*fim3, fim3*	phase	structural	SSM[e]
	multiple virulence factors	*bvgS*	phase	regulatory	SSM
Borrelia burgdorferi	lipoprotein	*vlsE*	antigenic	structural	recombination[f]
Campylobacter coli[g]	flagella	*flhA*	phase	regulatory	SSM
Campylobacter fetus[g]	SLP	*sapA*	antigenic	structural	recombination[f]
Campylobacter jejuni[g]	LOS modification[h]	*wlaN, cgtA*	phase	enzyme	SSM
	flagella	*mafl*	phase	unknown function	SSM
Escherichia coli	fimbriae (type 1, CS18)	*fim, fot* operons	phase[i]	structural, regulatory[i]	CSSR[j]
	fimbriae (Pap, S, F1845, Clp)	*pap* and family of *pap*-like operons (*sfa, daa, clp*)	phase[i]	structural, regulatory[i]	DNA methylation[k]
	outer membrane protein	*agn*43 (*flu*)	phase	structural	DNA methylation[k]
Haemophilus influenzae[g]	LOS modification[h] ChoP Neu5Ac Other	*lic1A* *lic3A* *lic2A, lgtC*	phase	enzyme	SSM
	DNA modification	*mod*	phase	enzyme	SSM
	fimbriae LKP	*hifA, hifB*	phase	structural	SSM
Helicobacter pylori[g]	DNA R/M	*mod*	phase	enzyme	SSM
	LPS modification (Lewis antigen)[h]	*futA, futB, futC*	phase	enzymes	SSM
	flagella	*fliP*	phase	structural	SSM
	membrane lipid composition[h]	*pldA*	phase	enzyme	SSM
Moraxella catarrhalis	adhesin	*uspA1*	phase	structural	SSM
N. gonorrhoeae[g]	type IV pilin modfication[h]	*pgtA*	phase	enzyme	SSM
	siderophore receptor	*fetA*	phase	structural	SSM
N. meningitidis[g]	outer membrane protein	*porA*	phase	structural	SSM
	outer membrane protein	*opc*	phase	structural	SSM
	hemoglobin receptors	*hpuAB, hmbR*	phase	structural	SSM
	capsule	*siaD*	phase	enzyme	SSM
	capsule	*siaA*	phase	enzyme	Transposition/
N. gonorrhoeae and	opacity proteins	*opa*	phase	structural	SSM

(Continued)

bacterial species	affected moiety or phenotype[b]	gene(s) or operon regulated	variation of regulated gene(s)[c]	class(es) of regulated gene or operon	molecular mechanism
N. meningitidis	type IV pilin	*pilE, pilS*	phase and antigenic	structural	recombination
	adhesin, type IV pilus associated	*pilC*	phase	structural	SSM
	LOS modification	*lgtA, C, lgtD*	phase	enzyme	SSM
Commensal *Neisseria* spp.	LPS modification[h] (ChoP)	*licA*	phase	enzyme	SSM
Proteus mirabilis	fimbriae (MR/P)	*mrp* operon	phase[i]	structural, regulatory[i]	CSSR
Salmonella enterica serotype	fimbriae (Pef)	*pef* operon	phase[i]	structural, regulatory[i]	DNA methylation
Typhiurium	flagella	*fljBA, fliC*	antigenic	structural regulatory	CSSR
Staphylococcus epidermis	adhesin (polysaccharide)	*ica*	phase	structural	transposition
Streptococcus pneumonieae	capsule	*cap3A*	structural	recombination	379
	metabolism	*spxB*	phase	enzyme	SSM
	DNA R/M	DNA methylase	phase	enzyme	SSM
Streptococcus pyogenes	surface protein	*sclB*	phase	structural	SSM

Table 9.7 Bacterial species in which phase and/or antigenic variation occurs. Phase-variable expression, modification, or antigenic variation of the moiety is usually a direct effect of phase-variable expression of the associated gene. Classified as an on/off phase variation or altered antigenic properties of a constantly expressed moiety. Slipped strand misparing (SSM) results in variable numbers of DNA sequence repeat units that can affect transcription or translation. General recombination. Modification of the moiety that results in antigenic variation is caused by phase-variable expression of the gene, which affects (one of) the enzymatic steps leading to the modification. Expression of the entire operon, which consists of genes encoding proteins of multiple classes, phase varies. CSSR causes inversion of a DNA element that contains the main promoter for the corresponding operon. DNA methylation by the DNA maintenance methylase Dam. Reversible insertion–excision of IS (like) element. (From van der Woude, M.W. and Baumler, A.J., *Clin. Microbiol. Rev.*, 17, 581–611, 2004.)

SUBVERTING B LYMPHOCYTES (B CELLS)

As discussed in the *Antigen Presentation* section, the production of high-affinity, class switched antibodies by plasma cells and the generation of memory B cells require help from follicular helper T cells previously activated by licenced DCs. Thus, antibody synthesis and generation of memory B cells can be blocked upstream of the B cell itself. However, B cells play important roles in both innate and adaptive immunity beyond antibody synthesis and B cell memory, and they can be the direct target of some pathogens (**Table 9.8**). Certain bacterial and viral pathogens and parasites infect B cells both acutely and chronically. Infection can have several effects on B cells including, inducing IL-10 and TGF-β-secreting suppressor

pathogen	infection of B cells	diversion of B cell activation	diversion of antibody production	induction of regulatory B cells	induction of B cell death	induction of B cell survival
parasites	NR	• T. cruzi • P. falciparum	• T. cruzi • P. falciparum	• L. major	• T. cruzi • T. bruce • P. chabaudi	NR
viruses	• CMV • measles virus • EBV • MMTV • JCV • HCV • influenza virus • norovirus	• measles virus • EBV • MMTV • HCV • HIV-1	• measles virus • EBV • HCV • HIV-1 • influenza virus	• CMV • HBV • HIV-1 • polyoma virus	• influenza virus	• EBV • HCV • HIV-1
bacteria	• B. abortus • M. catarrhalis • S. Typhimurium • L. monocytogenes • S. flexner	• M. catarrhalis • S. aureus • N. gonorrhoeae • Y. pseudotuberculosis	• E. muris • B. burgdorferi • M. catarrhalis • B. anthracis • S. aureus	• B. abortus • S. Typhimurium • C. abortus	• L. monocytogenes • F. tularensis • S. flexneri • H. pylori	• H. pylori • S. Typhimurium

Table 9.8 Pathogen manipulation of B-cell function. *B. abortus, Brucella abortus; B. anthracis, Bacillus anthracis; B. burgdorferi, Borrelia burgdorferi; C. abortus, Chlamydia abortus; CMV, cytomegalovirus; EBV, Epstein–Barr virus; E. muris, Ehrlichia muris; F. tularensis, Francisella tularensis; HBV, hepatitis B virus; HCV, hepatitis C virus; H. pylori, Helicobacter pylori; JCV, polyoma JC virus; L. major, Leishmania major; L. monocytogenes, Listeria monocytogenes; M. catarrhalis, Moraxella catarrhalis; MMTV, mouse mammary tumour retrovirus; N. gonorrhoeae, Neisseria gonorrhoeae; NR, no report; P. chabaudi, Plasmodium chabaudi; P. falciparum, Plasmodium falciparum; S. aureus, Staphylococcus aureus; S. flexneri, Shigella flexneri; S. Typhimurium, Salmonella enterica subsp. enterica serovar Typhimurium; T. brucei, Trypanosoma brucei; T. cruzi, Trypanosoma cruzi; Y. pseudotuberculosis, Yersinia pseudotuberculosis.*

(regulatory) B cells, activating IgM-secreting memory B cells, suppressing B-cell function and killing B cells. Pathogens enter B cells by mechanisms described in Chapter 5, *Facilitated Cell Entry*.

Suppression of B-cell activity: HIV-1 Nef can enter B cells and block co-stimulatory signalling between CD40 and CD40L, the co-stimulatory molecules used by follicular helper T cells and B cells to transmit signals. Among bacteria, the lethal toxin of *Bacillus anthacis* inhibits B cell proliferation and antibody production by hydrolysing mitogen-activated protein kinases. Several gram-negative pathogens employ T3SSs to modulate B-cell functions. For example, the *Yersinia pseudotuberculosis* effector YopH inhibits BCR signalling and cytosolic pathway peptide presentation.

Polyclonal activation of B cells: *Trypanosoma cruzi* kills immature B cells, but it polyclonally activates more mature B cells in peripheral lymphoid tissues. The activation of B cells regardless of their antibody specificity and without the requirement for T-cell help defocuses the antibody response against the pathogen. Polyclonal B cell activation is also a mechanism employed by the parasite *Plasmodium falciparum* and several viruses such as the Epstein-Barr virus (EBV), the hepatitis C virus (HCV) and HIV-1 and bacteria such as *Borrelia burgdorferi* and *Moraxella catarrhalis*.

Induction of regulatory B cells: Various classes of pathogen induce regulatory B cells that secrete IL-10 and TGF-β. These include the parasite *Leishmania major* and several viruses such as cytomegalovirus (CMV), hepatitis B virus (HBV), and HIV-1 and the bacteria *Salmonella enterica* serovar Typhimurium, *Shigella flexneri* and *Chlamydophila abortus*. In gram-negative pathogens, T3SS effectors are thought to be responsible for this activity.

Decreasing B-cell survival: Pathogens can kill B cells at various stages of B-cell development. For example, pathogen-specific, IgG class-switched B cells can be killed by the parasite *Trypanosoma cruzi* by inducing up-regulation of Fas and Fas ligand (FasL) on B cells so that the B cells are killed by cell-to-cell receptor–ligand interaction. *T. brucei* induces apoptosis in transitional B cells. Transitional B cells are B cells that have completed development in the bone marrow, migrated to the spleen but are not yet fully mature naïve B cells in the peripheral blood and secondary lymphoid tissues. Influenza A virus can kill B cells that bear a BCR specific for the virus hemagglutinin. Several bacterial pathogens can kill B cells by apoptosis either with exotoxins such as listeriolysin produced by *Listeria monocytogenes* or by infecting B cells and activating apoptosis cascades as is the case with *Francisella tularensis*, *Shigella flexneri* and *Helicobacter pylori*. Interestingly, extracellular bacterial cells of *F. tularensis* and *S. flexneri* are also capable of killing B cells, perhaps by ligating PRRs or other receptors and inducing formation of the death-inducing signalling complex (DISC).

Increasing B-cell survival: In contrast to killing B cells, some pathogens increase the survival of B cells, although this property is most associated with viral proteins, notably the LMP2A protein of EBV, which activates SRC kinases, and the E2 envelope protein of HCV, which ligates CD81 which is part of the B-cell co-receptor complex. Ligation of CD81 activates anti-apoptotic pathways. The bacterial pathogens *H. pylori* and *S. enterica* serovar Typhimurium also can perform this function, at least *in vitro*. *H. pylori* maintains B-cell viability using the effector CagA of its T4SS to up-regulate anti-apoptotic proteins and *S. enterica* serovar Typhimurium employs its T3SS to block activation of the inflammasome.

SUBVERTING T LYMPHOCYTES (T CELLS)

The most infamous pathogen infecting T cells is HIV-1, which infects CD4$^+$ T cells, MOs and DCs. HIV-1 is a retrovirus that is a member of the Lentivirus genus. The Deltaretrovirus genus contains a family of human T-cell lymphotropic viruses, HTLV 1-4. HTLV-1 predominantly affects CD4$^+$ lymphocytes, whereas HTLV-2 predominantly affects CD8$^+$ lymphocytes. Both HTLV-1 and HTLV-2 immortalise T cells. In contrast, HIV-1 is a cytotolytic virus that progressively depletes CD4$^+$ T cells by killing them. The outcome of infection with these viruses is immunodeficiency that increases susceptibility, not only to frank pathogens, but also to opportunistic pathogens. The Ebola virus can bring about activation-induced cell death of T cells both indirectly and directly. Indirect T cell death results from the release of tumour necrosis factor-alpha (TNFα) by Ebola-infected monocytes. TNFα can cause activation and bystander death of T lymphocytes. Ebola virus can kill T cells directly by the binding of its envelope glycoprotein to TLR4 and likely to other receptors such DC-SIGN, L-SIGN, folate receptor-α and Tyro3 receptor tyrosine kinases. Binding of these receptors triggers the activation of several signalling pathways, including cell death signalling pathways.

Pathogenic bacteria may also target T lymphocytes. For example, the *H. pylori* Cag pathogenicity island induces Fas ligand expression on T cells, leading to apoptosis. The *Y. pseudotuberculosis* protein tyrosine phosphatase (PTP), YopH, inhibits T-cell activation and IL-12 production by inhibiting tyrosine phosphorylation of the T-cell signalling complex immunoreceptor tyrosine-based activation motifs (ITAMs). *Neisseria gonorrhoeae* opacity-associated proteins (Opa) bind to carcino embryonic antigen cell adhesion molecule 1 (CEACAM1) on the membrane of T lymphocytes. Following coincident TCR ligation, the cytoplasmic domain of CEACAM1, which contains two immunoreceptor tyrosine-based inhibitory motifs (ITIMs), becomes phosphorylated, preventing the normal tyrosine phosphorylation of the CD3 zeta-chain and ZAP-70 kinase and thus inhibiting activation of the T cell (see the discussion of receptor crosstalk in the section *Up-regulation of IL-10*).

Superantigens: Superantigens (SAs) are proteins produced by some pathogenic bacteria and viruses that stimulate CD4⁺ (helper) T cells in an antigen–non-specific manner (see Chapter 6). SAs bridge MHC class II molecules on professional APCs with the TCR of CD4⁺ T cells outside of the peptide-binding groove of MHC class II molecules and the paratope of the TCR. Each SA binds to a specific motif on the variable domain of the β chain of an α:β TCR, which may be present on as much as 20% of CD4⁺ T cells such that any particular SA bring about can activate all CD4⁺ T cells that share that particular motif. CD4⁺ T cells activated by SA bring about release massive amounts of cytokines that result in shock and suppression of the adaptive immune response. Bacteria that produce SA are *Staphylococcus aureus*, *Streptococcus pyogenes* and some other Lancefield group streptococci, *Yersinia pseudotuberculosis* and *Mycoplasma arthritidis*. *S. aureus* produces many SAs, including staphylococcal toxic shock syndrome toxin 1 (TSST1) and a family of enterotoxins comprising some 18 members that cause food poisoning. At least 12 SA have been identified in *S. pyogenes*, including the streptococcal pyrogenic exotoxin (SPE) family, the streptococcal superantigen (SSA), and the streptococcal mitogenic exotoxins 1 and 2. SAs have also been found in the rabies virus, EBV and cytomegalovirus. At least one of the *S. aureus* SAs, staphylococcal enterotoxin B (SEB), also binds the co-stimulatory molecule CD28. This finding suggests that a larger and more stable complex is formed at the immunological synapse than was previously thought. The role(s) of SAs in pathogenicity remain unclear; however, the ability of SAs to blur the focus of helper T cells may contribute to immune evasion.

KEY CONCEPTS

- Adaptive cellular and humoral immunity depends on the interactions between the licenced DC, the thymus-derived lymphocyte (T cell) and the bone marrow-derived lymphocyte (B cell).
- Many pathogens target the DC and, by doing so, impair both cellular and humoral immunity.
- Pathogens may target T and B cells directly and comprise adaptive immunity.
- Pathogens have many strategies to evade soluble effector molecules such as antibodies.

BIBLIOGRAPHY

Abbas AK, Lichtman AH, Pillai S (Eds). 2015. Antigen capture and presentation to Lymphocytes. In *Basic immunology*, 5rd ed. Elsevier, St. Louis, MO.

Alcami A. 2003. Viral mimicry of cytokines, chemokines and their receptors. Nature Reviews Immunology, 3(January): 37-50.Baena A, Porcelli SA. 2009. Evasion and subversion of antigen presentation by *Mycobacterium tuberculosis*. *Tissue Antigens*, 74(3): 189–204.

Van Avondt K et al. 2015. Bacterial immune evasion through manipulation of host inhibitory immune signaling. *PLoS Pathogens*, March 5. doi:10.1371/journal.ppat.1004644.

Bowie AG, Unterholzner L. Viral evasion and subversion of pattern-recognition receptor signalling. *Nature Reviews Immunology*, 8(December): 911–922.

Brodsky FM et al. 1999. Human pathogen subversion of antigen presentation. *Immunological Reviews*, 168: 199–215.

Brodsky IE, Medzhitov R. 2009. Targeting of immune signalling networks by bacterial pathogens. *Nature Cell Biology*, 11(5): 521–526.

Castañeda-Sánchez JI et al. 2017. Chapter 8, B lymphocyte as a target of bacterial iInfections. In: Ed. Isvoranu G (Ed.), *Lymphocyte Updates: Cancer, Autoimmunity and Infection*. IntechOpen, London, UK. https://www.intechopen.com/books/lymphocyte-updates-cancer-autoimmunity-and-infection.

Collette JR, Lorenz MC. 2011. Mechanisms of immune evasion in fungal pathogens. *Current Opinion in Microbiology*, 14: 668–675.

Deitsch KW et al. 2009. Common strategies for antigenic variation by bacterial, fungal and protozoan pathogens. *Nature Reviews Microbiology*, 7(7): 493–503.

Dinko B, Pradel G. 2018. Immune evasion by *Plasmodium falciparum* parasites: Converting a host protection mechanism for the parasite's benefit. *Advances in Infectious Diseases*, 6: 82–95.

Feng Z-P et al. 2006. Abundance of intrinsically unstructured proteins in *P. falciparum* and other apicomplexan parasite proteomes. *Molecular & Biochemical Parasitology*, 150: 256–267.

Fernandes RK et al. 2015. *Paracoccidioides brasiliensis* interferes on dendritic cells maturation by inhibiting PGE2 production. *PLoS One*. doi:10.1371/journal.pone.0120948.

Finlay BB, McFadden G. 2006. Anti-immunology: Evasion of the host immune system by bacterial and viral pathogens. *Cell*, 124(February 24): 767–782.

Fruh K et al. 1999. A comparison of viral immune escape strategies targeting the MHC class I assembly pathway. *Immunological Reviews*, 168: 157–166.

Geijtenbeek TBH, Gringhuis SI. 2013. An inside job for antibodies: Tagging pathogens for intracellular sensing. *Nature Immunology*, 14(4): 309–311.

Goldszmid RS et al. 2009. Host ER-parasitophorous vacuole interaction provides a route of entry for antigen cross-presentation in *Toxoplasma gondii*-infected dendritic cells. *Journal of Experimental Medicine*, 206(2): 399–410.

Guizetti J, Scherf A. 2013. Silence, activate, poise and switch! Mechanisms of antigenic variation in *Plasmodium falciparum*. *Cellular Microbiology*, 15(5): 718–726.

Hajishengallis G, Lambris JD. 2011. Microbial manipulation of receptor crosstalk in innate immunity. *Nature Reviews Immunology*, 11(3): 187–200.

Hegde NR et al. 2003. Viral inhibition of MHC class II antigen presentation. *Trends in Immunology*, 24(5): 278–285.

Hernández-Chávez MJ et al. 2017. Fungal strategies to evade the host immune recognition. *Journal of Fungi*, 3: 51. doi:10.3390/jof3040051.

Horn D. 2014. Antigenic variation in African trypanosomes. *Molecular & Biochemical Parasitology*, 195: 123–129.

Horst D et al. 2011. Viral evasion of T cell immunity: Ancient mechanisms offering new applications. *Current Opinion in Immunolology*, 23(1): 96–103.

Jain N, Fries BC. 2009. Antigenic and phenotypic variations in fungi. *Cellular Microbiology*, 11(12): 1716–1723.

Katze MG, Korth MJ, Law GL, Nathanson N. 2016. *Viral Pathogenesis: From Basics to Systems Biology*. 3rd ed. Academic Press, Elsevier, London, UK.

Khan N et al. 2012. Manipulation of costimulatory molecules by intracellular pathogens: Veni, vidi, vici!! *PLoS Pathogens*, 8(6): e1002676.

Kotwal GJ, Kulkarni AP. 2013. Antigenic variation in microbial evasion of immune responses. In: *eLS*. John Wiley & Sons, Chichester, UK. doi:10.1002/9780470015902.a0001207.pub3.

Lee SF. 1995. Active release of bound antibody by *Streptococcus mutans*. *Infection and Immunity*, 63(5): 1940–1946.

Loker ES, Hofkin BV. 2015. *Parasitology: A Conceptual Approach*. Garland Science, New York.

Maizels RM, McSorley HJ. 2016. Regulation of the host immune system by helminth parasites. *Journal of Allergy and Clinical Immunology*, 138: 666–675.

Morales RAV et al. 2015. Structural basis for epitope masking and strain specificity of a conserved epitope in an intrinsically disordered malaria vaccine candidate. *Scientific Reports*, 5: 10103. doi:10.1038/srep10103.

Murphy K, Weaver C. 2016. *Janeway's Immunobiology*. 9th ed. Garland Science, New York.

Nash AA, Dalziel RG, Fitzgerald JR. 2015. *Mim's Pathogenesis of Infectious Disease*. 6th ed. Elsevier, London, UK.

Nathanson N. 2007. *Viral Pathogenesis and Immunity*. 2nd ed. Academic Press, Elsevier, London, UK.

Nothelfer K et al. 2015. Pathogen manipulation of B cells: The best defence is a good offence. *Nature Reviews Microbiology*, 13(3): 173–184.

Palmer GH et al. 2016. Antigenic variation in bacterial pathogens. *Microbiology Spectrum*, 4(1). doi:10.1128/microbiolspec.VMBF-0005-2015.

Pérez Arellano JL et al. 2001. Evasion mechanisms of parasites. *Revista Ibérica de Parasitología*, 61(1–2): 4–16.

Pleass RJ, Woof JM. 2001. Fc receptors and immunity to parasites. *Trends in Parasitology*, 17(11): 545–551.

Recker M et al. 2011. Antigenic variation in *Plasmodium falciparum* malaria involves a highly structured switching pattern. *PLoS Pathogens*, 7(3): e1001306.

Rópolo AS, Touz MC. 2010. A lesson in survival, by *Giardia lamblia*. *The Scientific World Journal*, 10: 2019–2031.

Sansonetti PJ, Di Santo JP. 2007. Debugging how bacteria manipulate the immune response. *Immunity*, 26(2): 150–161.

Schmid-Hempel P. 2009. Immune defence, parasite evasion strategies and their relevance for 'macroscopic phenomena' such as virulence. *Philosophical Transactions of the Royal Society B*, 364: 85–98.

Sebina I, Pepper M. 2018. Humoral immune responses to infection: Common mechanisms and unique strategies to combat pathogen immune evasion tactics. *Current Opinion in Immunology*, 51: 46–54.

Stanisic DI et al. 2013. Escaping the immune system: How the malaria parasite makes vaccine development a challenge. *Trends in Parasitology*, 29(12): 612–622.

Steverding D. 2007. Antigenic variation in pathogenic micro-organisms: Similarities and differences. *African Journal of Microbiology Research*, 1(7): 104–112.

Stijlemans B et al. 2016. Immune Evasion Strategies of *Trypanosoma brucei* within the mammalian host: Progression to pathogenicity. *Frontiers in Immunology*, 7: 233. doi:10.3389/fimmu.2016.00233.

Wilson BA, Salyers AA, Whitt DD, Winkler ME. 2011. *Bacterial Pathogenesis: A Molecular Approach*. 3rd ed. ASM Press, Washington, DC.

Wilson M, McNab R, Henderson B. 2002. *Bacterial Disease Mechanisms: An Introduction to Cellular Microbiology*. Cambridge University Press, Cambridge, UK.

van der Woude MW, Baumler AJ. 2004. Phase and antigenic variation in bacteria. *Clinical Microbiology Reviews*, 17(3): 581–611.

Chapter 10: Persistent and Latent Infections

INTRODUCTION

In this chapter, we consider pathogens that persist or remain latent in immunologically intact hosts and explore whether they are different from pathogens that cause acute disease. Before beginning, it is useful to define the terms *latent* infection and *persistent* infection. Latent infection is defined as any infection in which there are no clinical signs or infectious agent detectable in blood or secretions. Persistent infection is defined as any infection where there may be long-lasting or lifelong latent infections with asymptomatic periods and recurring acute episodes of clinical disease. Persistent and latent pathogens employ the same strategies to subvert the innate and acquired immune system as do pathogens that cause acute infections. These mechanisms have been described earlier in the text (Chapters 8 and 9) and we will reconsider them in this chapter for certain persistent pathogen. The answer to the question as to whether persistent pathogens have unique methods of immune evasion, appears to be no. However, they may have a greater complement of evasins and/or employ them more effectively. Moreover, persistent infection can result in immune exhaustion or deviation by re-polarizing $CD4^+$ helper T (T_H) cells to interleukin-10 (IL-10) producing regulatory T cells (T_{REG}). However, the production of T_{REG} may benefit the host by dampening inflammation and limiting tissue injury.

In immunocompetent humans, the majority of infectious agents are eliminated by the host immune system and, in many cases, immunological memory ensures that the host is immune to reinfection. The extent of issue injury is determined by the balance between the pathogen and the host immune system and it is tempting to assume that an agent capable of persistence has tipped the balance in its favour. However, another view is that persistent infection represents a particular phase in the pathogenesis of certain microorganisms rather than an imbalance in the pathogen–host interaction. This perspective is supported by the finding of genes that are associated with persistent infections but not with acute infections. Microorganisms associated with the human host might be considered to fall on a scale of persistence. At one end of the scale are commensal microorganisms that persistently colonize the barrier epithelia and are generally in a state of dynamic equilibrium with the host. Further along the scale are persistent pathogens; then pathogens that exhibit intermittent colonization (carriage), such as *Haemophilus influenzae, Streptococcus pneumoniae*, and *Neisseria meningitides* in which certain genes encoding determinants of pathogenesis may be switched off during the carriage period. These pathogens cause acute disease once they invade. Finally, at the other end of the scale are extrinsic pathogens that cause acute infections. However, the apparent dynamic equilibrium of persistent pathogens with the host may continue forever, as will be seen later in this chapter.

From what we have learned about pathogenesis, it is reasonable to conclude that pathogens capable of persistence are more likely to be intracellular where they are largely protected from humoral elements of the immune system. Intracellular pathogens, certainly those that persist, need to take up residence in long-lived host cells, inhibit phagosome/endosome fusion with lysosomes, autophagy and avoid attack by natural killer (NK) cells and cytotoxic $CD8^+$ T cells. However, intracellular persistent pathogens need to move extracellularly to facilitate eventual transmission. Mathematic modelling of persistent infections show that persistent pathogens occupy both intra- and extracellular locations. Although, in many cases, persistent infections are asymptomatic, they may reactivate or lead to malignancy. We will begin by considering several extracellular persistent bacterial pathogens before moving inside the host cell to consider facultative and obligate intracellular persistent pathogens. Here we examine some bacteria, fungi, viruses and parasites that cause persistent or latent infections and revisit the mechanisms by which they evade the host immune system.

PERSISTENT BACTERIAL INFECTIONS

Introduction

Bacteria causing persistent infections are either those kept in check but are unable to be eliminated by the adaptive immune response, or those that are members of the commensal microbiota that are able to cause invasive disease in immunocompetent individuals. As stated above, there appear to be no unique evasion mechanisms employed by bacteria that cause persistent infections. **Table 10.1** lists mechanisms suggested to increase persistence of pathogenic bacteria, but they are no different to those used by bacteria that cause acute infections (see Chapters 8 and 9). It may be that persistent bacterial and other classes of persistent pathogens have a greater armamentarium of them.

Helicobacter pylori

H. pylori (Hp) is an indigenous bacterium whose habitat is the stomach of human beings. There is evidence that it has co-evolved with humans over many hundreds of thousands of years. In the vast majority of humans,

mechanism	example of pathogen	comments
antigenic variation	N. meningitides N. gonorrhea B. burgdorferi B. fragilis	bacteria vary surface antigens using a repertoire of variable genes within their genome or with efficient horizontal gene transfer mechanisms
colonization of a particular tissue or organ	S. enterica Typhi M. leprae H. pylori	S. enterica Typhi cart persist in the reticuloendothelial system for extended periods of time (particularly in the bone marrow); carriers are chronically infected in the gall bladder colonization of nervous system lives in stomach partially in layer beyond epithelial cell barrier
host mimicry	N. gonorrhea N. meningitidis T. pallidum	sialylation of LPS mimicry of host polysaccharides coating with host proteins?
resistance to immune effector mechanisms	B. pseudomallei M. tuberculosis	ability to resist killing in phagocytic cells ability to persist in macrophages
modification of the intracellular environment	S. enterica Typhi Brucella species M. tuberculosis Chlamydia species	ability to avoid fusion with the host lysosomal compartment and modify the intravacuole environment establishment of an obligate intracellular lifestyle
antiphagocyte defense	Pseudomonas aeruginosa M. tuberculosis	production of extracellular alginate or polysaccharide production of extracellular glycolipid
selective gene inactivation	M. leprae S. enterica Typhi	loss of genes and accumulation of pseudogenes compared to M. tuberculosis; reduction of replication rate and metabolic activity loss of attachment factors and proteins for intracellular lifestyle may promote uptake by a favored pathway promoting systemic spread and persistence

Table 10.1 Mechanisms for increasing persistence used by pathogenic bacteria. (From Young, D. et al., *Nat. Immunol.*, 3, 1026–1032, 2002.)

Hp exists as a commensal, producing an asymptomatic, superficial chronic inflammation of the gastric mucosa that can only be detected by histology. However, in a small percentage of individuals, colonization of Hp leads to the formation of peptic ulcers and gastric adenocarcinoma. All strains of Hp have a pore-forming, vacuolating toxin (VacA), albeit in many polymorphic forms. VacA creates pores in the membrane of gastric epithelial cells and loosens the tight junction between them, likely by targeting claudins. Both of these actions allow the release of urea and anions such as chloride and bicarbonate and perhaps nutrients for the bacterium that may contribute to its persistence.

Hp subverts the immune system in several ways. Hp lipopolysaccharide (LPS) is poorly recognized by the cell membrane pattern recognition receptor (PRR) Toll-like receptor 4 (TLR4), and flagella are not recognized by the PRR TLR5 (**Figure 10.1**). In addition, ligation of the membrane C-type lectin PRR, dendritic cell-specific intercellular adhesion molecule-3 grabbing non-integrin (DC-SIGN), by Hp produces anti-inflammatory signals. Although Hp is recognized by the endosomal PRRs TLR8 and TLR9, activation of these TLRs induces IL-10, an immunoregulatory cytokine, and anti-inflammatory pathways. Hp has

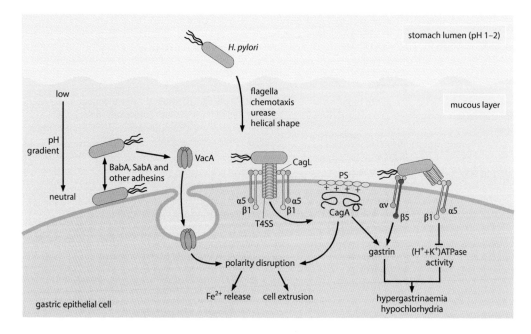

Figure 10.1 *Helicobacter pylori* colonization and persistence factors. During initial infection of the stomach lumen, urease-dependent ammonia production locally raises the pH, which promotes bacterial survival and solubilizes the mucous gel to facilitate bacterial motility. Chemotaxis (driven by pH and possibly other gradients) and the helical rod shape promote flagellar motility away from the acidic lumen to the preferred habitat of *H. pylori*, which is on and adjacent to gastric epithelial cells. SabA, BabA and other variably expressed adhesins might shift the balance from mucus-associated to cell-associated bacteria. Cell-associated bacteria alter gastric epithelial cell function *via* vacuolating cytotoxin (VacA), cytotoxin-associated gene A (CagA) and CagL, which all have multiple cellular targets. CagL interactions with the α5β1 cell surface receptor are mediated through the RGD motif of the protein, whereas interactions with the other cell surface receptor (αvβ5 integrin) are RGD-independent. The combined action of these three effectors leads to a number of changes in the gastric epithelial cell, including CagA- and VacA-dependent disruption of cell polarity, which can promote iron acquisition and cell extrusion; CagA- and CagL-dependent induction of chemokines and/or the gastric hormone gastrin; and CagL-dependent inhibition of acid secretion by the (H$^+$+K$^+$) ATPase and cellular proliferation, apoptosis and differentiation, which are mediated by all three effectors. In addition to CagL, CagA and CagY (not shown) have been demonstrated to bind α5β1 integrins, although the precise interaction surface is unknown. PS, phosphatidylserine; T4SS, type IV secretion system. (From Salamam, N.R. et al., *Nat. Rev. Microbiol.*, 11, 385–399, 2013.)

a type IV secretion system encoded by the *cag* pathogenicity island that delivers the cytotoxin, CagA, to the cytosol of the gastric epithelial cell (**Figure 10.2**). In addition, CagA activates anti-apoptosis pathways that may reduce the rate at which gastric epithelial cells are replaced. Furthermore, CagA and peptidoglycan injected *via* the T4SS can induce gastric epithelial cells to drive CD4$^+$ T cells to become T$_{REG}$ by mechanisms not currently understood. However, the finding of *cag*$^-$ clinical isolates suggests that the pathogenicity island, a large genetic element acquired through horizontal transmission, may not contribute to persistence of this bacterium.

VacA binds the LFA1 receptor on naïve T lymphocytes that inhibits IL-2 production and IL-2 receptor expression that are required for T cell activation, proliferation and subset decision. Additionally, VacA causes

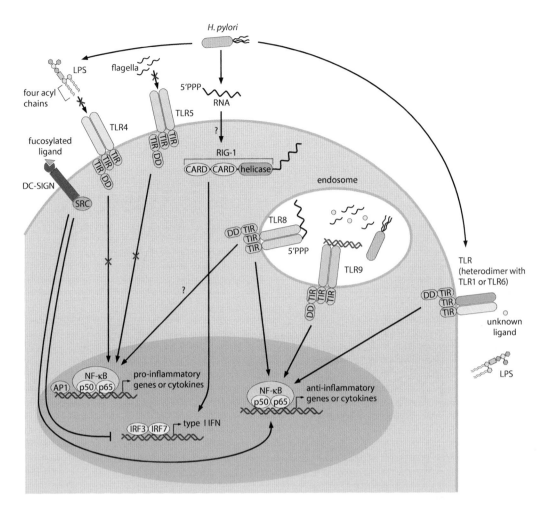

Figure 10.2 *Helicobacter pylori* subversion of innate immunity recognition. *H. pylori* has microbe-associated molecular patterns (MAMPs) that evade detection by Toll-like receptors (TLRs). *H. pylori* expresses tetra-acylated lipopolysaccharide (LPS), which is less biologically active than the hexa-acylated form that is typical of other gram-negative pathogens owing to specific lipid A modifications that prevent detection by TLR4. *H. pylori* flagella are not detected by TLR5 owing to mutations in the TLR5 binding site of flagellin. The DNA of the bacterium, as well as a currently uncharacterized MAMP (and possibly *H. pylori* LPS) are detected by TLR9 and TLR2, respectively; these TLRs predominantly activate anti-inflammatory signalling pathways and anti-inflammatory IL-10 expression. 5′ triphosphorylated RNA is detected by the RIG-like helicase receptor family (RLR) RIG-I, which activates the transcription factors IRF3 and IRF7 to induce type I interferon (i.e., IFN-α and IFN-β) expression. Bacterial RNA is also potentially detected by TLR8 in endosomes. The fucosylated DC-SIGN ligands of *H. pylori* suppress activation of the signalling pathways downstream of this C-type lectin receptor (CLR) and activate anti-inflammatory genes. CARD, caspase activation and recruitment domain; DC-SIGN, dendritic cell-specific intercellular adhesion molecule-3 grabbing non-integrin; DD, death domain; IL, interleukin; IRF, interferon regulatory factor; MYD88, myeloid differentiation primary response gene 88; NF-κB, nuclear factor-κB; TIR, Toll/IL-1 receptor domain. (From Salamam, N.R. et al., *Nat. Rev. Microbiol.*, 11, 385–399, 2013.)

dendritic cells (DCs) to produce IL-10 that drives naïve CD4+ T cells to become T$_{REG}$ (**Figure 10.3**). One feature of Hp that may be significant in its persistence is its high level of genetic diversity. Hp exhibits a large number of point mutations and recombination that permits escape from the pressure exerted by the immune system. This high degree of genetic

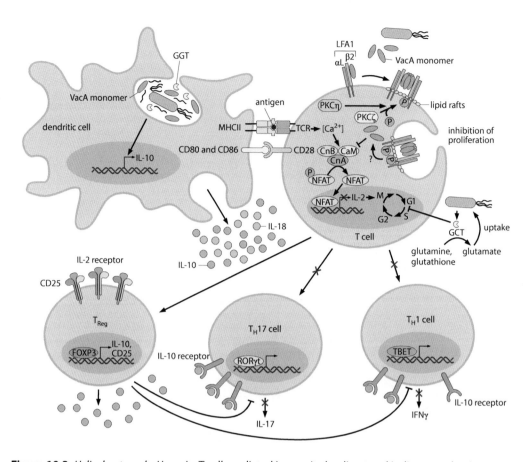

Figure 10.3 *Helicobacter pylori* impairs T cell-mediated immunity by direct and indirect mechanisms. All strains of *H. pylori* express the secreted virulence factors vacuolating cytotoxin (VacA) and gamma-glutamyl transpeptidase (GGT) to directly inhibit T-cell activation, proliferation and effector functions. VacA binds to the β2 integrin subunit of the heterodimeric transmembrane receptor lymphocyte function-associated antigen 1 (LFA1); the receptor complex is internalized following protein kinase C (PKC)–mediated serine/threonine phosphorylation (P) of the β2 integrin cytoplasmic tail. Cytoplasmic VacA prevents nuclear translocation of nuclear factor of activated T cells (NFAT) by inhibiting its dephosphorylation by the Ca^{2+}/calmodulin-dependent phosphatase calcineurin, and thereby blocks interleukin-2 (IL-2) production and subsequent T-cell activation and proliferation. GGT arrests T cells in the G1 phase of the cell cycle, preventing their proliferation. Both VacA and GGT also indirectly prevent T-cell immunity by re-programming DCs; VacA- and GGT-exposed DCs produce IL-10, and induce the FOXP3- and contact-dependent differentiation of T cells into $CD4^+CD25^+FOXP3^+$ T_{REG}s while simultaneously preventing T_H1 and T_H17 differentiation. T_{REG} cell differentiation further depends on dendritic cell (DC)–derived IL-18, which is processed upon activation of caspase 1, and binds to its receptor on naïve T cells. Depicted interactions at the T cell–DC synapse include major histocompatibility complex (MHC) class II binding to the T-cell receptor (TCR) and binding of co-stimulatory molecules CD80 and CD86 to CD28. DC-derived and/or T_{REG}-derived IL-10 further suppresses T_H1 and T_H17 effector functions. CaM, calmodulin; CnA, calcineurin A; GGT, γ-glutamyl-transpeptidase; RORγt, retinoid-related orphan receptor γt; TBET, T-box transcription factor; T_{REG}, regulato ry T cell. (From Salamam, N.R. et al., *Nat. Rev. Microbiol.*, 11, 385–399, 2013.)

diversity is consistent with other commensal bacteria. For example, *Streptococcus mitis*, a commensal in the human oral cavity, is antigenically and physiologically diverse, properties that aid in avoiding host immunity and promote re-colonization of a habitat or transfer to a new habitat in the mouth.

Treponema pallidum subspecies pallidum

Syphilis is a sexually transmitted infection caused by *T. pallidum*. Syphilis is divided into four stages, primary, secondary, latent and tertiary. The latent stage is divided into early latent and late latent phases depending on the frequency of infectious relapse. *T. pallidum* is a motile, microaerophilic, spiral-shaped, gram-negative spirochete that can be transmitted both vertically (mother to foetus) and horizontally (person to person). The spirochete is highly invasive and can transverse mucosae or damaged skin. Interestingly, its outer membrane lacks LPS and displays few microbe-associated molecular patterns (MAMPS) on the outer leaflet of its outer membrane. In fact, only two lipoproteins have been confirmed to be surface exposed. Like other spirochetes such as *Borrelia* and *Leptospira*, even the flagella are not exposed to the external environment but are located in the periplasmic space (**Figure 10.4**). These endoflagella give the spirochete its characteristic corkscrew motility. The genome of *T. pallidum* is small, containing a little over 1,000 genes that encode roughly the same number of proteins.

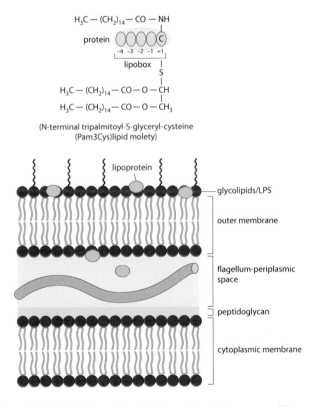

Figure 10.4 Lipopolysaccharide (LPS) is not found in *Borrelia* and *Treponema*. The periplasmic space contains the flagellum. The distribution of lipoproteins varies among spirochetes and they may be present in different cellular compartments: the outer membrane and the extracellular and the periplasmic spaces. For example, the pro-inflammatory lipoproteins of *T. pallidum* are located below its cell surface and thus do not interact directly with the immune system of the host, although they may be exposed to pattern recognition receptors (PRRs) during phagocytosis. Right upper corner: structure of spirochetal lipoproteins. (Adapted from Kelesidis, T., *Front. Immunol.*, 5, 2014.)

It is clear that humans infected with *T. pallidum* mount an innate and adaptive immune response against the organism. The digestion of spirochetes in the phagolysosomes of DCs generates peptides that are presented to both CD4$^+$ T$_H$1 and CD8$^+$ T cells (by cross-presentation). Both types of T cells release interferon-gamma (IFN-γ) that up-regulates macrophage activity so that they are more efficient at phagocytosing spirochetes in addition to driving them to release pro-inflammatory cytokines. In addition, CD4$^+$ T$_H$1 T cells help B cells make opsonizing and complement-fixing IgG antibodies that are able to assist phagocytosis by professional phagocytes. Nevertheless, the immune response is ineffective at eliminating the spirochete. During secondary and latent syphilis, high levels of opsonic antibodies are unable to control spirochetes in the bloodstream despite the fact that they recognize a large number of *T. pallidum* antigens. However, it is likely that many of these antigens are located in the periplasm rather than surface-exposed, so antibodies directed against them are likely ineffective in promoting complement-mediated and antibody-mediated opsonisation and lysis because the antibodies cannot bind their antigens. A family of outer-membrane proteins encoded by the *T. pallidum* repeat (*tpr*) gene family undergoes antigenic and phase variation and may be masked by the acquisition of host molecules. Interestingly, antibody-binding and antibody non–binding sub-populations of spirochetes have been detected in human infections. In addition, it has been posited that the spirochete may occupy sequestered sites such as the eyes, ovaries and testes that are not accessible to the immune system. **Figure 10.5** shows the role of known spirochete lipoproteins in regulation of host immunity. Perhaps what we have here during latency is a stalemate between the adaptive immune system and the spirochete. If this equilibrium maintains, then the individual remains asymptomatic; however, if the equilibrium tilts towards the spirochete, then the individual manifests tertiary syphilis that can affect multiple organ systems including the brain, nerves, eyes, heart, blood vessels, liver, bones, and joints.

Mycobacterium tuberculosis

M. tuberculosis (Mt) is an aerobic, facultatively intracellular bacterial pathogen that, based on skin test reaction, infects about one-third of the population of the planet. The bacterium is very efficiently transmitted by droplets inhaled into the respiratory tree; however, less than 10% of immunologically intact individuals develop disease. Mechanisms used by Mt to evade the immune system are shown in **Table 10.2** and include the arrest of phagosome–lysosome fusion (**Figure 10.6**), resistance against reactive nitrogen intermediates and nitric oxide (**Figure 10.7**) and interference with major histocompatibility complex (MHC) Class II antigen-presentation (**Table 10.3**). These have been discussed in Chapter 8.

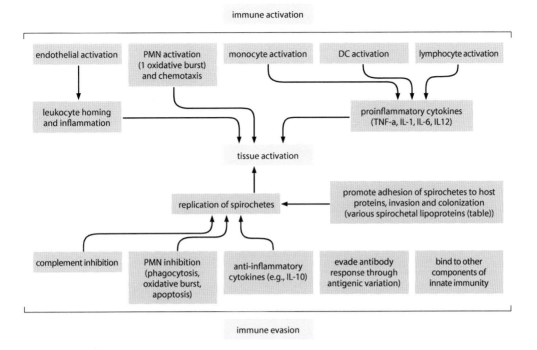

Figure 10.5 Activation and suppression of the immune response by *T. pallidum*. Lipoproteins in the outer leaflet of the spirochete play an important role in immune evasion through inhibition of complement, neutrophils, production of anti-inflammatory cytokines, evasion of antibody responses through antigenic variation, and binding to other components of innate immunity (e.g., apolipoproteins). In addition, spirochete lipoproteins may directly promote tissue invasion and colonization. (Adapted from Kelesidis, T., *Front. Immunol.*, 5, 2014.)

In the alveoli, the bacteria are initially taken up by macrophages and DCs by engaging endocytic receptors such as DC-SIGN and other C-type lectins. Mt has many lipoproteins and lipoglycans that are recognized by TLR2 on the cell membrane of DCs and macrophages, and TLR2 activation drives production of IL-10, which may inhibit development of antigen-specific CD4$^+$ T cells. Mt is able to impair the maturation of phagosomes and prevent their fusion with lysosomes so that the bacteria remain viable within this compartment. In addition, the mycobacterial type VII secretion system ESX-1 facilitates phagosome membrane rupture. Once Mt enter the cytosol, they are exposed to cytosolic PRRs such as NOD-like receptors. NOD-like receptors recognize peptidoglycan fragments and Mt produces an unusual muramyl-dipeptide that triggers type 1 interferons (IFN-α and IFN-β) instead of the type 2 interferon, IFN-γ. IFN-γ is critical to up-regulating the microbicidal activity of macrophages, whereas type 1 IFNs are ineffective in doing so.

At the focus of infection in the sub-pleura of the lung, pro-inflammatory cytokines and chemokines released from phagocytes following ligation of their signalling PRRs recruit additional cells such as neutrophils and monocytes to initiate the formation of a granuloma (see below).

	cell type	effector mechanism(s)	evasion mechanism(s)
innate immunity	neutrophils	degranulation	unknown
		production of ROIs	genetic resistance (mechanism unknown)
		NFTs	unknown
	macrophages	phagocytosis	arrest of phagosome-lysosome fusion
		production of RNIs and NO	genetic resistance (mechanism unknown)
	NK cells	cytolysis of infected cells/ intracellular bacteria	unknown
		activation of macrophages through IFN-γ	unknown
	complement system	membrane attack complex, opsonisation and phagocytosis by macrophages	unknown-possibly through binding Factor H
adaptive immunity	CD4+ T cells	cytolysis of infected cells/ intracellular bacteria	interference with MHC Class II antigen presentation
		activation of macrophages through IFN-γ and TNF-α	interference with MHC Class II antigen presentation
	CD8+ T cells	cytolysis of infected cells/ intracellular bacteria	unknown
		activation of macrophages through IFN-γ	unknown
	B cells	opsonisation and phagocytosis by macrophages	no evasion mechanism known. *M. tuberculosis* appears susceptible to Ab-opsonised phagocytosis

Table 10.2 Effector mechanisms of innate and adaptive immunity and their evasion by *M. tuberculosis*. Ab, antibody; IFN, interferon; MHC, major histocompatibility complex; NET, neutrophil extracellular trap; NO, nitric oxide; RNI, reactive nitrogen intermediate; ROI, reactive oxygen intermediate; TNF, tumor necrosis factor. (From Gupta, A. et al., *Immunobiology*, 217, 363–374, 2012.)

Macrophages and DCs that have taken up the pathogen carry it to other areas of the lung, to other organs and to secondary lymphoid tissues. Cell-mediated immunity is initiated in lymph nodes that drain the granuloma by DCs that arrive from the granuloma displaying mycobacterial peptides in the groove of MHC class II molecules. In the paracortex of the lymph nodes, the DCs present mycobacterial peptides to naïve CD4+ T cells. Effector CD4+ T_H1 T cells resulting from this interaction traffic from the draining lymphoid tissues to the granuloma where they release IFN-γ that activates macrophages to kill the phagocytosed mycobacteria. Interestingly, it takes about 6 weeks before the cell-mediated immune response ensues, perhaps because the phagocytosed mycobacteria inhibit migration of DCs to draining lymph nodes. Furthermore, mycobacteria-specific T_{REG} are induced which may affect priming of naïve T cells.

Although the cell-mediated immune response eliminates the mycobacteria in the majority of individuals, in some, the bacteria persist throughout life in a non-replicative state inside macrophages contained within the granuloma. The granuloma consists of a core of infected macrophages surrounded by lipid-containing macrophages known as foam cells and multinucleate giant cells formed by the fusion of macrophages.

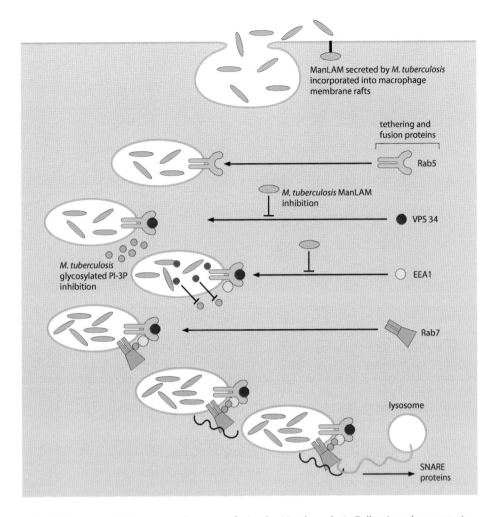

Figure 10.6 The arrest of phagosome–lysosome fusion by *M. tuberculosis*. Following phagocytosis. *M. tuberculosis* arrests phagosome–lysosomes fusion by interfering with the recruitment of the following molecules. EEA1, early endosome antigen 1: ManLAM, mannosylated liporabinomannan; PI-3P, phosphatydilinositol-3-phosphate; Rab, proteins from Ras superfamily of small GTPases; SNARE, soluble NSF (N-ethylmaleimide-sensitive factor) attachment protein receptors; VPS, vacuolar protein sorting. (From Gupta, A. et al., *Immunobiology*, 217, 363–374, 2012.)

These cells are surrounded by a wall of collagen outside which are CD4+ and CD8+ α:β T cells, γ:δ T cells and B cells. The organization of the granuloma is driven by IFN-γ secreted by CD4+ T$_H$1 T cells, CD8+ T cells and, likely, NK T cells. As the granuloma matures, it becomes vascularized. Mycobacteria in the granuloma are in a dormant state thought to be induced by the reduced oxygen tension and acidic environment of the granuloma and exposure to macrophage-produced nitric oxide and nutrient deprivation. Currently, three mechanisms are thought to control the metabolism and growth of mycobacteria in the granuloma. They are the DosR-DosS controlled dormancy regulon, resuscitation-promoting factors and toxin–antitoxin systems. The reasons that the cell-mediated immune response is unable to eliminate the mycobacteria have been mentioned above and include bacterial interference with MHC class II

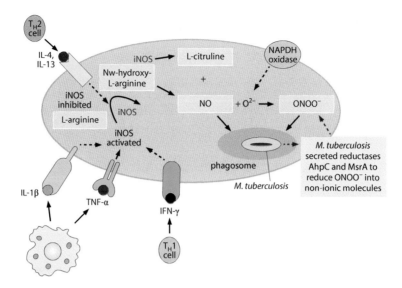

Figure 10.7 Synthesis, regulation and anti-mycobacterial action of NO and its evasion by *M. tuberculosis*. Cytokines TNF-α and IL-1-β, and IFN-γ, secreted by APCs and T cells respectively, activate the inducible form of NO synthase (iNOS) within macrophage phagosomes. iNOS catalyses the conversion of arginine to NO, which reacts with superoxide anion (O_2^-) to form peroxynitrite ($ONOO^-$), a powerful oxidant. This causes damage to bacterial DNA and kills *M. tuberculosis* within phagosomes. Cytokines released by T_H1 cells activate iNOS, whereas those from T_H2 cells inhibit it. *M. tuberculosis* has evolved mechanisms to resist nitrosative stress by using reductases to break down oxidizing agents such as peroxynitrite. (From Gupta, A. et al., *Immunobiology*, 217, 363–374, 2012.)

pathway step	possible interference
1. antigen processing	inhibition of proteolysis within phagosome
2. MHC Class II molecule formation	inhibition of gene expression (i.e., mRNA synthesis) for MHC Class II, including the normal up-regulation on activation with IFN-γ
3. MHC Class II-peptide co-location in late endosomes	disruption to assembly of MHC Class II complex, or their transport to endosomal vesicles
	prevention of co-localisation of MHC Class II with processed antigen
4. peptide loading	inhibition of peptide loading on to MHC Class II molecules in endosome
5. endosome transport to cell membrane	inhibition of endosomal transport of loaded peptide to cell membrane

Table 10.3 Antigen presentation pathway of MHC class II and its inhibition by *M. tuberculosis*. CLIP, class II-associated invariant chain peptides; LAM, lipoarabinomannan; RNA, ribonucleic acid. (From Gupta, A. et al., *Immunobiology*, 217, 363–374, 2012.)

expression, inhibition of macrophage activation in response to IFN-γ, induction of T_{REG} and induction of lipoxin A4, which is an anti-inflammatory mediator.

Bacteria within granulomas can reactivate to cause active disease termed reactivation tuberculosis. Several immune and mycobacterial mechanisms likely account for tuberculosis reactivation. Immune mechanisms include T-cell exhaustion, which is a well-recognized phenomenon seen in many chronic infections. T-cell exhaustion is a progressive loss of T-cell functions, culminating in cell death and reduced entry of effector T cells into the granuloma. Pathogen mechanisms include down-regulation of antigens that are the source of epitopes that are the targets of effector T cells and the presence of systems for recovery from dormancy.

The population structure of Mt in the lung, long considered to be clonal, is now known to exhibit significant genotype diversity within a host that may contribute to persistence and reactivation. It is possible that pioneer strains that establish in the lung in the early granuloma undergo micro-diversity in response to pressures exerted by the immune system and environment and generate phenotypes capable of reactivation.

Salmonella typhi serovar Typhi

The portal of entry of *Salmonella* is the Peyer's patches in the small intestine, although the bacterium can enter enterocytes as well. The bacteria are transcytosed by M cells that hand them off to resident DCs and macrophages that are located at the baso-lateral surface of M cells. Engagement of salmonella microbe-associated molecular patterns (MAMPS) by signalling PRRs results in the release of pro-inflammatory cytokines and chemokines that initiate an acute inflammatory response and recruit neutrophils, monocyte-macrophages and lymphocytes (**Figure 10.8**). However, the Vi capsule of salmonella may mask MAMP recognition by PRR while other MAMPS may be down-regulated. These mechanisms may assist the bacteria in preventing DC and macrophage activation. Salmonella has two type-3 secretion systems (T3SSs): T3SST-1, which is involved in facilitating entry into enterocytes, and T3SS-2, which is involved in the manipulation and arrest of phagosome maturation giving rise to what is termed a salmonella-containing vacuole. The fact that individuals infected with salmonella produce mucosal IgA and circulating IgG anti-salmonella antibodies and effector CD4$^+$and CD8$^+$ T cells indicates that fusion of phagosomes with lysosomes cannot be completely blocked in DCs because salmonella-derived peptides presented by MHC class I and class 2 molecules are required to produce effector T cells and plasma cells. Peyer's patches are secondary lymphoid organs that are inductive sites for mucosal IgA production, so DCs that phagocytose salmonella can present peptides in the context of MHC class II to naïve T cells to initiate a local mucosal IgA antibody response.

In the vast majority of individuals, salmonella is a self-limited infection confined to the lamina propria, however, a few individuals become long-term carriers of the bacterium. In chronically infected humans, the primary habitat of salmonella is the gall bladder, and there is evidence that the bacteria form biofilms on gall stones, which appears to be a predisposing factor for chronic biliary carriage. Salmonella may also inhabit the spleen and

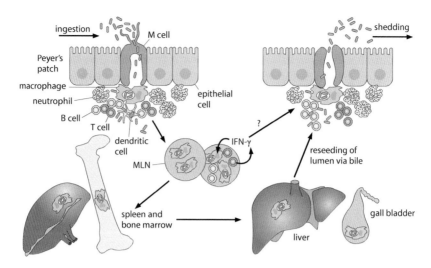

Figure 10.8 Persistent infection with *Salmonella enterica* serovar Typhi in humans. Bacteria are taken up by M cells of Peyer's patches. This is followed by inflammation and phagocytosis of bacteria by macrophages and neutrophils and recruitment of T and B cells. In systemic salmonellosis, such as typhoid fever, salmonella may target specific types of host cells, such as dendritic cells (DCs) and/or macrophages that favour dissemination through the lymphatics and blood stream to the mesenteric lymph nodes (MLNs) and to deeper tissues. Transport to the spleen, bone marrow, liver and gall bladder follow. Bacteria can persist in the MLNs, bone marrow and gall bladder for life; periodic reseeding of the mucosal surface *via* the bile ducts and/or the MLNs of the small intestine occur; and shedding can take place from the mucosal surface. Interferon-gamma (IFN-γ), secreted by CD4+T$_H$1, CD8+ T cells and NK cells has a role in maintaining persistence by controlling intracellular Salmonella replication. Interleukin (IL)-12, which can increase IFN-γ production, and the pro-inflammatory cytokine tumour-necrosis factor-alpha (TNF-α) also contribute to the control of persistent Salmonella. (From Monack, D.M. et al., *Nat. Rev. Microbiol.*, 2, 747–765, 2004.)

mesenteric lymph nodes. Salmonella in the gall bladder may be shed into the bile and, by this means, into the small intestine where they can repeat their infectious cycle, or they can be excreted in the faeces to infect new susceptible hosts. Little is known about the physiology of salmonella during chronic carriage (persistence). There is a suggestion that they may be dormant inside the salmonella-containing vacuole of macrophages. Virtually all research concerning persistence has been performed in constructed strains of mice with *Salmonella enterica* serovar Typhimurium, so whether data resulting from such studies are relevant to the persistence of *Salmonella enterica* serovar Typhi in humans remains to be determined.

PERSISTENT VIRUS INFECTIONS

Introduction

Persistent virus infections are of two types: (1) persistent lytic infections that show a low level of cell lysis, such as HIV and hepatitis, and (2) latent infections in which the virus exits the lytic cycle, persists in host cells, and does not produce new virions, such as herpesvirus. Persistent viruses share several properties such as parasitism of cells that allow long-term residence, perhaps by suppressing apoptosis, regulation of viral gene expression and evasion of the host immune system. Often the cellular residence of persistent viruses is different from the cells that are parasitized during the acute infection. The aim during acute infection is to produce a large number of virus particles in permissive cells so that the virus can disseminate to infect new susceptible hosts. Generally, persistent viruses are less transmissible than viruses causing acute infections and require tissue-to-tissue contact or the exchange of body fluids for transmission.

Herpesviruses

Herpesvirus are renowned for causing latent, persistent infections. Nine herpesvirus types infect humans and they are divided into α, β and γ herpesvirus groups: herpes simplex viruses 1 and 2, (HSV-1 and HSV-2), also known as HHV1 and HHV2; varicella-zoster virus (VZV), also known as HHV-3; Epstein-Barr virus (EBV) also known as HHV-4; human cytomegalovirus (HCMV) also known as HHV-5; human herpesvirus 6A and 6B (HHV-6A and HHV-6B); human herpesvirus 7 (HHV-7); and Kaposi's sarcoma-associated herpesvirus (KSHV), also known as HHV-8. The types of host cells infected during acute and persistent infections by these viruses are shown in **Table 10.4**.

In an active infection, herpesviruses replicate in the host cell nucleus but they do not produce virus particles during latency (persistent infection). The viral

subfamily	virus	primary target cell	site of latency	means of spread
Alphaherpesvirinae				
HHV-1	herpes simplex type 1	mucoepithelial cells	neuron	close contact (STD)
HHV-2	herpes simplex type 2	mucoepithelial cells	neuron	close contact (STD)
HHV-3	varicella-zoster virus	mucoepithelial and T cells	neuron	respiratory and close contact
Gammaherpesvirinae				
HHV-4	Epstein–Barr virus	B cells and epithelial cells	B cell	saliva (kissing disease)
HHV-8	kaposi sarcoma-related virus	lymphocytes and other cells	B cell	close contact (sexual), saliva?
Betaherpesvirinae				
HHV-5	cytomegalovirus	monocytes, granulocytes, lymphocytes, and epithelial cells	monocyte, myeloid stem cell, and ?	close contact (STD), transfusions, tissue transplant, and congenital
HHV-6	herpes lymphotropic virus	lymphocytes and ?	T cell and ?	saliva
HHV-7	HHV-7	like HHV-6	T cell and ?	saliva

Table 10.4 Properties distinguishing the herpesviruses. HHV, human herpesvirus; STD, sexually transmitted disease; "?", indicates that other cells may also be the primary target or site of latency. (From Murray, P. et al., *Medical Microbiology*, 8th ed., Elsevier, London, UK, p. 426, 2015.)

DNA exists as a circular episome associated with histones. An episome is DNA that is not integrated into the host cell chromosome. In addition to suppressing viral gene transcription, herpesviruses can prevent viral gene translation. CD8$^+$ T cells and type 1 and type 2 interferons are important in controlling herpesviruses, but these viruses can block interferon-induced inhibition of protein synthesis by the host cell. Herpesviruses can impair the production of virus-specific CD8$^+$ effector T cells by blocking the transporter associated with antigen processing (TAP) so that cytosolic viral peptides cannot enter the endoplasmic reticulum (ER) to be placed in the peptide binding groove of MHC class I molecules. In addition, herpesviruses can suppress MHC class I expression (**Figure 10.9**). Furthermore, those herpesviruses that are latent in neurons take advantage of the fact that these cells naturally express very low levels of MHC class I molecules.

EBV is latent in memory B cells and does not replicate but expresses a few proteins, one of which is the Epstein-Barr nuclear antigen 1 (EBNA-1). The purpose of EBNA-1 is to inhibit the function of the proteasome.

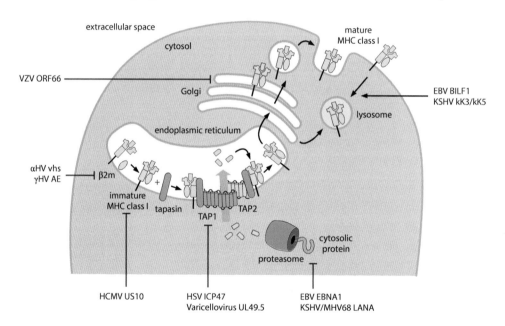

Figure 10.9 Herpesviruses encode a large variety of proteins that interfere with major histocompatibility (MHC) I antigen presentation to prevent elimination of the infected cell by cytotoxic T lymphocytes (CTLs). The Epstein-Barr Virus (EBV) EBNA1 and Kaposi's sarcoma-associated herpesvirus (KSHV) LANA avoid their own proteasomal degradation, thus limiting the generation of antigenic peptides. Viral host shut-off proteins, such as the UL41 homologues of alpha herpesviruses (αHV) like HSV, and the alkaline exonucleases of the gamma herpesviruses (γHV) EBV (BGLF5), and KSHV (SOX), degrade host mRNAs, thereby reducing protein levels of α Chain and β-2 microglobulin. The human cytomegalovirus (HCMV) US10, and HCMV US2 and US11 direct immature MHC class I molecules to the cytosol where they are degraded. Transport of peptides through TAP is inhibited by herpes simplex virus (HSV) ICP47, varicellovirus UL49.5 homologues, and HCMV US6. HCMV US3 interferes with the function of tapasin, thereby preventing proper peptide loading of MHC class I. The varicella-zoster virus (VZV) ORF66 retains mature MHC class I complexes in the cis/medial Golgi. EBV BILF1 and KSHV kK3 and kK5 increase the endocytosis and lysosomal degradation of cell surface MHC class I molecules. (From Curr. Opin. Immunol., 23, 96–103, 2011. Modified to display only human herpesviruses.

The proteasome serves to breakdown non-functional or misfolded proteins and is the mechanism by which proteins produced by cytosolic pathogens are broken down to peptides that can then be loaded onto MHC class I molecules and presented to CD8$^+$ T cells. Rarely, EBV-infected memory B cells become malignant and, in these malignant B cells, EBV down-regulates TAP-1 and TAP-2 so that cytosolic peptides cannot be transported into the ER to be loaded onto newly assembled MHC class I molecules. As the result of these attacks on the cytosolic pathway of antigen-presentation, virus-infected cells are able to evade effector CD8$^+$ T cells.

HCMV that is latent in monocytes and myeloid stem cells produces a homologue of IL-10 termed cmvIL-10. IL-10 is an immunoregulatory cytokine that down-regulates inflammation and immune responses by modulating the pro-inflammatory cytokines IFN-γ, IL-1β, GM-CSF, IL-6 and TNFα. In addition, IL-10 down-regulates MHC class I and class II molecules while up-regulating human leukocyte antigen (HLA)-G. (HLA)-G is a non-classical MHC class I molecule that binds to the killer-inhibitory receptors 1 and 2 (KIR1 and KIR2) expressed on the cell membrane of NK cells. In this way, virus-infected cells evade both CD8$^+$ cytotoxic T cells and NK cells.

HHV6A, HHV6B and HHV7 are lymphotrophic viruses that infect and replicate in activated CD4$^+$ T cells. HHV-6A and B establish latency in monocyte/macrophages and HHV6B may persist in epithelial cells in tonsillar crypts. HHV6A and HHV6B may become latent in the brain. As with other herpesviruses, HHV6 evades NK cell–mediated cytolysis. It does so by down-regulating the stress-induced molecules ULBP1, ULBP3 and MICB displayed by the infected cells. These molecules are recognized by NKG2D and B7-H6 activating receptors on NK cells. By this means, HHV6A, 6B and 7 are able to divert MHC class I molecules from their destination on the cell membrane of the infected cell to an endo-lysosomal compartment.

One quarter of the proteins encoded by HHV8 are involved in the regulation of various aspects of the human immune response to the virus. However, which of these play a role in establishing and maintaining latent infection is unclear. Major viral transcripts associated with latency are the latency-associated nuclear antigen (LANA), v-cyclin, v-FLIP, kaposin and 17 microRNAs (miRNAs). The main function of LANA is to maintain the viral episome; however, it has other important cellular regulatory functions as well. HHV8 uses v-cyclin and v-FLIP to suppress apoptosis and autophagy. Autophagy is the mechanism by which cells destroy damaged or redundant cellular components. HHV8 encodes a viral homologue of cellular Bcl-2 termed vBcl-2 to inhibit autophagy. Inhibition of autophagy by vBcl-2 is important in maintaining latent infections. HHV8 encodes four viral interferon regulatory factors (vIRF1–4) but only two (vIRF1 and vIRF3) have been detected in latently infected cells. HHV8 produces a viral version of human IL-6 (vIL-6) that is important at several stages of the HHV8 life cycle. Acting *via* autocrine and intracrine mechanisms vIL-6 may support the growth and survival of latently infected cells. Similar to other human herpesviruses, HHV8 has mechanisms by which to prevent MHC class I molecules displaying viral peptides on the surface of infected cells for presentation to effector CD8⁺ cytotoxic T cells. HHV8 encodes two proteins termed modulators of immune recognition (MIR) 1 and MIR2. These proteins interact with MHC class I molecules and target them for endocytosis and breakdown *via* the proteasome. In addition, HHV8 can inhibit MHC class I transcription. To circumvent NK cells targeting HHV8-infected cells that are missing MHC class I molecules, the virus down-regulates the NKG2D activating receptor.

Herpesviruses are able to modulate the role of CD4⁺ T cells in anti-viral immunity. The role of CD4⁺ T cells in anti-viral immunity extends beyond helping B cells make class-switched high-affinity anti-virus antibodies and in helping produce effector CD 8⁺ T cells and memory CD 8⁺ T cells. Many CD4⁺ T cells have cytotoxic functions and other direct antiviral activities (**Figure 10.10**). One way to inactivate CD4⁺ T cells is to interfere with the

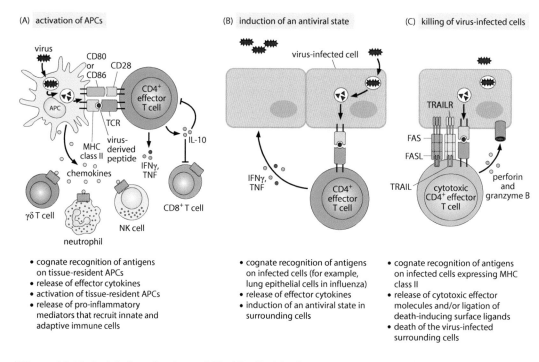

Figure 10.10 Antiviral mechanisms of CD4⁺ T cells. (A) After migrating to sites of infection, effector CD4⁺ T cells that recognize antigens on antigen-presenting cells (APCs) produce cytokines that help determine the character of the inflammatory responses in the tissue. Some cytokines of effector CD4⁺ T cells, such as interleukin-10 (IL-10), dampen inflammation and regulate immunopathology, whereas others, such as interferon-gamma (IFN-γ), are pro-inflammatory and activate macrophages, which in turn drive further inflammation. The production of IL-10 by effector CD4⁺ T cells can have a significant impact on the outcome of a viral infection. (B) IFN-γ, tumour necrosis factor (TNF) and other cytokines produced by CD4⁺ T cells help to coordinate an antiviral state in infected tissues. (C) Cytotoxic CD4⁺ T cells can directly lyse infected cells by FAS-dependent and perforin-dependent killing. FASL, FAS ligand; NK, natural killer; TCR, T cell receptor; TRAIL, TNF-related apoptosis-inducing ligand; TRAILR, TRAIL receptor. (From Swain, S.L. et al., *Nat. Rev. Immunol.*, 12, 136–148, 2012.)

exogenous pathway of antigen presentation that involves loading peptides into MHC class II molecules.

As we have seen the class II MHC transactivator (CIITA) is essential for transcriptional activity of the HLA class II promoter and is a target for viruses that modulate MHC class II. Herpesviruses target the CIITA promoter, for example, the EBV-encoded IE protein, BZLF1, vIRF3 from KSHV and VZV can reduce expression of MHC class II. IFN-γ, by increasing transcription of CIITA, can drive the expression of MHC class II by non-antigen-presenting cells. However, MHC class II expression can be inhibited by HCMV by suppression of HLA-DR synthesis. HLA-DR is central to the correct loading of peptides into nascent MHC class II molecules in late endosomes/phagosomes because it removes CLIP from the peptide-binding groove and selects appropriate peptides for loading.

When MHC class II molecules assemble in the ER, a polypeptide known as the invariant chain is placed in the peptide-binding groove to prevent peptide loading in this compartment. Suppression of the invariant polypeptide chain impairs loading of peptides into MHC class II in late endosomes/phagosomes. HSV-1 and BZLF1, an immediate-early viral gene of EBV, suppress the level of invariant chain.

Herpesviruses are able to modulate MHC class II:peptide on the cell membrane. For example, the BZLF2 protein of EBV blocks the interaction between the MHC class II molecules and the T-cell receptor (TCR). Alpha- and gamma-herpesviruses can affect both the endocytic and cytosolic pathways of antigen presentation by virtue of their ability to shut down global host protein synthesis following entry into the lytic cycle. Furthermore, many herpesvirus such as HCMV and EBV encode viral homologues of IL-10 (vIL-10) that can down-regulate antigen presentation.

Hepatitis B, C and D viruses

Hepatitis B virus (HBV) is an enveloped DNA virus; hepatitis C virus (HCV) is a positive strand, enveloped RNA virus; and hepatitis D (HDV) is a spherical RNA-enveloped viroid. HDV can replicate only in the presence of HBV, so it is either transmitted with HBV as a co-infection or acquired by an HBV-infected individual. HBV, HCV and HVD cause acute and chronic inflammatory liver disease that may result in hepatocellular carcinoma. There are a number of similarities between the immune evasion strategies of HBV and HCV but the kinetics of replication and chronicity of these viruses are quite different.

The vast majority of HBV infections in adults are self-limiting. In infected cells, HBV exists as a type of episomal DNA termed covalently closed circular DNA (cccDNA). There is a long lag phase before HBV begins exponential DNA replication in the liver and almost all hepatocytes are infected. The slow kinetics of virus replication likely allows a cellular immune response to develop in advance of it. In addition, HBV-infected cells express a high level of viral antigens, particularly the HBV core antigen (HBcAg) and peptides derived from the core antigen are presented by MHC class I molecules displayed on the hepatocyte cell membrane to effector CD8⁺ T cells. The effective display of viral peptides in the context of MHC class I molecules may account for the efficiency with which the immune system clears the virus in almost all cases. Mutation of HBV antigens leads to escape from antibodies and T cells, and both evasion strategies are observed in chronic HBV infection. In chronic HBV infection, there is evidence of T-cell deletion, anergy, exhaustion, and immunological ignorance. Transgenic mouse models have revealed two possible tolerogenic HBV proteins – pre-core and HBe antigen (HBeAg) – that may play a role in chronic infection by anergizing or deleting HBcAg/HBeAg cross-reactive T cells. Moreover, the blood of chronically infected individuals contains an extremely high level of the hepatitis B surface antigen (HBsAg), which could function as a high-dose tolerogen that could suppress the number of HBsAg-specific CD8⁺ T cells. Type 1 and type 2 interferons play an important role in the non-cytolytic elimination of HBV. Occult HBV infection (OBI) is an unusual presentation characterized by the absence of HBsAg and HBV DNA in blood, although HBV DNA is present in hepatocytes. Individuals with OBI show a long-lasting cell-mediated immune response with highly effective, broadly reactive HBV-reactive T cells and the formation of memory T cells.

In contrast to HBV infections, over 70% of adult-onset HCV infections are persistent. HCV infection induces both humoral and cell-mediated immunity (**Figure 10.11**). However, the appearance of neutralizing antibodies and effector T cells are delayed possibly as the result of impaired priming of naïve B cells and T cells. The neutralizing antibody response is essentially limited to the IgG1 subclass and can take as long as 20 weeks to develop. There is evidence that HCV can alter B cell function and development.

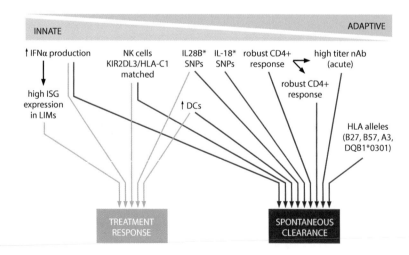

Figure 10.11 Immune response to hepatitis C infection. (From Terilli, R.R. and Cox, A.L., *Curr. HIV/AIDS Rep.*, 10, 51–58, 2013.)

The importance of neutralizing antibody in resolving HCV infection is a matter of conjecture because the virus can be eliminated before the onset of an antibody response and individuals with agamma-globulinemia are able to control the virus. Nevertheless, neutralizing antibodies do appear to be associated with the clearance of virus. In some cases, both CD4+ and CD8+ T cell responses to HCV may not appear for as long as 12 weeks after infection, although this delay does not appear to affect clearance of the virus. CD4+ T cells are required to eliminate the virus likely, because CD4+ T cells are required to activate cytotoxic CD8+ T cells that have received sub-optimal co-stimulation, as appears to be the case in HCV infection. Effector CD4+ T cells provide additional co-stimulation and IL-2 to CD8+ T cells. CD4+ T cells are also required for the generation of memory CD8+ T cells. Therefore, it is not surprising that, without help provided by CD4+ T cells, cytotoxic CD8+ T cells decline and this is associated with progression to chronic infection. HCV undergoes mutation during acute and chronic infection but effector T cells with TCRs specific for new HCV epitopes do not appear after the first 6 months of infection, suggesting arrest of T-cell development or T cell exhaustion. Another aspect of T cell biology in HCV infection is the presence of inhibitory receptors such as PD-1 (programmed death-1) on the membrane of HCV-specific CD8+ T cells. Mechanisms employed by HCV to evade the adaptive immune response are shown in **Figure 10.12**.

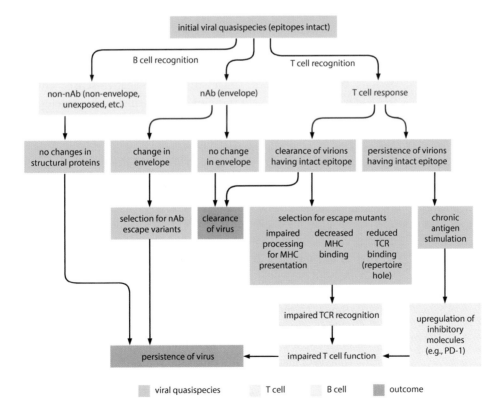

Figure 10.12 Viral evasion from the adaptive immune response. Immune pressure lead to clearance or select for mutations that successfully evade the immune response. nAb, neutralizing antibody; TCR, T cell receptor. (From Burke, K.P. and Cox, A.L., *Immunol. Res.*, 47, 216–227, 2010.)

HDV is a defective RNA virus that depends on HBVsAg to package its genome. There are two modes of infection. The first mode is co-infection of an HBV-negative individual with both HDV and HBV. Co-infection generally leads to resolution of both viruses. The second mode is superinfection of a chronic HBV carrier with HDV. In this case, most individuals continue to chronic HDV infection and hepatitis. Chronic infection with HDV intensifies HBV-related liver damage leading ultimately to cirrhosis of the liver and hepatocellular carcinoma. Cytotoxic CD8$^+$ T cells are the principal effectors of anti-viral immunity by lysing HDV-infected cells. The limited number of HDV epitopes available for presentation by MHC class 1 molecules likely impairs cytotoxic T cell activity. Cytotoxic CD8$^+$ T cells are a two-edged sword for although they kill virus-infected hepatocytes, this results in liver damage. Consistent with previous examples of cell-mediated anti-virus immunity, CD4$^+$ T$_H$1 T cells are required to support the production and maintenance of effector CD8$^+$ T cells and formation of memory CD8$^+$ T cells by providing IL-2 and additional co-stimulation. In common with other viruses that establish persistent infections, HDV can subvert type 1 interferons. HDV inhibits the activation of the IFN-α signalling pathways by interfering with the early steps of signal transduction. There are two forms of the HDV antigen (HDAg), termed large and small. They differ by only 19 amino acids but play different roles in the virus life cycle. Both modulate many aspects of the host cell, including apoptosis and cell growth, and may play a role in malignancy.

Measles virus

Measles virus (MV) is a single-stranded, negative-sense, enveloped RNA virus that encodes eight proteins, two of which – the envelope hemagglutinin (H) and fusion (F) protein – are ligands for CD46 expressed on epithelial cells and signalling lymphocyte-activation molecule (SLAM) receptors on the membrane of immune cells and, as such, are targets of neutralizing antibodies that can block infection. In most cases, MV causes a highly contagious acute infection that resolves in about two-thirds of susceptible persons in about 7–10 days after the late prodrome. Infection leads to life-long immunity. However, the MV suppresses all compartments of the immune response for 4–6 weeks during which time, in the remaining one-third of infected patients, the course is complicated by secondary infections such as pneumonia, encephalitis and otitis media that may lead to permanent hearing loss. Respiratory droplets transmit measles horizontally. The virus infects respiratory epithelial cells, but many types of cells are permissive, including monocyte/macrophages, lymphocytes, DCs, endothelial cells, neurons and astrocytes. MV infection leads to fusion of infected cells to form syncytia or giant cells and cytolysis. MV traffics from the respiratory tract to draining secondary lymphoid tissues within alveolar macrophages and DCs and from there to the blood. From the blood, the virus spreads to additional lymphoid tissues, and to epithelial and endothelial cells in various organs including the liver, brain, and skin. At a very low frequency, MV can become latent in the brain and reactivation causes subacute sclerosing panencephalitis (SSPE). Whether MV can persist at other sites in the body is unknown. Mechanisms by which MV becomes persistent are not fully understood but include control of interferons, receptor down-regulation and redistribution and limiting virus replication to levels below that which can be detected by cytosolic signalling PRRs (CSPPRs) such as RIG-1 and MDA5. Another mechanism of avoiding CSPPR is the inhibition of the phosphatase PP1 that prevents activation of RIG-I and MDA5.

Adenoviruses

Adenoviruses are non-enveloped double-stranded DNA (dsDNA) viruses. Types 1–7 cause most human adenovirus infections and include respiratory tract infections, pharyngoconjunctivitis, haemorrhagic cystitis and gastroenteritis. The virus causes lytic and latent infections. Human adenoviruses (HAdV) are transmitted by aerosol, close contact and by the faecal–oral route. They show tropism for mucosal epithelial cells but persist in lymphoid tissues such as tonsils, adenoids and Peyer's patches. Persistent shedding of the virus continues for a long period after apparent resolution of the primary infection. A limitation of the study of HAdV persistence has been the use of transformed human tumour cell lines, and it has been unclear whether findings obtained from these cells are extrapolatable to normal human cells. Recently, the mechanisms by which HAdV achieves persistence in non-transformed human cells have begun to be understood. As with other viruses, type 1 and type 2 interferons are important in controlling the primary infection and HAdV is able to subvert interferon signalling by the infected cell, which allows the virus to express the immediate-early gene *E1A*. *E1A* is required for expression of other immediate-early genes and reprogramming the cell to facilitate infection. E1A contains a repressive element termed RB that is responsible for repressing IFNs. Persistent HAdV infection is characterized by low virus expression, which likely limits viral peptide expression by MHC class I molecules to avoid recognition by cytotoxic CD8$^+$ T cells.

Human papilloma viruses

Human papilloma viruses (HPVs) are naked capsid DNA viruses that are transmitted vertically and horizontally by direct contact. There are over 100 HPV types that are divided into 16 groups. Some of these groups have tropism for skin, whereas others have tropism for mucosa. HPVs cause both benign epithelial proliferation (warts, oral, conjunctival and genital papilloma) and some types (HPV-16 and HPV-18) cause cervical carcinoma. HPV infects the basal layer of epithelia and virus replication drives division of the epithelial cells. Virus assembly occurs concomitantly with epithelial cell differentiation as the cells move from the basal layer to the epithelium surface where the virus is shed. HPV DNA persists in long-lived basal epithelial cells that are probably long-lived stem cells (**Figure 10.13**). As with other persistent viruses, there is little viral gene expression so that few viral peptides

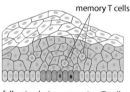

1. INFECTION

wound

long-lived stem/ stem-like cell   PV particle binds to basal lamina

2. LESION FORMATION

virus infects long-lived stem/ stem-like cell

DIVISION
cell division during wound repair populates basal layer with transiently amplifying (TA) cells containing HPV genomes

3. ESTABLISHED LESION

(A) FLAT WART; little or no viral gene expression in basal layer

TA   TA

slow division of infected stem/ stem-like cell maintains lesion   genome maintained in transiently amplifying (TA) basal cells prior to asymmetric cell division and differentiation

(B) NEOPLASIA (LSIL); viral gene expression and basal cell proliferation

expression of HIGH-RISK E6/E7 proteins increases rate of cell proliferation in basal and parabasal cells

4. IMMUNE REGRESSION AND LATENCY

memory T cells

following lesion regression, T-cells surveillance leads to suppression of viral gene expression and low level viral genome maintenance, possibly in the absence of viral proteins

Figure 10.13 Suggested mechanism for latent papilloma virus infection. (1) Access of viral particles to the basal lamina following a wound or micro-wound leading to infection of a long-lived basal cell such as a stem or 'stem-like' cell in which the viral genome can persist for months or years. (2) If viral titres are sufficiently high, then infection and viral gene expression can facilitate cell proliferation (red nuclei) and lesion-formation, possibly enhanced by the wound-healing environment, which favours cell division during wound repair. (3A) Once the basal layer has reformed, viral genomes may persist in a stem or 'stem-like' cell with little stimulation of cell proliferation. These cells divide occasionally to produce infected 'transiently amplifying' (TA) cells that populate the differentiating cell layers. In flat warts caused by low-risk human papilloma virus (HPV) types, the viral E6 and E7 proteins drive cell-cycle re-entry and genome amplification in the upper epithelial layers (green). (3B) Following high-risk HPV infection, E6/E7 expression in the basal and parabasal cell layers may persist, leading to the development of neoplasia. Such lesions can support late gene expression (green) and genome amplification (light blue) to a similar extent as flat warts, and may progress to high grade neoplasia (high-grade squamous intraepithelial lesion [HSIL]) and cancer. (4) That parasitophorous vacuole (PV) genomes can persist in the basal layer with limited viral gene expression means that latency may be the default state following lesion clearance as a result of a cell-mediated immune response. In such individuals, memory T cells present in the epithelium are thought to suppress viral gene expression, but they may not effectively eliminate the genome-containing basal cells that express viral proteins at a very low level.

can be loaded onto MHC class I molecules to display to HPV-specific CD8$^+$ cytotoxic T cells. HPV proteins E6 and E7 play an important role in subverting the host immune response. They are able to dysregulate DNA methylation, a mechanism used by host cells to arrest viral gene transcription and histone modification, which, for example, represses TLR9 transcription and NF-κB to shut down secretion of pro-inflammatory cytokines by infected cells. HPV16 E7 is able to up-regulate indoleamine 2,3-dioxygenase 1 (IDO1) expression in dermal DCs. IDO is immunosuppressive and drives the production of induced regulatory T cells. HPV E5 binds to MHC-I, CD1d, and the MHC class II associates invariant chain in the Golgi and ER that prevents their expression on the cell surface (**Figure 10.14**). In this way CD8$^+$, CD4$^+$ and NK T cells are rendered ineffective. Importantly, E6 and E7 interfere with tumour suppressor proteins p53 and Rb, leading to transformation to cancer cells. For this reason, E6 and E7 are known as oncoproteins.

Human polyomaviruses

BK virus (BKV) and JC virus (JCV) are omnipresent but rarely cause infections in immunologically intact individuals. The viruses are non-enveloped and have a small, circular, dsDNA genome. They are acquired during childhood, but the route of transmission is unknown. The main sites of persistence are the genitourinary tract, the central nervous system, and the hematopoietic system. In order to persist, the viruses minimize gene expression to prevent viral peptides from being loaded onto MHC class I molecules that would alert cytotoxic CD8$^+$ T cells. Very little is

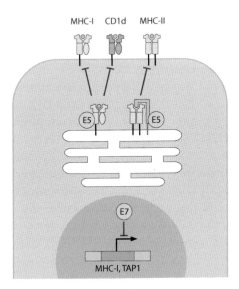

Figure 10.14 Human papilloma virus (HPV) E5 and E7 prevent surface expression of major histocompatibility (MHC) molecules. HPV E5 binds to MHC-I, CD1d, and the invariant chain of MHC-II in the Golgi and endoplasmic reticulum (ER), preventing their trafficking to the cell surface. High-risk HPV E7 represses transcription of MHC-I genes. (From Westrich, J.A. et al., *Virus Res.*, 231, 21–33, 2017.)

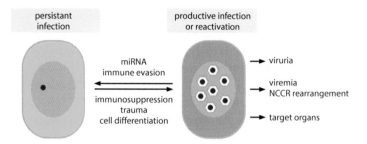

Figure 10.15 Polyomavirus persistence and reactivation. In healthy hosts, polyomaviruses persist in the host cell nucleus as episomal DNA. Viral microRNA (miRNA) and immune evasion strategies may contribute to persistence. Under certain immunosuppressed, trauma, or cell differentiation conditions, polyomaviruses can reactivate and undergo productive replication. Reactivation can ultimately lead to viruria, viraemia, potential noncoding control region (NCCR) rearrangement, as well as dissemination of the virus from persistent sites to other target organs. (From Imperiale, M.J. and Jiang, M., *Annu. Rev. Virol.*, 3, 517–532, 2016.)

known about the mechanisms of persistence of human polyoma viruses (**Figure 10.15**), and much of what is known has come from studies of mouse polyoma virus and simian virus 40 (SV40).

Human immunodeficiency virus

Human immunodeficiency virus (HIV) is a retrovirus that is a member of the genus *Lentivirus*. The virus is an enveloped, positive-strand RNA virus that has a reverse transcriptase, which allows the virus to make a DNA copy of the viral genome that integrates into the genome of the host cell. There are two types of HIV, HIV-1 and HIV-2. HIV-2 is largely limited to West Africa. Transmission occurs both vertically and horizontally. In horizontal transmission, HIV breaches the barrier epithelia *via* the use of infected needles by which the virus is introduced directly into the bloodstream, *via* inflammation or ulcers in the rectal or oral mucosa or *via* uptake by the specialized antigen sampling cells, M cells, of the mucosa-associated lymphoid tissues. The virus infects CD4$^+$ T cells, macrophages and DCs. Engagement of the primary host cell membrane receptor CD4, and one of several co-receptors, primarily CCR5 or CXCR4, allows fusion of the viral envelope with the host cell membrane. DCs can also capture the virus using the endocytic receptor DC-SIGN. After cell entry, the virus un-coats in the cytoplasm and the RNA genome of the virus is reverse transcribed to create a double-stranded cDNA that traffics to the nucleus and integrates into the infected cell genome at various sites. Initially, the virus utilizes the co-receptor CCR5 that allows tropism for DCs, macrophages and CD4$^+$ T cells, whereas later in infection, the virus utilizes CXC4 as co-receptor, limiting infection to CD4$^+$ T cells. The virus requires an activated T cell to replicate likely because, in a resting cell, reverse transcription and nuclear import are inefficient and the virus genome does not integrate. In resting T cells, HIV exists as a provirus. Despite robust innate and acquired immune responses against HIV, the virus resists elimination. The high rate of mutation of retroviruses gives rise to escape mutants with new epitopes that evade existing neutralizing anti-virus antibodies and cytotoxic CD8$^+$ T cells. The strategy of down-regulating MHC class I molecules is also employed by HIV. In this case, the virus interferes with the traffic of MHC class I molecules to the cell surface and down-regulates expression of CD4. To avoid NK cells, HIV down-regulates activating receptors on the infected cell surface, leaving a preponderance of inhibitory receptors. HIV is able to induce apoptosis of immune effector cells without affecting the cells in which it resides.

The introduction of highly active anti-recto viral therapy (HAART), which consists of cocktails of drugs each with a different mechanism of action (target) to prevent the emergence of drug resistance by the virus, has revolutionized the clinical outcome of infection by HIV. However, HAART cannot eliminate latent HIV, and replication-competent provirus exists within long-lived memory CD4$^+$ T cells that undergo clonal expansion. That this is the case is supported by the finding that virus isolated after initiation of HAART is oligoclonal and that replication-competent provirus is integrated in the same position in the genome of these CD4$^+$ T cells. Furthermore, these T cells can proliferate without producing virus, only to do so once the T cells are stimulated. Only a small proportion of latently infected CD4$^+$ T cells contain replication-competent provirus, most contain defective provirus. T_H1 CD4$^+$ T cells contain a greater proportion of replication-competent provirus than do other CD4$^+$ T cell subsets. Several mechanisms contribute to maintaining HIV latency, including viral integration sites, chromatin environment, down-regulated transcription factors, impaired activity of Tat, and cellular miRNA interference with viral protein translation.

Human T-cell lymphotropic virus type 1

Human T-cell lymphotropic virus type 1 (HTLV-1) is a retrovirus that is transmitted vertically and horizontally. Horizontal transmission occurs by sexual contact or intravenous injection with needles of syringes containing infected blood. HTLV-1 produces persistent infection and, rarely, causes adult T-cell leukaemia, HTLV-associated myelopathy/tropical spastic paraparesis, uveitis and infective dermatitis. HTLV-1 shares features of its life cycle and immune evasion strategies with HIV. HTLV-1 can infect T_H1, T_H2 and T_{REG} T cell subsets. Once the provirus is integrated into the genome of the infected T cell, HTLV-1 replicates as the result of clonal expansion of these cells rather than by continuing to infect additional T cells. Neutralizing antibodies and NK cells do not appear to contribute to HTLV-1 immunity and the expression of MHC class 1 molecules is not markedly reduced on infected T cells. The ineffectiveness of NK cells likely reflects the absence of activating receptors on infected cells. Cytotoxic CD8$^+$ T cells are the principal effector cells involved in HTLV-1 immunity. Nevertheless, cytotoxic T cells are unable to clear the virus seemingly because virus-infected T cells have low immunogenicity because no viral genes are transcribed except for HTLV-1 basic leucine zipper factor (HBZ). *HBZ* reduces the immunogenicity of infected T cells by inhibiting IFN-γ production, a pro-inflammatory cytokine, and increasing sensitivity to TGF-β, a regulatory cytokine. Furthermore, *HBZ* enhances expression of the transcription factor FoxP3 that drives T cells to become regulatory cells and promotes T-cell proliferation. In addition to *HBZ*, the viral gene *TAX* has an important role in viral persistence and pathogenesis. *TAX* activates many cell-signalling pathways, reprograms the cell cycle, interferes with checkpoint control and inhibits DNA repair.

PERSISTENT PARASITE INFECTIONS

Introduction

Similar to persistent viruses, parasites regulate the immune response by inducing T_{REG} and by inducing the production of IL-10 and TGF-β, two immunosuppressive cytokines. However, parasites also modulate B cells, macrophages, and innate immunity, and helminths alter the intestinal microbiota and its metabolites. Infections with *Ascaris lumbricoides*, *Trichuris trichiura*, hookworm and schistosomiasis appear to ameliorate allergy. Live helminths have been ingested in clinical trials to determine whether they can mitigate allergy and autoimmunity.

Helminths

Helminths are multicellular worms that include cestodes, nematodes and trematodes. Their life cycles are diverse and they enter their human host *via* the skin, mucosae or by employing the mosquito as an intermediate vector. As mentioned above a general property of helminths is their ability to down-regulate both the innate and adaptive human immune system. This down-regulation extends beyond helminth-specific immunity to general immune suppression.

Adaptive immunity to helminths is mediated by T_H2 CD4$^+$ T cells. The T_H2 pathway involves IgE antibodies, mast cells, basophils and eosinophils. These mediator-releasing cells have high-affinity Fc receptors for IgE. When cell membrane-bound IgE antibodies are cross linked by helminth antigens these cells rupture and release mediators the functions of which are contraction of smooth muscle and increased secretion of mucins in an attempt to dislodge parasites attached to the mucosal epithelium. The cytokine IL-4 is essential for the differentiation of naïve CD4$^+$ T cells to effector T_H2 CD4$^+$ T cells. IL-4 is produced by T_H2 CD4$^+$ T cells and mast cells and the cytokine reinforces T_H2 polarization. In addition, IL-4 is anti-inflammatory and suppresses production of the pro-inflammatory cytokines IL-1β and TNF-α thereby limiting inflammation. Furthermore, IL-4 directs the production of alternative macrophages termed type 2 macrophages that counteract inflammation by producing IL-10 and TGF-β (**Figure 10.16**). IL-4 drives class switching of B cells to produce both IgE and IgG4 antibodies. However,

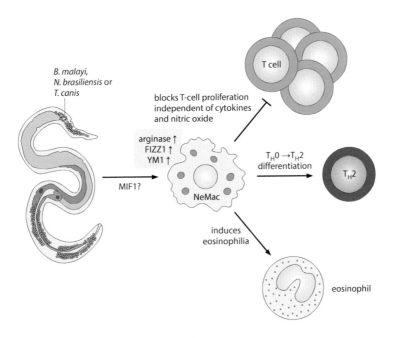

Figure 10.16 Role of alternatively activated macrophages in helminth infections. Helminth infections induce alternatively activated or type-2 macrophages (termed nematode-elicited macrophages, NeMacs, in this figure). Among their properties are production of IL-10, prostaglandins or nitric oxide, induction of T helper 2 (T$_H$2)-cell differentiation of naïve T-cells, attraction of eosinophils and production of arginase and the genes, *Yfi.4 f* and *FIZZ1*.

IL-10 inhibits class switching of IgE in favour of IgG4. IgG4 antibodies cannot activate mast cells, basophils and eosinophils, cannot activate the complement cascade or function to opsonize parasites. Thus, the relative levels of IgE and IgG4 reflect protection from, or susceptibility to infection, respectively. The roles of IgE and IgG4 in acute and chronic infection with helminths is shown in **Figure 10.17**. Consistent with other classes of pathogens discussed in this chapter helminths also interfere with antigen processing, modulation of antigen-presenting cells, and interference with and mimicking cytokines (**Figure 10.18**).

Plasmodium

The genus *Plasmodium* causes the disease malaria. Plasmodia are protozoan parasites of erythrocytes that require two hosts to complete their life cycle. Sexual reproduction occurs in the *Anopheles* mosquito and asexual reproduction occurs in humans or animals. Malaria in humans is caused by four species of the genus, namely, *P. falciparum, P. malariae, P. ovale, P. vivax* and, rarely, *P. knowlesi*. Persistence of the parasite in the vertebrate host increases the opportunity of the mosquito vector to acquire male and female gametocytes to spread the infection. After deposition in the dermis, sporozoites employ two proteins termed sporozoite microneme protein essential for cell traversal (SPECT)-1 and to facilitate passage through cells SPECT-2 on their way to the liver. The parasites are recognized by immune

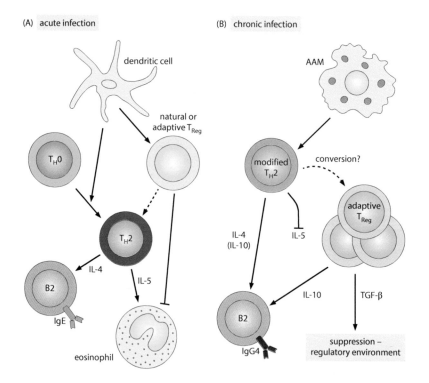

Figure 10.17 Roles of IgE and IgG4 in helminth infections. (A) Infection with a parasitic helminth leads to the activation of dendritic cells (DCs) in the peripheral tissues. Licensed DCs initially activate naïve T cells to become T helper 2 (T_H2) cells that drive B cells to produce IgE. (B) With increasing levels of parasite antigens, alternatively activated macrophages promote the development of modified T_H2 cells that drive production of IgG4 by B cells.

cells in the liver by the cytosolic PRR MDA5. The innate immune response involves DCs, macrophages, NK cells, NK T cells and γ:d T cells and cytokines such as TNF-α, IL-1β, IL-6, IL-13 and, particularly, IFN-γ. The adaptive immune system recognizes the malaria parasite but develops slowly and is unable to eliminate it and prevent reinfection. The acquired immune response is species- and strain-specific, so that responses to strains other than the dominant strain in mosquitoes take time to develop. Neutralizing antibodies are important in attempting to block sporozoites from entering hepatocytes. Cytotoxic CD8$^+$ T cells kill infected hepatocytes. Because erythrocytes do not express MHC class I molecules, the mechanism by which effector CD8$^+$ T cells kill infected erythrocytes is likely by Fas–FasL interaction given that infected erythrocytes are known to express Fas. Neutralizing antibodies are also important in attempting to block merozoites from entering erythrocytes by blocking red cell membrane receptors or opsonizing parasites for removal by phagocytosis.

Antigenic variation is the principal mechanism used by the parasite to evade the host immune response. *Plasmodium* exhibits a high level of genetic diversity and antigenic polymorphism. Variable antigens are expressed on the surface of the parasite or on the surface of the infected erythrocyte and are termed variant surface antigens (VSAs). VSAs are often species-specific.

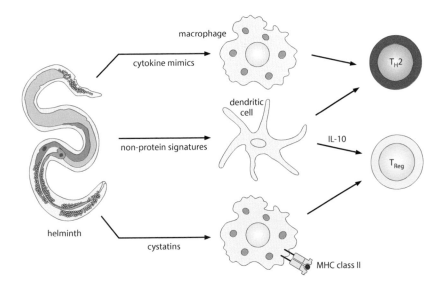

Figure 10.18 Parasite-derived immune modulators. Cytokine mimics, such as macrophage-migration inhibition factor (MIF) and transforming growth factor-13 (TGF-13) can modulate macrophage function and lead to the induction of T helper 2 (T_H2) cells and immune suppression. Non-protein signature molecules such as phosphatidylserine and phosphorylcholine and glycans can interact with dendrite cells and result in the induction of both T_H2 and regulatory T cells (T_{REG}). Helminths also produce prostaglandin D_2 (PGD_2), and PGE_2, which can have similar effects on accessory cells (not shown). Direct interference with antigen presentation by parasite-derived cystatins prevents T-cell activation both at the level of T-cell receptor engagement by MHC:peptide complexes by inhibiting proteases involved in antigen presentation, and by enhancing the production of interleukin-10 (IL-10) by accessory cells.

In addition to antigenic variation, there is a suggestion that the malaria parasite may be able to modulate the function of DCs, T cells and B cells, but this remains to be confirmed in human infections. Antigenic variation at different stages of the life cycle of *Plasmodium* is shown in **Figure 10.19**.

Leishmania

The genus *Leishmania* currently contains about 15 species that infect humans. They are obligate intracellular parasites that infect cells of the mononuclear phagocyte system. *Leishmania* species are transmitted from animals to humans and from human to human by the intermediate vector, the female sand fly. Promastigotes in the saliva of the sand fly are injected into the skin of the host and are engulfed by phagocytes in which they become amastigotes and replicate. Most infections are asymptomatic, but the parasite can cause skin ulcers and can spread from the skin to cause visceral or mucocutaneous leishmaniasis. NK cells, NK T cells and the complement cascade are important in clearing promasitgotes. Promastigotes employ several strategies to avoid the complement system such as preventing the insertion of the membrane attack complex (MAC), and inactivating C3, C5, and C9 by phosphorylation. T_H1 CD4$^+$ T cells and NK cells are central to controlling the infection once the parasites take up residence in macrophages. This is because both T_H1 CD4$^+$ T cells and NK cells are potent sources of IFN-γ, which up-regulates

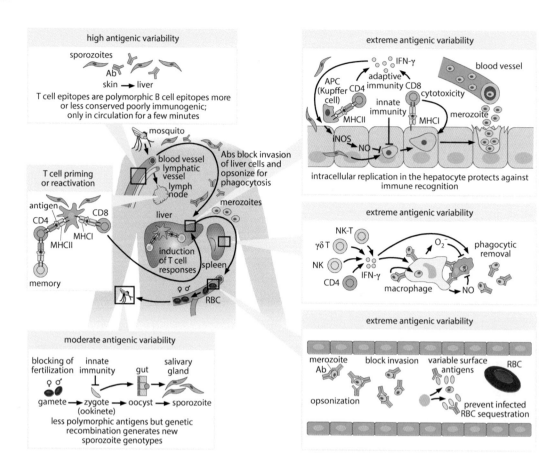

Figure 10.19 Antigenic variation in plasmodium. Sporozoites, injected into the skin by the biting mosquito, drain to the lymph nodes, where they prime T and B cells, or to the liver, where they invade hepatocytes. Antibodies (Ab) trap sporozoites in the skin or prevent their invasion of liver cells. Interferon-gamma (IFN-γ)–producing CD4$^+$ and CD8$^+$ T cells inhibit parasite development into merozoites inside the hepatocyte. However, this immune response is frequently insufficient, and merozoites emerging from the liver invade red blood cells, replicate, burst out of the infected erythrocyte and invade new erythrocytes. Merozoite-specific antibodies agglutinate and opsonize the parasite and can inhibit the invasion of red blood cells through receptor blockade. Antibodies to variant surface proteins also opsonize and agglutinate infected red blood cells (RBCs) (cytoadherence) and prevent their sequestration in small blood vessels. IFN-γ-producing lymphocytes activate macrophages and enhance the phagocytosis of opsonized merozoites and infected red blood cells (iRBCs). Complement-fixing antibodies to gametocyte and gamete antigens lyse parasites inside the mosquito gut or prevent the fertilization and development of the zygote. Sporozoite, liver-stage and gametocyte and gamete antigens are somewhat polymorphic, whereas merozoite antigens and variant surface antigens are highly polymorphic. APC, antigen-presenting cell. (From Riley, E.M. and Stewart, A.A., *Nat. Med.*, 19, 168–178, 2013.)

the antimicrobial activity of macrophages. *Leishmania* arrests fusion of the phagosome and lysosomes once engulfed by macrophages and also inhibits antigen presentation by inhibiting MHC class II:peptide loading and presentation on the cell membrane. *Leishmania* also impairs the function of the co-stimulatory molecules B7-1, B7-2, and CD40 and interferes with cell signalling cascades, affecting the production of reactive oxygen species (ROS) and the release of cytokines and chemokines. That the parasite number remains constant suggests that there is a balance between replication of the parasite and their destruction by the immune system.

Trypanosoma cruzi

The intermediate vector of *T. cruzi* (Tc) is the reduviid bug that carries the tryptomastigote of Tc in its faeces so that wounds created by biting and feeding of the bug are exposed to typtomastigotes. From the skin, the parasite migrates to various other tissues that include the heart, liver, brain secondary lymphoid organs, skeletal muscles, neurons in the intestine and the oesophagus as well as other tissues (**Figure 10.20**). The parasite invades host cells and becomes an amastigote that divides and eventually ruptures the infected cells and spreads to new uninfected cells. Infection with *T. cruzi* begins with an acute phase characterized by parasitaemia that is controlled by the immune

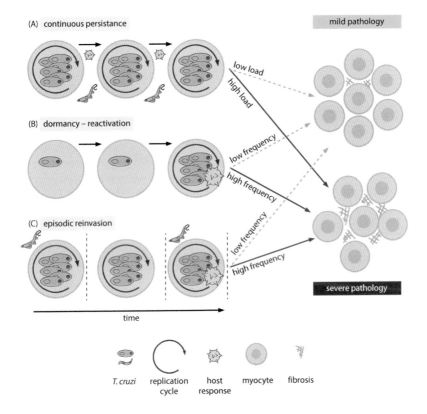

Figure 10.20 *T. cruzi* pathogenesis. In chronic infections *T. cruzi* predominantly parasitizes myocytes. These infected cells are typically scarce and focally distributed; they can be within cardiac, skeletal, or smooth muscle tissues, such as those from the vasculature or the gastrointestinal (GI) tract. The heart is the most common site of pathology. Parasite persistence within an individual host may occur through different modes: (A) Continuous persistence describes an ever-present, low-abundance parasite load that is sustained as a locally contained equilibrium between intracellular parasite replication and host immune responses. (B) Although they have not been proven to exist, dormant forms of *T. curzi* may reside within tissues and evade host immunity. As seen for other pathogens, reactivation into typical replication cycles could occur on an intermittent basis. (C) Due to the ability of *T. curzi* to invade multiple tissues and migrate between them, an organ may be subject to discrete episodes of infection by reinvasion. These three modes are not mutually exclusive and may overlap to different degrees at different times. Over time, the cumulative parasite load is likely to dictate the frequency and intensity of local inflammatory responses, which, depending on their quality, result in differing degrees of pathology. (From Lewis, M.D. and Kelly, J.M., *Trends Paristol.*, 32, 899–911, 2016.)

response. However, the immune response induced during the acute infection does not eliminate the parasite and chronic infection ensues. Roughly, two-thirds of individuals who progress to chronic infection are asymptomatic. However, in the remaining one-third of individuals, the chronic form of the disease affects the peripheral autonomous nervous system in the gastrointestinal tract and heart and the heart muscle. Chronic infection may give rise to autoimmune disease as the result of molecular mimicry. As with other chronic/persistent infections, the spectrum of disease depends on complex interactions between the parasite and the immune system. As mentioned previously, IFN-γ is important in the development of resistance to *T. cruzi*, as it is in many other intracellular infections. NK cells are likely the source of IFN-γ in the early stages of *T. cruzi* infection. DCs, macrophages, NK T cells, γ:δ T cells and B cells all play a role in immunity to *T. cruzi*.

Major mechanisms involved in *T. cruzi* survival and control during the initial phase of infection are shown in **Figure 10.21**. The ability of *T. cruzi* to evade host immunity may vary from strain to strain and reduviid bugs have been show to harbour different combinations of strains. The failure to clear the parasite may be related to the large repertoire of highly polymorphic and immunogenic surface proteins that are expressed by the parasite. *O*-glycosylated glycoproteins terminated with sialic acid are the principal molecules on the surface of the parasite and are encoded by over 800 genes. These glycoproteins are centrally involved in the parasite–host interaction and in modulation of host immunity. Immune evasion by *T. cruzi* tryptomastigotes

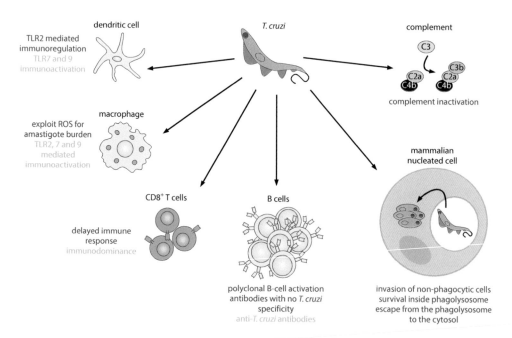

Figure 10.21 Major mechanisms involved in *T. cruzi* survival and control during the initial phase of infection. Mechanisms associated with the control of parasite load are highlighted in green, whereas those involved with parasite modulation of the host immune system and/or with increased parasite load are highlighted in red. (From Cardoso, M.S. et al., *Front. Immunol.*, 6, 659, 2015.)

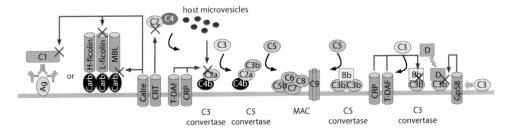

Figure 10.22 *T. cruzi* complement evasion mechanisms. To avoid lysis, *T. cruzi* relies on molecules, such as calreticulin and gp58/68 (Gp58), which block the initial steps of classic/lectin or alternative pathways, respectively, and CRIT, T-DAF, CRP, and host-derived microvesicles that disrupt or block C3 convertase assembly. Ag, antigen; Calre, calreticulin; Carb, carbohydrate. (From Cardoso, M.S. et al., *Front. Immunol.*, 6, 659, 2015.)

is directed towards subverting the complement cascade (**Figure 10.22**) and inhibiting professional phagocytes (amastigotes). *T. cruzi* avoids destruction by macrophages by escaping from the phagolysosome and entering the cytoplasm where it replicates. As seen with *Leishmania*, *T. cruzi* can affect transcription of cytokines secreted by infected macrophages and a *T. cruzi* cysteine-protease prevents macrophage expression of pro-inflammatory cytokines. *T. cruzi* is effective in suppressing both cellular and humoral immunity. *T. cruzi* induces anergy and clonal deletion of T cells. The parasite also induces polyclonal activation of B cells, which means that every B cell is activated, not just those specific for *T. cruzi* antigens. As the result of polyclonal activation, the antibody response is un-focused. Because *T. cruzi* can infect the thymus, it has been suggested that *T. cruzi* antigens may be involved in negative selection of *T. cruzi*-specific T cells or the generation of T_{REG}, both of which may impact the effectiveness of the adaptive immune response to the parasite.

Toxoplasma gondii

T. gondii is an intracellular parasite whose reservoir is the domestic cat. Cats acquire the parasite from eating infected rodents. Unsporulated oocysts are excreted in cat faeces and take between 1 and 5 days to sporulate in the environment. Humans are infected by hand-to-mouth transmission with oocysts from contaminated cat litter or soil, by drinking contaminated water or by eating under-cooked meat containing cysts. In addition, the parasite may be transmitted by blood transfusion, transplantation and across the placenta. Tachyzoites rupture the cytoplasmic membrane of infected cells and continue to infect nucleated cells until they are cleared by the immune system, leaving behind encysted bradyzoites, primarily in brain, eye, skeletal and cardiac muscle. The cyst wall is mainly host-derived, which masks the bradyzoites from the host immune response and is important in the ability of the parasite to persist in the host.

Once they cross the intestinal epithelium, tachyzoites are recognized by PRRs and endocytic receptors on DCs and macrophages. However, the parasite avoids the phagocytic (endocytic) pathway and enters the cell, forming a parasitophorous vacuole (PV) that does not fuse with lysosomes. Moreover, the parasite is also able to suppress maturation of DCs and macrophages. Therefore, these antigen-presenting cells are for the most part unable to process *T. gondii* antigens for presentation to naïve T cells. Nevertheless, ligation of *T. cruzi* microbe-associated molecular patterns by PRRs of DCs, macrophages and recruited neutrophils leads to the production of IL-12. IL-12 stimulates NK cells and NK T cells and later effector T_H1 CD4$^+$ T cells to produce IFN-γ. IFN-γ stimulation enables DCs to overcome suppression of antigen processing. In addition, IFN-γ inhibits intracellular parasite replication and may kill intracellular *T. gondii* by driving production of ROS and reactive nitrogen species (RNS) and by tryptophan depletion. Activated DCs migrate to draining lymph nodes to present *T. gondii* peptides to naïve T cells and drive them to become *T. gondii*-specific T_H1 CD4$^+$ T cells. However, in doing so, DCs may act as a Trojan horse carrying parasites to lymph nodes and aiding in rapid dissemination of the parasite. Considering the importance of IFN-γ in controlling *T. gondii*, it is not surprising that the parasite is able to subvert IFN-γ signalling, down-regulate iNOS and inhibit expression of MHC class I and class II on the infected cell membrane. As stated above, recognition of *T. cruzi* MAMPs by DC PRRs does, however, stimulate release of IL-12 by these antigen-presenting cells that, in turn, causes release of IFN-γ by NK cells and NK T cells. The parasite also triggers the release of TGF-β and IL-10, both of which are immunoregulatory cytokines. Therefore, there is a balance between pro- and anti-inflammatory host responses that may benefit both parasite and host. Although NK cells and T_H1 CD4$^+$ T cells are prominent in host resistance during the acute stages of the infection, CD8$^+$ T cells are the major producer of IFN-γ in chronic infection. *T. gondii*-peptides for loading onto MHC class I molecules for presentation to naïve CD8$^+$ T cells may be delivered to the cytosol by diffusion across the PV and/or by cross-presentation. In either case, effector *T. gondii*-specific CD8$^+$ cytotoxic T cells are induced. Whether cytotoxicity of CD8$^+$ T cells is as important as their production of IFN-γ is unclear.

Myeloid-derived suppressor cells in chronic infections

When neutrophils, monocyte/macrophages and DCs receive strong signals from ligation of MAMPs by PRRs, they become highly phagocytic, generate ROS and RNS (neutrophils and macrophages), secrete pro-inflammatory cytokines and up-regulate expression of MHC class I and class II and costimulatory molecules. However, during persistent infection, the signals received by myeloid cells are weak and chronic and delivered by growth factors and inflammatory mediators. Myeloid cells receiving such signals take on a suppressive phenotype. They are poorly phagocytic, produce elevated background levels of ROS and RNS and high levels of arginase, prostaglandin E_2 (PGE_2) and immunosuppressive cytokines such as IL-10 and TGF-β. These cells are termed myeloid-derived suppressor cells and have been mentioned previously in the section on *Helminths*.

The physiological role of suppressive myeloid cells is to return the immune system to rest once pathogens are eliminated by shutting down inflammation and effector T cells and generating T_{REG}. However, suppressor myeloid cells are implicated in the pathogenesis of chronic bacterial, viral, protozoa, helminth and fungal infections (**Table 10.5**). For example, monocytic

microbial organism	context of M-MDSC investigation	major outcome; immunosuppressive effect
viruses		
immunodeficiency virus [human immunodeficiency virus (HIV), simian immunodeficiency Virus, LP-BM5]	M-MDSC and total MDSC	host detrimental; suppress T-cell and B-cell responses, express inducible nitric oxide synthase (iNOS), and produce reactive oxygen species (ROS), ARG-1, IL-10, induce Treg
cytomegalovirus (CMV)	M-MDSC-like	host detrimental; impair T-cell expansion, slowing viral clearance
hepatitis C virus (HCV)	M-MDSC and total MDSC	host detrimental; suppress CD4 T-cell and NK cell function, increase Treg
hepatitis B virus (HBV)	M-MDSC and total MDSC	host detrimental; express IL-10, suppress T-cell function, promote disease chronicity
viral coinfection (HIV/CMV, HCV/HIV)		host detrimental; impair T-cell function, accelerate disease progression
bacteria		
Staphylococcus aureus	M-MDSC and PMN-MDSC	host detrimental; suppress T-cell function, express ARG-1, iNOS, IL-10, exacerbate disease, promote disease chronicity
Francisella tularensis	Total MDSC	host detrimental; reduced phagocytosis, reduced survival
Mycobacteria spp.	M-MDSC and total MDSC	host beneficial/detrimental; suppress T-cell function; express ARG-1 and iNOS, impaired pathogen killing; TNF-dependent suppression of CD4 T cells
Klebsiella pneumonia	M-MDSC and PMN-MDSC	host beneficial/detrimental; pro-resolving, express ARG-1, IL-10/impair phagocytosis/killing
Helicobacter pylori	M-MDSC	host detrimental; suppress protective TH1 development.
Polymicrobial sepsis	M-MDSC and total MDSC	host beneficial/detrimental; suppress T-cell function, express nitric oxide and pro-inflammatory cytokines (early) and ARG-1, IL-10, and TGF-β (late)
Escherichia coli	M-MDSC	host detrimental; suppress T-cell activation, innate immunity, impair bacterial uptake and increase disease severity, infection susceptibility
protozoa		
Leishmania spp.	M-MDSC and total MDSC	host beneficial/detrimental; species-specificity, suppress CD4 T-cell proliferation, improved killing of parasites
Trypanosoma cruzi	M-MDSC and PMN-MDSC	host beneficial/detrimental; dependent on MDSC subset, express ROS, NO, suppress CD8 T-cell proliferation
Toxoplasma gondii	Total MDSC	host protective; express NO, control parasite replication
helminths		
Schistosoma spp.	Total MDSC	not evaluated; express ROS, suppress T-cell responses
Echinnococcus granulosus	Total MDSC	not evaluated; association with increased Treg and impaired T-cell L-selectin
Nippostrongylus brasiliensis	M-MDSC and PMN-MDSC	host beneficial/detrimental; dependent on MDSC subset, express TH2 cytokines, reduce parasite burden (PMN-MDSC)
Heligmosomoides polygyrus bakeri	Total MDSC	host detrimental; suppress CD4 T-cell proliferation, increase parasite burden, and promote chronic infection

Table 10.5 Immunosuppressive effects of monocytic myeloid-derived suppressor cells (M-MDSC) on infectious disease outcome. (From Dorhol, A. and Du Plessis, N., *Front. Immunol.*, 8, Article 1895, 2018.)

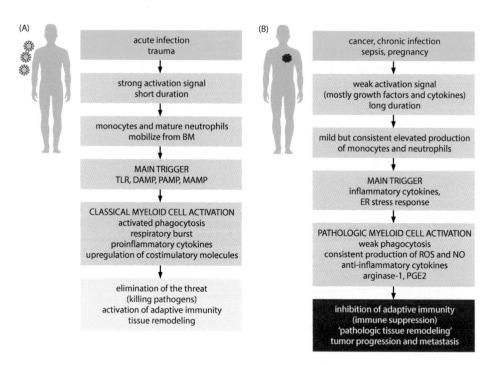

Figure 10.23 Monocyte and neutrophil activation in acute and chronic infection. (A) In the presence of strong activation signals coming from pathogens in the form of Toll-like receptor (TLR) ligands, damage-associated molecular patterns (DAMPs) and/or pathogen-associated molecular patterns (PAMPs) or microbe-associated molecular patterns (MAMPs), monocytes and neutrophils are activated as shown in the figure. (B) In the presence of weak activation signals mediated mostly by growth factors and cytokines, myeloid cell populations undergo modest but continuous expansion. Pro-inflammatory cytokines and endoplasmic reticulum (ER) stress responses contribute to pathologic myeloid cell activation that manifests as weak phagocytic activity, increased production of reactive oxygen species (ROS) and nitric oxide (NO), and expression of arginase-1 (not expressed in human monocytes or M-MDSCs) and prostaglandin E_2 (PGE$_2$). This results in immune suppression. (From Dorhol, A. and Du Plessis, N., *Nat. Immunol.*, 19, 108–119, 2018.)

myeloid suppressor cells (M-MSC) are found in the lungs in *Mycobacterium tuberculosis, Francisella tularensis* and influenza A infections; in the liver in HBV infection; and systemically in HIV infection. A comparison of monocyte and neutrophil activation in acute and chronic infections is shown in **Figure 10.23**. The functional characteristics of monocytic myeloid-derived suppressor cells (M-MDSC) and their interactions with immune cells during chronic infections are shown in **Figure 10.24**.

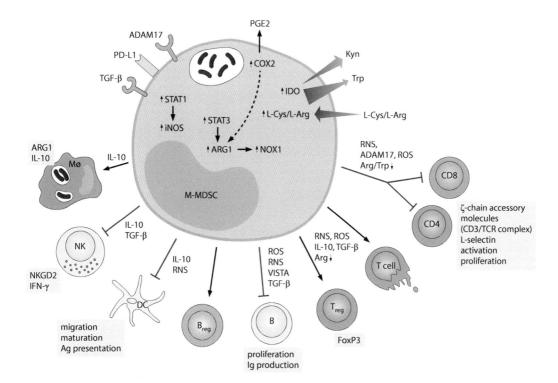

Figure 10.24 Features of monocytic myeloid-derived suppressor cells (M-MDSCs) and their interactions with immune cells during infection. M-MDSCs express membrane-bound inhibitory receptors and up-regulate enzymatic pathways (iNOS, ARG1, COX2, IDO) conferring suppressive activity towards multiple myeloid and lymphoid cell subsets. The key function of M-MDSC is suppression of T-cell immunity. M-MDSC restrict proliferation and release of cytokines by effector CD4 and CD8 lymphocytes and induce apoptotic cell death in these cells. In addition, these myeloid regulatory cells induce regulatory T and B cells while limiting antibody release and proliferation of conventional B cells. M-MDSCs alter activity of NK cells and antigen-presenting cells (APCs) and induce polarization of macrophages towards a regulatory phenotype. Color-coded arrows indicate induction/activation (green) or suppression (red), and molecules employed by M-MDSC for such effects are highlighted. Size- and color-coded arrows indicate gradient fluxes for selected essential amino acids. Boxes indicate cellular functions or pathways modulated by M-MDSC. ADAM17, ADAM metallopeptidase domain 17; ARG1, arginase 1; CD, cluster of differentiation; COX2, cyclooxygenase 2; DC, dendritic cell; IDO1, indoleamine dioxygenase 1; IFN-γ, interferon-gamma; IL-10, interleukin 10; iNOS, inducible nitric oxide synthase; Kyn, kynurenine; L-Arg, L-arginine; L-Cys, L-cysteine; MΦ, macrophage; NK, natural killer cell; NKGD2, killer cell lectin like receptor K1; NOX1, NADPH oxidase 1; PGE$_2$, prostaglandin E$_2$, PD-L1, programmed-death ligand 1; RNS, reactive nitrogen species; ROS, reactive oxygen species; STAT, signal transducer and activator of transcription; TGF- β, transforming growth factor beta; Trp, tryptophan; VISTA, V-domain Ig suppressor of T-cell activation. (From Dorhol, A. and Du Plessis, N., *Front. Immunol.*, 8, Article 1895, 2018.)

KEY CONCEPTS

- All classes of pathogens, with perhaps the exception of fungi, have members capable of persistent infection.

- All classes of persistent pathogens utilize the same mechanisms of immune evasion.

- These mechanisms are the same as those employed by pathogens causing acute infections.

BIBLIOGRAPHY

Algood HMS, Timothy L. Cover TL. 2006. *Helicobacter pylori* persistence: An overview of interactions between *H. pylori* and host immune defenses. *Clinical Microbiology Reviews*, 19(4): 597–613.

Blaser MJ, Atherton JC. 2004. *Helicobacter pylori* persistence: Biology and disease. *Journal of Clinical Investigation*, 113: 321–333.

Boxus M, Willems L. 2009. Mechanisms of HTLV-1 persistence and transformation. *British Journal of Cancer*, 101: 1497–1501.

Brodsky FM. 1999. Stealth, sabotage and exploitation. *Immunological Reviews*, 168(1): 5–11.

Brodsky IE, Medzhitov R. 2009. Targeting of immune signalling networks by bacterial pathogens. *Nature Cell Biology*, 11(5): 521–526.

Burke KP, Cox AL. 2010. Hepatitis C virus evasion of adaptive immune responses: A model for viral persistence. *Immunological Research*, 47(1–3): 216–227.

Cardillo F et al. 2015. Immunity and immune modulation in *Trypanosoma cruzi* infection. *FEMS Pathogens and Disease*, 73(9): doi:10.1093/femspd/ftv082.

Cardoso MS et al. 2016. Evasion of the immune response by *Trypanosoma cruzi* during acute infection. *Frontiers in Immunology*, 6: 659. doi:10.3389/fimmu.2015.00659.

Doorbar J. 2013. Latent papillomavirus infections and their regulation. *Current Opinion in Virology*, 3(4): 416–421.

Dorhol A, Du Plessis N. 2018. Monocytic myeloid-derived suppressor cells in chronic infections. *Frontiers in Immunology*, 8: Article 1895. doi:10.3389/fimmu.2017.01895.

Dougan G, Baker S. 2014. *Salmonella enterica* Serovar Typhi and the pathogenesis of typhoid fever. *Annual Review of Microbiology*, 68: 317–336.

Eswarappa SM. 2009. Location of pathogenic bacteria during persistent infections: Insights from an analysis using game theory. *PLoS One*, 4(4): e5383.

Fafi-Kremer S et al. 2012. Neutralizing antibodies and pathogenesis of hepatitis C virus infection. *Viruses*, 4: 2016–2030. doi:10.3390/v4102016.

Fisher RA et al. 2017. Persistent bacterial infections and persister cells. *Nature Reviews Microbiology*, 15(8): 453–464.

Godkin A, Smith KA. 2016. Chronic infections with viruses or parasites: Breaking bad to make good. *Immunology*, 150: 389–396.

Grant SS, Hung DT. 2013. Persistent bacterial infections, antibiotic tolerance, and the oxidative stress response. *Virulence*, 4(4): 273–283.

Gupta A et al. 2012. *Mycobacterium tuberculosis*: Immune evasion, latency and reactivation. *Immunobiology*, 217: 363–374.

Horst D et al. 2011. Viral evasion of T cell immunity: Ancient mechanisms offering new applications. *Current Opinion in Immunology*, 23: 96–103.

Iannello A et al. 2006. Viral strategies for evading antiviral cellular immune responses of the host. *Journal of Leukocyte Biology*, 79: 16–35.

Imperiale MJ, Jiang M. 2016. Polyomavirus persistence. *Annual Review of Virology*, 3: 517–532.

Jonjic S et al. 2008. Immune evasion of natural killer cells by viruses. *Currunt Opinion in Immunology*, 20(1): 30–38.

Kane M, Golovkina T. 2010. Common threads in persistent viral infections. *Journal of Virology*, 84(9): 4116–4123.

Katze MG, Korth MJ, Law GL, Nathanson N. 2016. *Viral Pathogenesis: From Basics to Systems Biology*. 3rd ed. Academic Press, Elsevier, London, UK.

Kelesidis T. 2014. The cross-talk between spirochetal lipoproteins and immunity. *Frontiers in Immunology*, 5(June): Article 310. doi:10.3389/fimmu.2014.00310.

Kwon KJ, Siliciano RF. 2017. HIV persistence: Clonal expansion of cells in the latent reservoir. *Journal of Clinical Investigation*, 127(7): 2536–2538.

Lewis MD, Kelly JM. 2016. Putting infection dynamics at the heart of Chagas disease. *Trends in Parasitology*, 32(11): 899–911.

Loker ES, Hofkin BV. 2015. *Parasitology: A Conceptual Approach*. Garland Science, New York.

Ma Y et al. 2016. Viral evasion of natural killer cell activation. *Viruses*, 8: 95. doi:10.3390/v8040095.

Maisels RM, Yasdanbakhsh M. 2003. Immune regulation by helminth parasites: Cellular and molecular mechanisms. *Nature Reviews Immunology*, 3(9): 733–744.

Monack DM et al. 2004. Persistent bacterial infections: The interface of the pathogen and the host immune system. *Nature Reviews Microbiology*, 2(9): 747–765.

Müller A et al. 2011. *H. pylori* exploits and manipulates innate and adaptive immune cell signaling pathways to establish persistent infection. *Cell Communication and Signaling*, 9: 25.

Murray P et al. 2015. *Medical Microbiology*. 8th ed., p. 426. Elsevier, London, UK.

Mzingwane ML, Tiemessen CT. 2017. Mechanisms of HIV persistence in HIV reservoirs. *Reviews in Medical Virology*, 27: e1924.

Nash AA, Dalziel RG, Fitzgerald JR. 2015. *Mim's Pathogenesis of Infectious Disease*. 6th ed. Elsevier, London, UK.

Nathanson N. 2007. *Viral Pathogenesis and Immunity*. 2nd ed. Academic Press, Elsevier, London, UK.

Radolf JD et al. 2016. *Treponema pallidum*, the syphilis spirochete: Making a living as a stealth pathogen. *Nature Reviews Microbiology*, 14(December): 744–759.

Rhen M et al. 2004. The basis of persistent bacterial infections. *Trends in Microbiology*, 11(2): 80–86.

Riley EM, Stewart AA. 2013. Immune mechanisms in malaria: New insights in vaccine development. *Nature Medicine*, 19(2): 168–178.

Røder G et al. Viral proteins interfering with antigen presentation target the major histocompatibility complex class I peptide-loading complex. *Journal of Virology*, 82(17): 8246–8252.

Salama NR et al. 2013. Nina R. Life in the human stomach: Persistence strategies of the bacterial pathogen *Helicobacter pylori*. *Nature Reviews Microbiology*, 11: 385–399.

Stanisic DI et al. 2013. Escaping the immune system: How the malaria parasite makes vaccine development a challenge. *Trends in Parasitology*, 29(12): 612–622.

Stevceva L (Ed.). 2018. *Vaccines for Latent Viral Infections*. Bentham Science, Sharjah, U.A.E.

Sutherland CJ. 2016. Persistent parasitism: The adaptive biology of malariae and ovale malaria. *Trends in Parasitology*, 32(10): 808–819.

Swain SL et al. 2012. Expanding roles for CD4$^+$ T cells in immunity to viruses. *Nature Reviews Immunology*, 12: 136–148.

Terilli RR, Cox AL. 2013. Immunity and hepatitis C: A review. *Current HIV/AIDS Reports*, 10: 51–58.

Veglia F et al. 2018. Myeloid-derived suppressor cells coming of age. *Nature Immunology*, 108: 19(2): 108–119.

Westrich JA et al. 2017. Evasion of host immune defenses by human papillomavirus. *Virus Research*, 231: 21–33.

Wiertz EJ et al. 2007. Herpesvirus interference with major histocompatibility complex class II-restricted T-cell activation. *Journal of Virology*, 81(9): 4389–4396.

Wilson BA, Salyers AA, Whitt DD, Winkler ME. 2011. *Bacterial Pathogenesis: A Molecular Approach*. 3rd ed. ASM Press, Washington, DC.

Wilson M, McNab R, Henderson B. 2002. *Bacterial Disease Mechanisms: An Introduction to Cellular Microbiology*. Cambridge University Press, Cambridge, UK.

Yewdell JW, Bennink JR. 1999. Mechanisms of viral interference with MHC class I antigen processing and presentation. *Annual Review of Cell and Developmental Biology*, 15: 579–606.

Young D et al. 2002. Chronic bacterial infections: Living with unwanted guests. *Nature Immunology*, 3(11): 1026–1032.

Zuo J, Rowe M. 2012. Herpesviruses placating the unwilling host: Manipulation of the MHC class II antigen presentation pathway. *Viruses*, 4: 1335–1353.

Index

Note: Page numbers in italic and bold refer to figures and tables, respectively.

A

AB$_5$ exotoxins, 178–180
AB exotoxins, 181
abiotic surfaces, 94–95
ablumenal surface, 92, 124
acquired pellicle, 61
actin-based protrusion, 245–247, *246*
actin-binding proteins, 142
actin filaments, 141, *142*
actin-related complex 2/3 (Arp 2/3) activation, 243, *243*
activating naïve T cells, 265
acute phase response, 189
adenoviruses, 317
adhesin–receptor interactions
    bacteria, 98–102
    fungi, 102
    parasites, 106–107
    protein–carbohydrate, 97–98
    viruses, 102–105
adhesins, 61, 87
    anchorless, 119
    cellular, 94
adhesins–receptor interactions, 97–107
adhesion, 194–195, **195**
adhesion to host surfaces
    adhesin–receptor interactions, 97–107
    barrier epithelia, 88–94
    ECM, 94–97
    endothelium, 124
    galectins, 122–123
    ICAM, 94–97
    protein–protein interactions, 107–121
aerolysins, 174
alimentary canal, microbiotas, 34
    large intestine, 44, 46–56
    mouth, 34–39
    oesophagus, 40, *41*
    small intestine, 41–44, *44, 45*
    stomach, 41, *42, 43*
α-helical PFTs, 172–173
AMPs, *see* antimicrobial peptides (AMPs)
anchorless adhesins, 119
animal models, 8–10
anthrax toxin (AT), 180
antibiotics, 18

antibodies, pathogen inaccessibility to, 281
antibody, evasion, 280–281
antibody enzymatic degradation, 280
antigenic determinant, 282
antigenic shift/drift, 285
antigenic variation, 281, 283–285, 324
antigen modulation, 281–283
antigen presentation, 263–264
    activating naïve T cells, 265
    follicular helper CD4$^+$ T cells, 267–270
    MAMPs by PRRs, 264–265
    MHC class I/class II pathways, 270–280
    pathway, **306**
antigen-presenting cells (APCs), 263
antimicrobial peptides (AMPs)
    anti-parasite effects, 203
    bacterial evasion, 204–205, *206*
    Candida albicans mechanisms, 205, *207*
    cytoplasmic membrane, 202–203
    deficiencies, inflammatory/infectious diseases, 203–204, **204**
    description, 201
    mechanism, 202, *202*
    parasite evasion, 205
    viruses, 205
anti-parasite effects, AMPs, 203
APCs (antigen-presenting cells), 263
apical complex, 134
apical membrane antigen (AMA) 1, 134–135
apicomplexans, *136*
Aspergillus fumigatus, 198
AT (anthrax toxin), 180
ATG5, *see* autophagy-related 5 protein (ATG5)
autophagy
    ATG5, 249
    autophagosomal maturation, 251
    DNA viruses, 253, **255**
    eukaryotic pathogens, 253
    fungi, 253, **255**
    herpesviruses, 250
    Listeria monocytogenes, LLO, 251, *252*
    microbial adaptations, 249–250, *250*
    protozoa, 253, **256**
    RNA viruses, 253, **254**
autophagy-related 5 protein (ATG5), 249, 251, *252*
autotransporters, 111

B

Bacillus anthracis, 2
bacteria; *see also* gram-negative bacteria;
    gram-positive bacteria
    adhesin–receptor interactions, 98–102
    protein–protein interactions, 117, 119, **120**
bacterial exotoxins, 167–168
    membrane-acting toxins, 168–170
bacterial phospholipases, 196–197
bacterial proteases
    affecting PARS, **191**
    pathogenesis, potential rules in, **192**
bacterial vaccines, 5
bacteriocins, 66
Bacteroides thetaiotaomicron, 51
barrier epithelia, 87
    blood and lymphatic vessels, 91–92
    blood-brain barrier, 92–93
    foeto-placental interface, 93–94
    mucous membranes, 89–91
    skin, 88
B-cell activity, suppression, 289
B-cell follicles, 265
B-cell receptors (BCRs), 267
B cells, polyclonal activation, 289
B-cell survival, 289–290
BCRs (B-cell receptors), 267
β-barrel pore-forming toxins, 173
    aerolysin, 174
    CDCs, 174
    haemolysins, 173–174
binary toxins, 176
biofilms, 59–60
    bladder, 69, *70*
    chronic wounds harbour, 71–73
    dental plaque, 59–60, *64*, 79
    by filamentous fungi, 76
    flow cell system, *82*, **82**
    in human infections, 67–76
    urinary catheter, 68, *69*
    by viruses, 77–78
biofilms structure and properties
    development and climax community, 63–65
    dispersal, 67
    formation, 61–63
    mucosae *versus* skin, 60–61
    quorum sensing, 65–66
bioreactor system, *81*
bladder biofilms, 69, *70*
blood and lymphatic vessels, 91–92, 124
blood-brain barrier, 92–93
blood–cerebrospinal fluid (CSF) barrier, *130*
blood-clotting cascades, activation, 190, *190*
bradykinin, activation, 190, *190*
breast-feeding, 50, *50*

bronchiolar lavage method, 31
budding, *Orientia tsutsugamushi*, *246*, 247

C

calcium bridging, 121
calcium-type lectins (CTLs), 221–223, 225
CAM, *see* cell adhesion molecules (CAM)
Candida albicans, 102, **113**, 188–189
capillary
    versus capillary in brain, *93*
    pericyte, *92*
capsules, 121, 214, 285
captive breeding, 8
Casadevall's damage–response framework, 4, *4*
catenins, 146
caveola-mediated endocytosis, 138–139
CD4⁺ T cells, antiviral mechanisms, *312*
CDCs (cholesterol-dependent cytotoxins), 174
CDTs (cytolethal distending toxins), 248
cell adhesion molecules (CAM), 94
    down-regulation, 279–280
cell entry
    endocytosis pathways, exploitation, 145
    intact mucosal epithelium, 132–144
    intact skin, 130–132
    microtubule reorganisation, 158, 160–161
    transcytosis, 161
    zipper mechanism, 145–158
cell internalisation, 188–189
cell lines, 12–14, *14*
cell-mediated immune response, 304
cell wall glycopolymers, 121
CF, *see* cystic fibrosis (CF)
chaperone–usher pilus systems, 99, **100**
chemoattractant molecules/chemoattraction,
    191, 216–218
chemokines
    mechanisms, pathogens, 218–219, **219**
    virus-encoded modulators, **220–221**
chemostat, 79
Chlamydia/Chlamydophila, 232–233, *234*, 270
cholera toxin (CT), 178
cholesterol-dependent cytotoxins (CDCs), 174
chronic HBV infection, 314
chronic otitis media, 73–74
chronic wounds harbour biofilms, 71–73
clathrin-independent endocytosis (CIE), 140, *141*
clathrin-mediated endocytosis (CME),
    139–140, *141*
climax community, 64
Clostridium botulinum neurotoxins, 181
Clostridium tetani neurotoxins, 181
CME (clathrin-mediated endocytosis), 139–140, *141*
coadhesion (coaggregation), 63–64

coaggregation aggregation, 87
cognate interaction, 263, 267
collagen, **114–115**
    hug, 113
colonoscopy, 44
commensalism, 17
community state types (CSTs), 22, *24*
complement receptor 3 (CR3), 279
complement system
    capsules, bacteria/fungi, 214
    cell wall structure, 214
    classical/lectin/alternative pathways,
        207–209, *208*
    fluid phase, 214–215
    microbial pathogens, 209, *209*
    proteases, 213–214
    RCAs, *see* regulators of complement activation
        (RCAs)
conformational/discontinuous epitopes, 282
conjunctiva, microbiota, 31, *32*
continuous endothelium, 92
continuous fenestrated endothelium, 92
coprophagy, 10
corneocytes, 88, 94–95
coronal sulcus (CS) specimen, 27, *30*, *31*
cortical actin, 142
Corynebacterium diphtheriae DT, 181
co-stimulatory molecules, manipulation, 275, **275**
covalently closed circular DNA (cccDNA), 313
Coxiella burnetii, 233, 235, *236*
CR3 (complement receptor 3), 279
cross-presentation process, 273
cryptic epitopes (cryptotopes), 283
CS1 pili/group 4 pili, 98, 101
CS (coronal sulcus) specimen, 27, *30*, *31*
CSTs (community state types), 22, *24*
CT (cholera toxin), 178
CTLs (calcium-type lectins), 221–223, 225
culture-independent analysis, *36*, *37*
curli/group 3 pili, 101
cysteine proteases, 189
cystic fibrosis (CF), 74–76, *77*
cytoadhesion, 106
cytokine storm, 184
cytolethal distending toxins (CDTs), 248
cytolysis, 245, *246*
cytosol, escape to, 238–240
cytotoxic CD8$^+$ T cells, 316

D

DAEC (diffusely adhering *Escherichia coli*), 99
danger-associated molecular patterns
        (DAMPs), 221
DC (dendritic cell), 265

decidua, 93
defensins, 203
degradative enzymes, 187
deltaretrovirus, 290
dendritic cell (DC), 265
dental plaque biofilm, 59-60, *64*, 79
Derjagiun, Landau, Verwey and Overbeek (DLVO)
        theory, 95, *97*
DIC (disseminated intravascular coagulation), 184
diffusely adhering *Escherichia coli* (DAEC), 99
discontinuous endothelium, 92
disseminated intravascular coagulation
        (DIC), 184
DLVO (Derjagiun, Landau, Verwey and Overbeek)
        theory, 95, *97*
dock, lock and latch (DLL), 109, *109*
dynamin, 139-140
dysbiosis, 18

E

early endosome (EEs), 144, *144*
ebola virus, 290
eccrine glands, 20
ECM, *see* extracellular matrix (ECM)
ecosystem, 17
EEs (early endosome), 144, *144*
EHEC (enterohaemorrhagic *Escherichia coli*), 153
Encephalitozoon cuniculi, 237
endocytosis, 138-141
    pathways, 145
endogenous microbiotas, 18
endosome/lysosome system, *144*
endothelium, 91-92
endotoxins, 183-184
endotracheal tube (ETT), 69, *70*
enterohaemorrhagic *Escherichia coli*
        (EHEC), 153
enteropathogenic *Escherichia coli* (EPEC), 153,
        154, 161
enterotoxins, 168
enzymatic degradation, 131-133
enzyme-linked immunosorbent assay (ELISA), 12
EPEC (enteropathogenic *Escherichia coli*), 153,
        154, 161
epitope masking, 281-283, *282*
Epstein-Barr nuclear antigen 1 (EBNA-1), 310
Escherichia coli pathogenic types, **153**
ETT (endotracheal tube), 69, *70*
eukaryotes, 187
eukaryotic signalling pathways, *197*
exotoxins, 4
    bacterial, 167-170
    intracellular, 176-181
    membrane-damaging, 171-176
    parasite, 182

extracellular degradative enzymes
  microbe/parasite glycosidases, 194–196
  microbe/parasite phospholipases, 196–198
  microbial proteases, potential roles, 188–193
  overview, 187–188
  proteases, 188
extracellular glycolytic enzymes, **121**
extracellular matrix (ECM), 94–97
  properties, **108**
  proteins, 13, **115**
extracellular polymeric substance (EPS) matrix, *63*
extrusion, *Chlamydia* spp., *246*, 247

F

faecal microbiota, 47, 50–51
Falkow's molecular postulates, 2–3
fibronectin-binding MSCRAMMs, 108–109, 111,
      113, 115
filamin, 139
fimbriae (pili), 285
flagella, 285
f-MLP (formyl-methionyl-leucyl-
      phenylalanine), 191
foeto-placental interface, 93–94
follicular helper CD4⁺ T cells, 267–270
formula-feeding, 50, *50*
formyl-methionyl-leucyl-phenylalanine
      (f-MLP), 191
Francisella tularensis, 238, *239*
fumonisins, 182
fungal evasion, AMPs, 205, *207*
fungal phospholipases, 198
fungal proteases, 188, *189*, 190
fungal toxins, 182
fungi
  adhesin–receptor interactions, 102
  biofilms formation, 76
  protein–protein interactions, 120
  zipper mechanism, 156, 158, **159**

G

GAGs (glycosaminoglycans), 117, 218
galectins, 107, *122*, 122–123
Gardnerella vaginalis, 25
gastrointestinal tract, **56**, 90, **90**
gelatinous plaque, 59
gingival sulcus, 35
glabrous skin, 130
glucosaminoglycans, **118**
glucosidases, production, **196**
glycocalyx, 90
glycosaminoglycans (GAGs), 117, 218
glycosidases, 187
glycosphingolipids, 103

G-protein–coupled receptors (GPCRs),
      217–218, **219**
gram-negative bacteria, 111, 145, 184
  collagen-binding, 113
  fibronectin-binding, **112**
  LPS, 183, *183*
  pili, 98–99
gram-positive bacteria, 65, 109, 115
  fibronectin-binding, **110–111**
  pili, **102**
granuloma, 304

H

HAART (highly active anti-recto viral therapy), 321
haemolysins, 173–174
HBV (hepatitis B virus), 313–316
HCMV (human cytomegalovirus), 272
HCV (hepatitis C virus), 313–316
HDV (hepatitis D virus), 313–316
heart valves, 71
heat-labile enterotoxin (LT), 178
heat-stable exotoxins (STs), 170
Helicobacter pylori, 41, 296–300
  colonization and persistence factors, *298*
  innate immunity recognition, *299*
  T cell-mediated immunity, *300*
helminths, 322–323, *323*, *324*
hemagglutinin (HA), 285
Henle-Koch postulates, 2–6
hepatitis B virus (HBV), 313–316
hepatitis C virus (HCV), 313–316
hepatitis D virus (HDV), 313–316
herpesvirus, 250, *271*, 309–313
highly active anti-recto viral therapy (HAART), 321
Histoplasma capsulatum, 235–236
HIV (human immunodeficiency virus), 320–321
host barriers, *130*
host immune system, modulation, *188*
HPVs, *see* human papilloma viruses (HPVs)
HTLV-1 basic leucine zipper (HBZ) factor,
      321–322
human cytomegalovirus (HCMV), 272
human immunodeficiency virus (HIV), 320–321
human infections, biofilms in, 67
  bladder, 69, *70*
  CF, 74–76, *77*
  chronic wounds, 71–73
  ETT, 69, *70*
  heart valves, 71
  otitis media, 73–74
  peripheral and central i.v. catheters, 67
  peritoneal dialysis catheters, 70, *71*
  prosthetic joint, 71
  urinary catheter, 68, *69*
human infectious agents, 8

human papilloma viruses (HPVs), 317–319,
    318, *319*
human pathogens, **274**
human polyomaviruses, 319–320
human T-cell lymphotropic virus type 1 (HTLV-1),
    321–322
human telomerase reverse transcriptase
    complementary DNA (hTERT
    cDNA), 12
hydrochloric acid, 41

I

IDPs (intrinsically disordered proteins), 282
IgG (immunoglobulin G) molecule, 191, 280
IL-10, *see* interleukin-10 (IL-10)
IL-12, 224; *see also* interleukin-10 (IL-10)
    inhibition, 276–278, **278**
immune exclusion, 280
immunoglobulin G (IgG) molecule, 191, 280
immunoglobulins, 191, 193
    deglycosylation, 194
    Fc receptors, 280–281
immunoreceptor tyrosine-base activation motifs
    (ITAMs), 225, 227
immunosuppressive mediators, synergistic
    induction, **277**
infectious diseases, 2–6
influenza virus neuraminidase, 195
informed consent, 7
injectosomes, 148
innate/adaptive immunity, effector
    mechanisms, **304**
inside-out signalling, 278–279, **279**
intact mucosal epithelium
    actin cytoskeleton and endosomal trafficking,
        141–144
    endocytosis, 138–141
    entry *via* M cells, 132–133
    enzymatic degradation, 133
    MJ, 134–135
    paracytosis, 135–138
    polar tube formation, 134
integrin signalling, 279
intercellular junctions, *136*
interferon-gamma (IFN-γ), 271–272
inter-generic aggregation, 87
intergeneric coaggregation, 63–64
interleukin-10 (IL-10), 223, 270, 311, 323
    up-regulation, 276–280
intracellular bacterial pathogens, survival
        strategies, *229*
    arresting, phagosome/endosome, 229–232
    diverting, endosomal/phagosomal pathways,
        232–233
    endolysosome/phagolysosome, 233–235

intracellular exotoxins
    AB₅ exotoxins, 178–180
    AB exotoxins, 181
    AB toxins, 176–177
intracellular pathogens, 296
intracellular pathogens, host cells
    actin-mediated cell-to-cell spread, 245–247
    cell cycle, 248
    cytolysis, 245
    to exit, 245, *246*
    extrusion, 247
    programmed cell death, 247–248
    reprogramming, 249
intrinsically disordered proteins (IDPs), 282
invasin, 146, *146*
invasion, 129
ITAMs (immunoreceptor tyrosine-base activation
    motifs), 225, 227

K

Koch, R., 2

L

Lactobacillus species, 22
lactoferricin, 203
LAM (lipoarabinomannan), 279
latent infections, 295, 308
Legionella pneumophila, 233, *235*
Leishmania donovani, 237, 273
Leishmania major, 123, 277
Leishmania species, 325–326
licencing process, 223, 265
ligands, manipulation, 276
lipases, 187; *see also* phospholipases
lipoarabinomannan (LAM), 279
lipoglycans, 106–107
lipooligosaccharide (LOS), 183–184
lipophosphoglycan (LPG), 237
lipopolysaccharide (LPS), 183–184, *301*
Listeria, molecular pathways, *147*
Listeria monocytogenes, 238, *239*, *244*
    adhesins of, 147
listeriolysin (LLO), 251, *252*
LOS (lipooligosaccharide), 183–184
LPG (lipophosphoglycan), 237
LPS (lipopolysaccharide), 183–184, *301*
lymphatic vessels, 91–92, *92*
lysophospholipids, 197

M

macrophage receptor with collagenous structure
    (MARCO), 225, *226*, 227
macrophages, 304

macropinocytosis, 138, 156
major histocompatibility complex (MHC), 263
Malassezia species, 22, *23*
MAMPs, *see* microbe-associated molecular patterns (MAMPs)
mannose-binding lectin (MBL) pathway, 207–209, *208*
MARTX, *see* multifunctional auto-processing repeats-in-toxin toxins (MARTX)
maternal-fetal interface, *93*
Matrigel membrane, 13
MBL pathway, *see* mannose-binding lectin (MBL) pathway
measles virus (MV), 316–317
megasomes, 273
membrane-acting toxins
  SAs, 168–170
  STs, 170
membrane-damaging exotoxins, 171–172
  α-helical PFTs, 172–173
  β-barrel PFTs, 173–174
  MARTX exotoxins, 175, *176*
  RTX exotoxins, 174–175
membrane vesicles (MVs), 283
methanogens, 51
methicillin-resistant *Staphylococcus aureus* (MRSA) biofilm, 67, *69*
MHC class I/class II pathways, *264*
  bacterial subversion, 270–271, 273
  parasite evasion, 273
  viral subversion, 270–273, **272**
microaspiration, 33
microbe-associated molecular patterns (MAMPs), 221, 223–225, 250, 264, 307
microbe–host interactions, 12–14
microbes
  and parasite glycosidases, 194–195
  and parasite phospholipases, 196–198
  pathogenesis, 1
microbial communities properties, **62**
microbial proteases, potential roles
  bradykinin-generating/blood-clotting cascades, 190, *190*
  chemoattractant molecules, 191
  immunoglobulins, 191, 193
  plasma protease inhibitors, inactivation, 189
  protease-activated receptor, 190
  tissue destruction/cell internalisation, 188–189
microbial strain, 10
microbial surface components recognising adhesive matrix molecules (MSCRAMMs), 108
  collagen- and laminin-binding, 113
  fibronectin-binding, 108–109, 111, 113, 115
  VN-binding, 116–117

microbiome, 17
  human gut components, *55*
  mid-vaginal profile, *24*
  mother's gut in, *47*
  urinary tract, **28**, *29*
microbiotas, 17
  alimentary canal, 34–56
  anatomical sites, stool and mucosa, *46*
  under breast-/formula-feeding, *50*
  conjunctiva, 31, *32*
  CS, 27
  culture-independent analysis, 36, *37*
  phylum/order-like phylogroups to, *52*
  respiratory tract, 31, *33*, 34, *34*
  skin, *20*, 20–22
  urinary tract, 25–31
  vagina, 22, 24–25, **26**, *27*
microfold (M) cells, 90–91, 132–133
microplate static biofilm model, 79
microtubule reorganisation, 158, 160–161
M-MDSC, *see* monocytic myeloid-derived suppressor cells (M-MDSC)
modulators of immune recognition (MIR), 312
molecular mimicry, 281
molecular techniques, 18–19
monocytic myeloid-derived suppressor cells (M-MDSC), 330–332, **331**, *332*, *333*
moonlighting proteins, 119, **120**
moving junction (MJ), 134–135
MSCRAMMs, *see* microbial surface components recognising adhesive matrix molecules (MSCRAMMs)
mucins, types, **90**
mucosae *versus* skin, 60–61
mucosal biofilm formation, *82*
mucosal surfaces, 10–12
mucous membranes, 89–91
multifunctional auto-processing repeats-in-toxin toxins (MARTX), 175
  exotoxins, 175, *176*
multivalent adhesin molecule 7 (MAM7), 111
mutualism, 17
MV (measles virus), 316–317
mycobacteria, 304–305
Mycobacterium tuberculosis (MT), 2, 230, *231*, 302–307
mycotoxins, 182

N

naïve T cells, 265–267, *266*, *269*
natural cytotoxicity receptors (NCRs), 256–258, **258**
natural killer (NK) cells
  CD8+ cytotoxic T cells, 253
  KIRs/KLRs, 256

NCRs, 256–258, **258**
NKG2D receptor, 257, **257**
NCRs, *see* natural cytotoxicity receptors (NCRs)
Neisseria gonorrhoeae, 117, 148, 290
neuraminidase (NA), 285
neutralisation, 280
neutrophil granules, 215–216, *216*
NK cells, *see* natural killer (NK) cells
non-fenestrated continuous endothelium, 92
normalisation, need for, 12
nucleic acid sequence, 3

O

occult HBV infection (OBI), 314
O-linked glycosylation, 194
oncoproteins, 319
opportunistic pathogens, 2
opsonophagocytosis, 214
oral microbiota, 35–36
oxytocin, 11

P

PAMPs (pathogen-associated molecular
        patterns), 221
pandemic influenza, 285
paracytosis, 135–138
parasites
    adhesin–receptor interactions, 106–107
    -derived immune modulators, *325*
    evasion, AMPs, 205
    exotoxins, 182
    phospholipases, 198
PARs (protease-activated receptors), 190
pathogen, 1, 9–10, 307
pathogen-associated molecular patterns
        (PAMPs), 221
pathogenesis, infectious diseases
    animal models in, 8–10
    experimental models, 10–12
    human experimentation in, 8
pathogenic(s), 1
    bacteria, **297**
    fungi, 198
    microorganisms, *5*
    prokaryotes, 187–197
    research, humans/animals/cell lines in, 6–8
pathogenicity, 3–4
    determinants, 5
    experimental models, 6–14
pathogen survival, host cells, 228
    bacteria, 229–235
    cell entry, 228
    fungi, 235–236
    parasites, 237, **238**

pattern recognition receptors (PRRs), 265
    β-1-3-glucan, 225
    B lymphocytes, 221
    IPS-1, RIG-I and MDA5, 224
    and ligands, 221, **222**
    pathogens, 221–222, 224
    TLR-TLR and TLR-CTL interactions, 223–224
peritoneal dialysis catheters, 70, *71*
persistent bacterial infections, 296
    Helicobacter pylori, 296–300, *299*, *300*
    Mycobacterium tuberculosis, 302–307
    Salmonella typhi serovar Typhi, 307–308, *308*
    Treponema pallidum subspecies pallidum,
        301–302, *303*
persistent/latent infections
    bacterial infections, 296–308
    lytic infections, 308
    overview, 295–296
    parasite infections, 322–332
    virus infections, 308–322
persistent parasite infections
    Helminths, 322–323, *323*, *324*
    Leishmania, 325–326
    M-MDSC, 330–332, *332*
    Plasmodium, 323–325, *326*
    Toxoplasma gondii, 329–330
    Trypanosoma cruzi, 327, *327*–329, *328*, *329*
persistent virus infections, 309–310
    adenoviruses, 317
    hepatitis B, C/D viruses, 313–316, *314*, *315*
    herpesvirus, **309**, 309–313, *310*
    HIV, 320–321
    HPVs, 317–319, *318*, *319*
    HTLV-1, 321–322
    human polyomaviruses, 319
    MV, 316–317
pertussis toxin (PT), 180
Peyer's patches, 307
phagocytosis
    cell membrane endocytic receptors, 215
    chemoattraction, 216–218
    chemokines, 218–221
    macrophages/DCs, 215
    neutrophils, 215–216, *216*
    phagosome maturation, 215–216, *217*
    vesicles, 216
phagosome–lysosome fusion, arrest, *305*
phagosome maturation, 215–216, *217*
phase variation, 281, 283–285
    moieties, **286–287**
phosphatidylinositol 3-phosphate (PI3P), 229–230
phospholipases, 196
pili/fimbriae, 98
pilocarpin, 11
plasma protease inhibitors, inactivation, 189
Plasmodium falciparum, 120, 182, 277

Plasmodium falciparum erythrocyte membrane protein 1 (PfEMP1), 106
Plasmodium species, 106, 189, 323–325, *326*
polar tube formation, 134
pore-forming toxins (PFTs), **171**
post hoc analysis, 7
primary minimum, 96
programmed cell death, 247–248
prosthetic joints, 71
protease-activated receptors (PARs), 190
proteases, 188
proteasomes, 263
protected brush method, 31
protein–carbohydrate interactions, 97–98, *98*
protein kinase B, 196
protein–protein interactions
    adhesive matrix molecules, 107–115
    anchorless adhesins, 119
    bacteria, 117, 119, **120**
    capsules, 121
    cell wall glycopolymers, 121
    fibrinogen-binding MSCRAMMs, 115
    fungi, 120
    parasites, 118
    proteoglycan-binding adhesins, 117
    protozoa and multicellular parasites, 120–121
    viruses, 117–118
    VN-binding MSCRAMMs, 116–117
proteoglycan-binding adhesins, 117
protozoan parasites, 195
PRRs, *see* pattern recognition receptors (PRRs)
Pseudomonas aeruginosa exotoxin A (ExoA), 177

Q

quorum sensing, 65–66

R

RCAs, *see* regulators of complement activation (RCAs)
reactivation tuberculosis, 306
recycling pathway, 161
regulators of complement activation (RCAs), 210
    classical/alternative pathways, **210**
    protein and targets, **211–213**
    SCRs, 210–211
    viruses, 211
regulatory B cells, induction, 289
regulatory receptors, manipulation, **275**, 276
repeats in toxin (RTX) exotoxins, 174–175
respiratory tract microbiotas, 31, *33*, 34, *34*
RHO-family GTPase-mediated modelling, *143*
rhoptry neck proteins (RONs), 134–135
ribotoxic stress, 178

risk–benefit ratio, 8
rodent models, limitations/disadvantages, 9–10
RONs (rhoptry neck proteins), 134–135
RTX (repeats in toxin) exotoxins, 174–175

S

Salmonella-containing vacuole, 307
Salmonella enterica serovar Typhimurium, 230–231, *231*, *232*, 307–308, *308*
Salmonella T3SS-1: SipC, *152*
SAs (superantigens), 168–170, 291
SCRs (short consensus repeats), 210–211
sebaceous glands, 20, 130–131
secondary lymphoid tissues, 265, 276
secondary minimum, 96
secretion systems, types, **149–150**
secretory component (SC), 11
secretory immunoglobulin A (SIgA), 10–11, 191–192
segmental gene conversion, 284
sequence-based methods, 3
serine proteases, 189, 207–208
SEs (staphylococcal enterotoxins), 168
Shiga and Shiga-like toxins, 178–180
Shigella effectors, 151
Shigella flexneri, *152*, 160
short consensus repeats (SCRs), 210–211
SIgA (secretory immunoglobulin A), 10–11, 191–192
16S ribosomal RNA, 21
skin, 88
    bacteria on, *21*
    microbiota, *20*, 20–22
sorting endosomes, 144
specific pathogen-free (SPF) primates, 8–9
spirochetes, 301–302
sporozoite microneme protein, 323
staphylococcal enterotoxins (SEs), 168
Staphylococcus aureus, 281
    α-haemolysin, *173*
Streptococcus mutans, 35, 283
Streptococcus pyogenes, 189, 194
substrata, 88, *96*
subverting B lymphocytes (B Cells), 287–289, **288**
subverting T lymphocytes (T Cells), 290–291
superantigen-like (SSL) proteins, 218
superantigens (SAs), 168–170, 291
symbiosis, 17

T

T3SSs, *see* type 3 secretion systems (T3SSs)
T4SS, *see* type 4 secretion system (T4SS)
T-cell exhaustion, 307
tissue culture cells, 13

tissue destruction, 188–189
Toll-like receptors (TLRs), 137, 221, **222**, 223–224
Toxoplasma gondii, 237, 329–330
transactivation process, 279
transcriptional control, 284
transcytosis, 132, 161
transgenic mouse models, 314
transitional B cells, 289
transurethral catheterisation, 25–26
TREM1 (triggering receptor expressed on myeloid
        cells), 225
Treponema pallidum subspecies pallidum,
        301–302, *303*
Trichomonas vaginalis, 195
trigger mechanism, *145*, 148–155
Trypanosoma cruzi, 240, 289, *327*, 327–329, *328*, *329*
type 1 secretion system (T1SS), 174
type 2 macrophages, 322
type 3 injectisome families, **151**
type 3 secretion systems (T3SSs), 148, 205, 283, 307
type 4 secretion system (T4SS), 137, 154–155
type IV pili, 99, **101**

U

UPEC (uropathogenic *Escherichia coli*), 99, 154
urinary catheter biofilms, 68, *69*
urinary tract, microbiotas, 25–31, *29*, *30*
urogenital protozoan, 195
uropathogenic *Escherichia coli* (UPEC), 99, 154

V

vaccines in the United States, 5, **6**
vaccinia virus, 243–244, *244*
vacuolating toxin (VacA), 297
vaginal microbiota, 22, 24–25, **26**, *27*
vagitypes, 22
variant surface antigens (VSAs), 324
viable but nonculturable (VBNC), 67
viral neuraminidase, 195
virulence, 3–4
viruses
        adhesin–receptor interactions, 102–105
        biofilm formation, 77–78
        endocytic pathways, *158*
        interactions, intracellular vacuoles, 240–242,
            241, **242**
        protein–protein interactions, 117–118
        receptors on epithelial cells, **157**
        zipper mechanism, 156
vitronectin-binding MSCRAMMs, 116–117

Z

zearalenone, 182
zipper mechanism, 145–148
        fungi, 156, 158, **159**
        trigger mechanism, *145*, 148–155
        viruses, 156, **157**, *158*
zymogens, 188, 207